Linear Regression Analysis

Theory and Computing

Linear Regression Analysis

Theory and Computing

Xin Yan
University of Missouri–Kansas City, USA

Xiao Gang Su
University of Central Florida, USA

World Scientific

NEW JERSEY • LONDON • SINGAPORE • BEIJING • SHANGHAI • HONG KONG • TAIPEI • CHENNAI

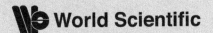

Published by

World Scientific Publishing Co. Pte. Ltd.
5 Toh Tuck Link, Singapore 596224
USA office: 27 Warren Street, Suite 401-402, Hackensack, NJ 07601
UK office: 57 Shelton Street, Covent Garden, London WC2H 9HE

Library of Congress Cataloging-in-Publication Data
Yan, Xin
Linear regression analysis : theory and computing / by Xin Yan & Xiaogang Su.
p. cm.
Includes bibliographical references and index.
ISBN-13: 978-981-283-410-2 (hardcover : alk. paper)
ISBN-10: 981-283-410-9 (hardcover : alk. paper)
1. Regression analysis. I. Su, Xiaogang II. Title.
QA278.2.Y36 2009
519.5'36--dc22
2009012000

British Library Cataloguing-in-Publication Data
A catalogue record for this book is available from the British Library.

Printed in Singapore.

Preface

In statistics, regression analysis consists of techniques for modeling the relationship between a dependent variable (also called response variable) and one or more independent variables (also known as explanatory variables or predictors). In regression, the dependent variable is modeled as a function of independent variables, corresponding regression parameters (coefficients), and a random error term which represents variation in the dependent variable unexplained by the function of the dependent variables and coefficients. In linear regression the dependent variable is modeled as a linear function of a set of regression parameters and a random error. The parameters need to be estimated so that the model gives the " best fit " to the data. The parameters are estimated based on predefined criterion. The most commonly used criterion is the least squares method, but other criteria have also been used that will result in different estimators of the regression parameters. The statistical properties of the estimator derived using different criteria will be different from the estimator using the least squares principle. In this book the least squares principle will be utilized to derive estimates of the regression parameters. If a regression model adequately reflects the true relationship between the response variable and independent variables, this model can be used for predicting dependent variable, identifying important independent variables, and establishing desired causal relationship between the response variable and independent variables.

To perform regression analysis, an investigator often assembles data on underlying variables of interest and employs regression model to estimate the quantitative causal effect of the independent variables to the response variable. The investigator also typically assesses the " statistical significance " of the estimated relationship between the independent variables and dependent variable, that is, the degree of confidence on how the true relationship is close to the estimated statistical relationship.

Regression analysis is a process used to estimate a function which predicts value of response variable in terms of values of other independent variables. If the regression function is determined only through a set of parameters the type of regression is the parametric regression. Many methods have been developed to determine various parametric relationships between response variable and independent variables. These methods typically depend on the form of parametric regression function and the distribution of the error term in a regression model. For example, linear regression, logistic regression, Poisson regression, and probit regression, etc. These particular regression models assume different regression functions and error terms from corresponding underline distributions. A generalization of linear regression models has been formalized in the " generalized linear model " and it requires to specify a link function which provides the relationship between the linear predictor and the mean of the distribution function.

The regression model often relies heavily on the underlying assumptions being satisfied. Regression analysis has been criticized as being misused for these purposes in many cases where the appropriate assumptions cannot be verified to hold. One important factor for such criticism is due to the fact that a regression model is easier to be criticized than to find a method to fit a regression model (Cook and Weisberg (1982)). However, checking model assumptions should never be oversighted in regression analysis.

By saying much about regression model we would like to go back to the purpose of this book. The goal of the book is to provide a comprehensive, one-semester textbook in the area of regression analysis. The book includes carefully selected topics and will not assume to serve as a complete reference book in the area of regression analysis, but rather as an easy-to-read textbook to provide readers, particularly the graduate students majoring in either statistics or biostatistics, or those who use regression analysis substantially in their subject fields, the fundamental theories on regression analysis, methods for regression model diagnosis, and computing techniques in regression. In addition to carefully selected classical topics for regression analysis, we also include some recent developments in the area of regression analysis such as the least absolute shrinkage and selection operator (LASSO) proposed by Tibshirani (1996) and Bayes averaging method.

The topics on regression analysis covered in this book are distributed among 9 chapters. Chapter 1 briefly introduces the basic concept of regression and defines the linear regression model. Chapters 2 and 3 cover the simple linear regression and multiple linear regression. Although the simple linear regression is a special case of the multiple linear regression, we present it without using matrix and give detailed derivations that highlight the fundamental concepts in linear regression. The presentation of multiple regression focus on the concept of vector space, linear projection, and linear hypothesis test. The theory of matrix is used extensively for the proofs of the statistical properties of linear regression model. Chapters 4 through 6 discuss the diagnosis of linear regression model. These chapters cover outlier detection, influential observations identification, collinearity, confounding, regression on dummy variables, checking for equal variance assumption, graphical display of residual diagnosis, and variable transformation technique in linear regression analysis. Chapters 7 and 8 provide further discussions on the generalizations of the ordinary least squares estimation in linear regression. In these two chapters we discuss how to extend the regression model to situation where the equal variance assumption on the error term fails. To model the regression data with unequal variance the generalized least squares method is introduced. In Chapter 7, two shrinkage estimators, the ridge regression and the LASSO are introduced and discussed. A brief discussion on the least squares method for nonlinear regression is also included. Chapter 8 briefly introduces the generalized linear models. In particular, the Poisson Regression for count data and the logistic regression for binary data are discussed. Chapter 9 briefly discussed the Bayesian linear regression models. The Bayes averaging method is introduced and discussed.

The purpose of including these topics in the book is to foster a better understanding of regression modeling. Although these topics and techniques are presented largely for regression, the ideas behind these topics and the techniques are also applicable in other areas of statistical modeling. The topics presented in the book cover fundamental theories in linear regression analysis and we think that they are the most useful and relevant to the future research into this area. A thorough understanding of the basic theories, model diagnosis, and computing techniques in linear regression analysis is necessary for those who would like to learn statistics either as a discipline or as a substantial tool in their subject field. To this end, we provide detailed proofs of fundamental theories related to linear regression modeling, diagnosis, and computing so that readers can understand the methods in regression analysis and actually model the data using the methods presented in the book.

To enable the book serves the intended purpose as a graduate textbook for regression analysis, in addition to detailed proofs, we also include many examples to illustrate relevant computing techniques in regression analysis and diagnosis. We hope that this would increase the readability and help to understand the regression methods for students who expect a thorough understanding of regression methods and know how to use these methods to solve for practical problems. In addition, we tried to avoid an oversized-textbook so that it can be taught in one semester. We do not intend to write a complete reference book for regression analysis because it will require a significantly larger volume of the book and may not be suitable for a textbook of regression course. In our practice we realize that graduate students often feel overwhelming when try to read an oversized textbook. Therefore, we focus on presenting fundamental theories and detailed derivations that can highlight the most important methods and techniques in linear regression analysis.

Most computational examples of regression analysis and diagnosis in the book use one of popular software package the Statistical Analysis System (SAS), although readers are not discouraged to use other statistical software packages in their subject area. Including illustrative SAS programs for the regression analysis and diagnosis in the book is to help readers to become familiar with various computing techniques that are necessary to regression analysis. In addition, the SAS Output Delivery System (ODS) is introduced to enable readers to generate output tables and figures in a desired format. These illustrative programs are often arranged in the end of each chapter with brief explanations. In addition to the SAS, we also briefly introduce the software R which is a freeware. R has many user-written functions for implementation of various statistical methods including regression. These functions are similar to the built-in functions in the commercial software package S-PLUS. We provide some programs in R to produce desired regression diagnosis graphs. Readers are encouraged to learn how to use the software R to perform regression analysis, diagnosis, and producing graphs.

X. Yan and X. G. Su

Contents

Preface v

List of Figures xv

List of Tables xvii

1. Introduction 1

1.1 Regression Model 1
1.2 Goals of Regression Analysis 4
1.3 Statistical Computing in Regression Analysis 5

2. Simple Linear Regression 9

2.1 Introduction 9
2.2 Least Squares Estimation 10
2.3 Statistical Properties of the Least Squares Estimation . . 13
2.4 Maximum Likelihood Estimation 18
2.5 Confidence Interval on Regression Mean and Regression Prediction 19
2.6 Statistical Inference on Regression Parameters 21
2.7 Residual Analysis and Model Diagnosis 25
2.8 Example 28

3. Multiple Linear Regression 41

3.1 Vector Space and Projection 41
3.1.1 Vector Space 41
3.1.2 Linearly Independent Vectors 44
3.1.3 Dot Product and Projection 44

3.2 Matrix Form of Multiple Linear Regression 48
3.3 Quadratic Form of Random Variables 49
3.4 Idempotent Matrices 50
3.5 Multivariate Normal Distribution 54
3.6 Quadratic Form of the Multivariate Normal Variables . . 56
3.7 Least Squares Estimates of the Multiple Regression Parameters 58
3.8 Matrix Form of the Simple Linear Regression 62
3.9 Test for Full Model and Reduced Model 64
3.10 Test for General Linear Hypothesis 66
3.11 The Least Squares Estimates of Multiple Regression Parameters Under Linear Restrictions 67
3.12 Confidence Intervals of Mean and Prediction in Multiple Regression 69
3.13 Simultaneous Test for Regression Parameters 70
3.14 Bonferroni Confidence Region for Regression Parameters . 71
3.15 Interaction and Confounding 72
3.15.1 Interaction 73
3.15.2 Confounding 75
3.16 Regression with Dummy Variables 77
3.17 Collinearity in Multiple Linear Regression 81
3.17.1 Collinearity 81
3.17.2 Variance Inflation 85
3.18 Linear Model in Centered Form 87
3.19 Numerical Computation of LSE via QR Decomposition . 92
3.19.1 Orthogonalization 92
3.19.2 QR Decomposition and LSE 94
3.20 Analysis of Regression Residual 96
3.20.1 Purpose of the Residual Analysis 96
3.20.2 Residual Plot 97
3.20.3 Studentized Residuals 103
3.20.4 PRESS Residual 103
3.20.5 Identify Outlier Using PRESS Residual 106
3.20.6 Test for Mean Shift Outlier 108
3.21 Check for Normality of the Error Term in Multiple Regression 115
3.22 Example 115

4. Detection of Outliers and Influential Observations in Multiple Linear Regression 129

4.1 Model Diagnosis for Multiple Linear Regression 130
4.1.1 Simple Criteria for Model Comparison 130
4.1.2 Bias in Error Estimate from Under-specified Model 131
4.1.3 Cross Validation 132
4.2 Detection of Outliers in Multiple Linear Regression 133
4.3 Detection of Influential Observations in Multiple Linear Regression 134
4.3.1 Influential Observation 134
4.3.2 Notes on Outlier and Influential Observation . . . 136
4.3.3 Residual Mean Square Error for Over-fitted Regression Model 137
4.4 Test for Mean-shift Outliers 139
4.5 Graphical Display of Regression Diagnosis 142
4.5.1 Partial Residual Plot 142
4.5.2 Component-plus-residual Plot 146
4.5.3 Augmented Partial Residual Plot 147
4.6 Test for Inferential Observations 147
4.7 Example 150

5. Model Selection 157

5.1 Effect of Underfitting and Overfitting 157
5.2 All Possible Regressions 165
5.2.1 Some Naive Criteria 165
5.2.2 PRESS and GCV 166
5.2.3 Mallow's C_P 167
5.2.4 AIC, AIC_C, and BIC 169
5.3 Stepwise Selection 171
5.3.1 Backward Elimination 171
5.3.2 Forward Addition 172
5.3.3 Stepwise Search 172
5.4 Examples 173
5.5 Other Related Issues 179
5.5.1 Variance Importance or Relevance 180
5.5.2 PCA and SIR 186

6. Model Diagnostics 195
6.1 Test Heteroscedasticity 197
6.1.1 Heteroscedasticity 197
6.1.2 Likelihood Ratio Test, Wald, and Lagrange Multiplier Test 198
6.1.3 Tests for Heteroscedasticity 201
6.2 Detection of Regression Functional Form 204
6.2.1 Box-Cox Power Transformation 205
6.2.2 Additive Models 207
6.2.3 ACE and AVAS 210
6.2.4 Example 211

7. Extensions of Least Squares 219
7.1 Non-Full-Rank Linear Regression Models 219
7.1.1 Generalized Inverse 221
7.1.2 Statistical Inference on Null-Full-Rank Regression Models 223
7.2 Generalized Least Squares 229
7.2.1 Estimation of (β, σ^2) 230
7.2.2 Statistical Inference 231
7.2.3 Misspecification of the Error Variance Structure . 232
7.2.4 Typical Error Variance Structures 233
7.2.5 Example 236
7.3 Ridge Regression and LASSO 238
7.3.1 Ridge Shrinkage Estimator 239
7.3.2 Connection with PCA 243
7.3.3 LASSO and Other Extensions 246
7.3.4 Example 250
7.4 Parametric Nonlinear Regression 259
7.4.1 Least Squares Estimation in Nonlinear Regression 261
7.4.2 Example 263

8. Generalized Linear Models 269
8.1 Introduction: A Motivating Example 269
8.2 Components of GLM 272
8.2.1 Exponential Family 272
8.2.2 Linear Predictor and Link Functions 273
8.3 Maximum Likelihood Estimation of GLM 274

8.3.1 Likelihood Equations 274
8.3.2 Fisher's Information Matrix 275
8.3.3 Optimization of the Likelihood 276
8.4 Statistical Inference and Other Issues in GLM 278
8.4.1 Wald, Likelihood Ratio, and Score Test 278
8.4.2 Other Model Fitting Issues 281
8.5 Logistic Regression for Binary Data 282
8.5.1 Interpreting the Logistic Model 282
8.5.2 Estimation of the Logistic Model 284
8.5.3 Example 285
8.6 Poisson Regression for Count Data 287
8.6.1 The Loglinear Model 287
8.6.2 Example 288

9. Bayesian Linear Regression 297

9.1 Bayesian Linear Models 297
9.1.1 Bayesian Inference in General 297
9.1.2 Conjugate Normal-Gamma Priors 299
9.1.3 Inference in Bayesian Linear Model 302
9.1.4 Bayesian Inference via MCMC 303
9.1.5 Prediction 306
9.1.6 Example 307
9.2 Bayesian Model Averaging 309

Bibliography 317

Index 325

List of Figures

2.1 (a) Regression Line and Scatter Plot. (b) Residual Plot, (c) 95% Confidence Band for Regression Mean. (d) 95% Confidence Band for Regression Prediction. 33
2.2 Q-Q Plot for Regression Model density=$\beta_0 + \beta_1$ stiffness $+ \varepsilon$. 34

3.1 Response Curves of Y Versus X_1 at Different Values of X_2 in Models (3.33) and (3.32). 74
3.2 Regression on Dummy Variables 82
3.3 Various Shapes of Residual Plots 98
3.4 Residual Plots of the Regression Model of the US Population . 100

4.1 Partial Residual Plots for Regression Models 145

5.1 (a) Plot of the Average and Standard Deviation of SSPE Prediction from 100 Simulation Runs; (b) A Hypothetical Plot of Prediction Error Versus Model Complexity in General Regression Problems. 163
5.2 Variable Importance for the Quasar Data: Random Forests vs. RELIEF. 181
5.3 A Simulation Example on Variable Importance: Random Forests vs. RELIEF. 185

6.1 Box-Cox Power Transformation of Y in the Navy Manpower Data: Plot of the Log-likelihood Score vs. λ. 212
6.2 Detecting the Nonlinear Effect of x_1 and x_2 via Additive Models in the Navy Manpower Data. 213
6.3 AVAS Results for the Navy Manpower Data: (a) Plot of $g(Y)$; (b) Plot of $f_1(X_1)$; (c) Plot of $f_2(X_2)$. 214

6.4 AVAS Results for the Navy Manpower Data: (a) 3-D Scatter Plot of the Original Data; (b) 3-D Scatter Plot of $g(Y)$ vs. $f_1(X_1)$ and $f_2(X_2)$ after Transformation. 215

7.1 (a) The Scatter Plot of DBP Versus Age; (b) Residual versus Age; (c) Absolute Values of Residuals $|\hat{\varepsilon}_i|$ Versus Age; (d) Squared Residuals $\hat{\varepsilon}_i^2$ Versus Age. 237
7.2 Estimation in Constrained Least Squares in the Two-dimensional Case: Contours of the Residual Sum of Squares and the Constraint Functions in (a) Ridge Regression; (b) LASSO. 240
7.3 Paired Scatter Plots for Longley's (1967) Economic Data. . . . 251
7.4 Ridge Regression for Longley's (1967) Economic Data: (a) Plot of Parameter Estimates Versus s; (b) Plot of GCV Values Versus s. 254
7.5 LASSO Procedure for Longley's (1967) Economic Data: (a) Plot of Standardized Parameter Estimates Versus Fraction of the L_1 Shrinkage; (b) Plot of Cross-validated SSE Versus Fraction. . . 257
7.6 Plot of the Logistic Function With $C = 1$: (a) When B Varies With $A = 1$; (b) When A Varies With $B = 1$. 260
7.7 Parametric Nonlinear Regression with the US Population Growth Data Reproduced from Fox (2002): (a) Scatter Plot of the Data Superimposed by the LS Fitted Straight Line; (b) Plot of the Residuals From the Linear Model Versus Year; (c) Scatter Plot of the Data Superimposed by the Fitted Nonlinear Curve; (d) Plot of the Residuals From the Nonlinear Fit Versus Year. 264

8.1 (a) Scatterplot of the CHD Data, Superimposed With the Straight Line from Least Squares Fit; (b) Plot of the Percentage of Subjects With CHD Within Each Age Group, Superimposed by LOWESS Smoothed Curve. 270
8.2 Plot of the Log-likelihood Function: Comparison of Wald, LRT, and Score Tests on $H_0 : \beta = b$. 280

9.1 Posterior Marginal Densities of $(\boldsymbol{\beta}, \sigma^2)$: Bayesian Linear Regression for Shingle Data. 308
9.2 Posterior Densities for Predicting y_p: the Shingle Data. 309
9.3 The 13 Selected Models by BMA: the Quasar Data. 313
9.4 Posterior Distributions of the Coefficients Generated by BMA: the Quasar Data. 314

List of Tables

1.1 Smoking and Mortality Data 1

2.1 Parent's Height and Children's Height 9
2.2 Degrees of Freedom in Partition of Total Variance 22
2.3 Distributions of Partition of Total Variance 23
2.4 ANOVA Table 1 23
2.5 Confidence Intervals on Parameter Estimates 29
2.6 ANOVA Table 2 30
2.7 Regression Table 30
2.8 Parameter Estimates of Simple Linear Regression 30
2.9 Table for Fitted Values and Residuals 31
2.10 Data for Two Parallel Regression Lines 36
2.11 Chemical Reaction Data 38
2.12 Weight-lifting Test Data 39

3.1 Two Independent Vectors 83
3.2 Two Highly Correlated Vectors 83
3.3 Correlation Matrix for Variables $x_1, x_2, \cdots, x_5$ 88
3.4 Parameter Estimates and Variance Inflation 88
3.5 Correlation Matrix after Deleting Variable x_1 88
3.6 Variance Inflation after Deleting x_1 88
3.7 United States Population Data (in Millions) 97
3.8 Parameter Estimates for Model Population=Year 99
3.9 Parameter Estimates for Model Population=Year+Year^2 99
3.10 Parameter Estimates for Regression Model Population=β_0 +β_1 Year+ β_2 Year^2+z 101
3.11 Coal-cleansing Data 109

3.12 Parameter Estimates for Regression Model for Coal-Cleansing Data . . . 109
3.13 Residuals . . . 110
3.14 Manpower Data . . . 111
3.15 Simultaneous Outlier Detection . . . 113
3.16 Detection of Multiple Mean Shift Outliers . . . 114
3.17 Stand Characteristics of Pine Tree Data . . . 116
3.18 Parameter Estimates and Confidence Intervals Using x_1, x_2 and x_3 . . . 117
3.19 Parameter Estimates and Confidence Intervals after Deleting x_3 117
3.20 Confidence Intervals on Regression Mean and Prediction Without Deletion . . . 118
3.21 Confidence Intervals on Regression Mean and Prediction After Deleting x_3 . . . 118
3.22 Regression Model for Centralized Data . . . 119
3.23 Test for Equal Slope Among 3 Groups . . . 121
3.24 Regression by Group . . . 121
3.25 Data Set for Calculation of Confidence Interval on Regression Prediction . . . 123
3.26 Propellant Grain Data . . . 124
3.27 Data Set for Testing Linear Hypothesis . . . 124
3.28 Algae Data . . . 126

4.1 Coal-cleansing Data . . . 141
4.2 Parameter Estimates for $y = b_0 + b_1x_1 + b_2x_2 + b_3x_3$ . . . 141
4.3 Residuals, Leverage, and t_i . . . 142
4.4 Navy Manpower Data . . . 144
4.5 Residuals for Model With Regressors x_1 and x_2 . . . 146
4.6 Residuals for Model With Regressors $x_1^{\frac{1}{3}}$ and x_2 . . . 146
4.7 Sales Data for Asphalt Shingles . . . 151
4.8 Values of PRESS Statistic for Various Regression Models . . . 152
4.9 PRESS Residuals and Leverages for Models Including x_1, x_2 (press12, h12) and Including x_1, x_2, x_3 (press123, h123) . . . 153
4.10 Data Set for Multiple Mean Shift Outliers . . . 154
4.11 Fish Biomass Data . . . 155

5.1 The Quasar Data from Ex. 4.9 in Mendenhall and Sinich (2003) 174
5.2 All Possible Regression Selection for the Quasar Data. . . . 175
5.3 Backward Elimination Procedure for the Quasar Data. . . . 190

5.4 Forward Addition Procedure for the Quasar Data. 191
5.5 The Clerical Data from Ex. 6.4 in Mendenhall and Sinich (2003) 192
5.6 Backward Elimination Results for the Clerical Data. 193
5.7 Results of a Simulated Example for Inspecting Overall Significance in Stepwise Selection 193
5.8 Female Teachers Effectiveness Data 194
5.9 All-subset-regression for Female Teacher Effectiveness Data . . 194

6.1 Results of Efficiency Tests 216

7.1 Layouts for Data Collected in One-way ANOVA Experiments. 220
7.2 Analysis of the Diastolic Blood Pressure Data: (a) Ordinary Least Squares (OLS) Estimates; (b) Weighted Least Squares (WLS) Fitting with Weights Derived from Regressing Absolute Values $|\hat{\varepsilon}_i|$ of the Residuals On Age. 238
7.3 The Macroeconomic Data Set in Longley (1967). 249
7.4 Variable Description for Longley's (1967) Macroeconomic Data. 250
7.5 Least Squares Results for Longley's (1967) Economic Data. . . 252
7.6 Ridge Regression and LASSO for Longley's (1967) Data. 255
7.7 Nonlinear Regression Results of the US Population Data from PROC NLIN and PROC AUTOREG. 265

8.1 Frequency Table of AGE Group by CHD. 271
8.2 Variable Description for the Kyphosis Data: Logistic Regression Example. 285
8.3 Analysis Results for the Kyphosis Data from PROC LOGISTIC. 286
8.4 Variable Description for the Log-Linear Regression Example. . 288
8.5 Analysis Results for the School Absence Data from PROC GENMOD. 290
8.6 Variable Description for the Prostate Cancer Data. 293
8.7 Table of Parameter Estimates and the Estimated Covariance Matrix for Model I: the Prostate Cancer Data. 294
8.8 Table of Parameter Estimates and the Estimated Covariance Matrix for Model II: the Prostate Cancer Data. 295
8.9 The Maximized Loglikelihood Scores for Several Fitted Models with the Prostate Cancer Data. 295

9.1 Bayesian Linear Regression Results With the Shingle Data. . . 307
9.2 BMA Results for the Quasar Data. 315

Chapter 1

Introduction

1.1 Regression Model

Researchers are often interested in the relationships between one variable and several other variables. For example, does smoking cause lung cancer? Following Table 1.1 summarizes a study carried out by government statisticians in England. The data concern 25 occupational groups and are condensed from data on thousands of individual men. One variable is smoking ratio which is a measure of the number of cigarettes smoked per day by men in each occupation relative to the number smoked by all men of the same age. Another variable is the standardized mortality ratio. To answer the question that does smoking cause cancer we may like to know the relationship between the derived mortality ratio and smoking ratio. This falls into the scope of regression analysis. Data from a scientific

Table 1.1 Smoking and Mortality Data

Smoking	77	112	137	113	117	110	94	125	116	133
Mortality	84	96	116	144	123	139	128	113	155	146
Smoking	102	115	111	105	93	87	88	91	102	100
Mortality	101	128	118	115	113	79	104	85	88	120
Smoking	91	76	104	66	107					
Mortality	104	60	129	51	86					

experiment often lead to ask whether there is a causal relationship between two or more variables. Regression analysis is the statistical method for investigating such relationship. It is probably one of the oldest topics in the area of mathematical statistics dating back to about two hundred years ago. The earliest form of the linear regression was the least squares method, which was published by Legendre in 1805, and by Gauss in 1809. The term "least squares" is from Legendre's term. Legendre and Gauss both applied the method to the problem of determining, from astronomical observations, the orbits of bodies about the sun. Euler had worked on the same problem (1748) without success. Gauss published a further development of the theory of least squares in 1821, including a version of the today's well-known Gauss-Markov theorem, which is a fundamental theorem in the area of the general linear models.

What is a statistical model? A statistical model is a simple description of a state or process. "A model is neither a hypothesis nor a theory. Unlike scientific hypotheses, a model is not verifiable directly by an experiment. For all models of true or false, the validation of a model is not that it is "true" but that it generates good testable hypotheses relevant to important problems." (R. Levins, Am. Scientist 54: 421-31, 1966)

Linear regression requires that model is linear in regression parameters. Regression analysis is the method to discover the relationship between one or more response variables (also called dependent variables, explained variables, predicted variables, or regressands, usually denoted by y) and the predictors (also called independent variables, explanatory variables, control variables, or regressors, usually denoted by $x_1, x_2, \cdots, x_p$).

There are three types of regression. The first is the simple linear regression. The simple linear regression is for modeling the linear relationship between two variables. One of them is the dependent variable y and another is the independent variable x. For example, the simple linear regression can model the relationship between muscle strength (y) and lean body mass (x). The simple regression model is often written as the following form

$$y = \beta_0 + \beta_1 x + \varepsilon, \tag{1.1}$$

where y is the dependent variable, β_0 is y intercept, β_1 is the gradient or the slope of the regression line, x is the independent variable, and ε is the random error. It is usually assumed that error ε is normally distributed with $E(\varepsilon) = 0$ and a constant variance $\text{Var}(\varepsilon) = \sigma^2$ in the simple linear regression.

The second type of regression is the multiple linear regression which is a linear regression model with one dependent variable and more than one independent variables. The multiple linear regression assumes that the response variable is a linear function of the model parameters and there are more than one independent variables in the model. The general form of the multiple linear regression model is as follows:

$$y = \beta_0 + \beta_1 x_1 + \cdots + \beta_p x_p + \varepsilon, \tag{1.2}$$

where y is dependent variable, $\beta_0, \beta_1, \beta_2, \cdots, \beta_p$ are regression coefficients, and $x_1, x_2, \cdots, x_n$ are independent variables in the model. In the classical regression setting it is usually assumed that the error term ε follows the normal distribution with $E(\varepsilon) = 0$ and a constant variance $\text{Var}(\varepsilon) = \sigma^2$.

Simple linear regression is to investigate the linear relationship between one dependent variable and one independent variable, while the multiple linear regression focuses on the linear relationship between one dependent variable and more than one independent variables. The multiple linear regression involves more issues than the simple linear regression such as collinearity, variance inflation, graphical display of regression diagnosis, and detection of regression outlier and influential observation.

The third type of regression is nonlinear regression, which assumes that the relationship between dependent variable and independent variables is not linear in regression parameters. Example of nonlinear regression model (growth model) may be written as

$$y = \frac{\alpha}{1 + e^{\beta t}} + \varepsilon, \tag{1.3}$$

where y is the growth of a particular organism as a function of time t, α and β are model parameters, and ε is the random error. Nonlinear regression model is more complicated than linear regression model in terms of estimation of model parameters, model selection, model diagnosis, variable selection, outlier detection, or influential observation identification. General theory of the nonlinear regression is beyond the scope of this book and will not be discussed in detail. However, in addition to the linear regression model we will discuss some generalized linear models. In particular, we will introduce and discuss two important generalized linear models, logistic regression model for binary data and log-linear regression model for count data in Chapter 8.

1.2 Goals of Regression Analysis

Regression analysis is one of the most commonly used statistical methods in practice. Applications of regression analysis can be found in many scientific fields including medicine, biology, agriculture, economics, engineering, sociology, geology, etc. The purposes of regression analysis are three-folds:

(1) Establish a casual relationship between response variable y and regressors $x_1, x_2, \cdots, x_n$.
(2) Predict y based on a set of values of $x_1, x_2, \cdots, x_n$.
(3) Screen variables $x_1, x_2, \cdots, x_n$ to identify which variables are more important than others to explain the response variable y so that the causal relationship can be determined more efficiently and accurately.

An analyst often follows, but not limited, the following procedures in the regression analysis.

(1) The first and most important step is to understand the real-life problem which is often fallen into a specific scientific field. Carefully determine whether the scientific question falls into scope of regression analysis.
(2) Define a regression model which may be written as

$$\text{Response variable} = \text{a function of regressors} + \text{random error},$$

or simply in a mathematical format

$$y = f(x_1, x_2, \cdots, x_p) + \varepsilon.$$

You may utilize a well-accepted model in a specific scientific field or try to define your own model based upon a sound scientific judgement by yourself or expert scientists in the subject area. Defining a model is often a joint effort among statisticians and experts in a scientific discipline. Choosing an appropriate model is the first step for statistical modeling and often involve further refinement procedures.
(3) Make distributional assumptions on the random error ε in the regression model. These assumptions need to be verified or tested by the data collected from experiment. The assumptions are often the basis upon which the regression model may be solved and statistical inference is drawn.
(4) Collect data y and $x_1, x_2, \cdots, x_p$. This data collection step usually involves substantial work that includes, but not limited to, experimental design, sample size determination, database design, data cleaning,

and derivations of analysis variables that will be used in the statistical analysis. In many real-life applications, this is a crucial step that often involve significant amount of work.

(5) According to the software used in the analysis, create data sets in an appropriate format that are easy to be read into a chosen software. In addition, it often needs to create more specific analysis data sets for planned or exploratory statistical analysis.

(6) Carefully evaluate whether or not the selected model is appropriate for answering the desired scientific questions. Various diagnosis methods may be used to evaluate the performance of the selected statistical model. It should be kept in mind that the model diagnosis is for the judgment of whether the selected statistical model is a sound model that can answer the desired scientific questions.

(7) If the model is deemed to be appropriate according to a well accepted model diagnosis criteria, it may be used to answer the desired scientific questions; otherwise, the model is subject to refinement or modification. Several iterations of model selection, model diagnosis, and model refinement may be necessary and very common in practice.

1.3 Statistical Computing in Regression Analysis

After a linear regression model is chosen and a database is created, the next step is statistical computing. The purposes of the statistical computing are to solve for the actual model parameters and to conduct model diagnosis. Various user-friendly statistical softwares have been developed to make the regression analysis easier and more efficient.

Statistical Analysis System (SAS) developed by SAS Institute, Inc. is one of the popular softwares which can be used to perform regression analysis. The SAS System is an integrated system of software products that enables users to perform data entry and data management, to produce statistical graphics, to conduct wide range of statistical analyses, to retrieve data from data warehouse (extract, transform, load) platform, and to provide dynamic interface to other software, etc. One great feature of SAS is that many standard statistical methods have been integrated into various SAS procedures that enable analysts easily find desired solutions without writing source code from the original algorithms of statistical methods. The SAS "macro" language allows user to write subroutines to perform particular user-defined statistical analysis. SAS compiles and runs on UNIX platform and Windows operating system.

Software S-PLUS developed by Insightful Inc. is another one of the most popular softwares that have been used substantially by analysts in various scientific fields. This software is a rigorous computing tool covering a broad range of methods in statistics. Various built-in S-PLUS functions have been developed that enable users to perform statistical analysis and generate analysis graphics conveniently and efficiently. The S-PLUS offers a wide collection of specialized modules that provide additional functionality to the S-PLUS in areas such as: volatility forecasting, optimization, clinical trials analysis, environmental statistics, and spatial data analysis, data mining. In addition, user can write S-PLUS programs or functions using the S-language to perform statistical analysis of specific needs. The S-PLUS compiles and runs on UNIX platform and Windows operating system.

Statistical Package for Social Sciences (SPSS) is also one of the most widely used softwares for the statistical analysis in the area of social sciences. It is one of the preferred softwares used by market researchers, health researchers, survey companies, government, education researchers, among others. In addition to statistical analysis, data management (case selection, file reshaping, creating derived data) and data documentation (a metadata dictionary is stored) are features of the SPSS. Many features of SPSS are accessible via pull-down menus or can be programmed with a proprietary 4GL command syntax language. Additionally, a "macro" language can be used to write command language subroutines to facilitate special needs of user-desired statistical analysis.

Another popular software that can be used for various statistical analyses is R. R is a language and environment for statistical computing and graphics. It is a GNU project similar to the S language and environment. R can be considered as a different implementation of S language. There are some important differences, but much code written for S language runs unaltered under R. The S language is often the vehicle of choice for research in statistical methodology, and R provides an open source route to participation in that activity. One of R's strengths is the ease with which well-designed publication-quality plots can be produced, including mathematical symbols and formulae where needed. Great care has been taken over the defaults for the design choices in graphics, but user retains full control. R is available as free software under the terms of the Free Software Foundation's GNU General Public License in source code form. It compiles and runs on a wide variety of UNIX platforms and similar system Linux, as well as Windows.

Regression analysis can be performed using various softwares such as SAS, S-PLUS, R, or SPSS. In this book we choose the software SAS to illustrate the computing techniques in regression analysis and diagnosis. Extensive examples are provided in the book to enable readers to become familiar with regression analysis and diagnosis using SAS. We also provide some examples of regression graphic plots using the software R. However, readers are not discouraged to use other softwares to perform regression analysis and diagnosis.

Chapter 2

Simple Linear Regression

2.1 Introduction

The term "regression" and the methods for investigating the relationships between two variables may date back to about 100 years ago. It was first introduced by Francis Galton in 1908, the renowned British biologist, when he was engaged in the study of heredity. One of his observations was that the children of tall parents to be taller than average but not as tall as their parents. This "regression toward mediocrity" gave these statistical methods their name. The term regression and its evolution primarily describe statistical relations between variables. In particular, the simple regression is the regression method to discuss the relationship between one dependent variable (y) and one independent variable (x). The following classical data set contains the information of parent's height and children's height.

Table 2.1 Parent's Height and Children's Height

Parent	64.5	65.5	66.5	67.5	68.5	69.5	70.5	71.5	72.5
Children	65.8	66.7	67.2	67.6	68.2	68.9	69.5	69.9	72.2

The mean height is 68.44 for children and 68.5 for parents. The regression line for the data of parents and children can be described as

$$\text{child height} = 21.52 + 0.69 \text{ parent height}.$$

The simple linear regression model is typically stated in the form

$$y = \beta_0 + \beta_1 x + \varepsilon,$$

where y is the dependent variable, β_0 is the y intercept, β_1 is the slope of the simple linear regression line, x is the independent variable, and ε is the random error. The dependent variable is also called response variable, and the independent variable is called explanatory or predictor variable. An explanatory variable explains causal changes in the response variables. A more general presentation of a regression model may be written as

$$y = E(y) + \epsilon,$$

where $E(y)$ is the mathematical expectation of the response variable. When $E(y)$ is a linear combination of exploratory variables $x_1, x_2, \cdots, x_k$ the regression is the linear regression. If $k = 1$ the regression is the simple linear regression. If $E(y)$ is a nonlinear function of $x_1, x_2, \cdots, x_k$ the regression is nonlinear. The classical assumptions on error term are $E(\varepsilon) = 0$ and a constant variance $\text{Var}(\varepsilon) = \sigma^2$. The typical experiment for the simple linear regression is that we observe n pairs of data $(x_1, y_1), (x_2, y_2), \cdots, (x_n, y_n)$ from a scientific experiment, and model in terms of the n pairs of the data can be written as

$$y_i = \beta_0 + \beta_1 x_i + \varepsilon_i \quad \text{for } i = 1, 2, \cdots, n,$$

with $E(\varepsilon_i) = 0$, a constant variance $\text{Var}(\varepsilon_i) = \sigma^2$, and all ε_i's are independent. Note that the actual value of σ^2 is usually unknown. The values of x_i's are measured "exactly", with no measurement error involved. After model is specified and data are collected, the next step is to find "good" estimates of β_0 and β_1 for the simple linear regression model that can best describe the data came from a scientific experiment. We will derive these estimates and discuss their statistical properties in the next section.

2.2 Least Squares Estimation

The least squares principle for the simple linear regression model is to find the estimates b_0 and b_1 such that the sum of the squared distance from actual response y_i and predicted response $\hat{y}_i = \beta_0 + \beta_1 x_i$ reaches the minimum among all possible choices of regression coefficients β_0 and β_1. i.e.,

$$(b_0, b_1) = \arg \min_{(\beta_0, \beta_1)} \sum_{i=1}^{n} [y_i - (\beta_0 + \beta_1 x_i)]^2.$$

The motivation behind the least squares method is to find parameter estimates by choosing the regression line that is the most "closest" line to all data points (x_i, y_i). Mathematically, the least squares estimates of the simple linear regression are given by solving the following system:

$$\frac{\partial}{\partial \beta_0} \sum_{i=1}^{n} [y_i - (\beta_0 + \beta_1 x_i)]^2 = 0 \tag{2.1}$$

$$\frac{\partial}{\partial \beta_1} \sum_{i=1}^{n} [y_i - (\beta_0 + \beta_1 x_i)]^2 = 0 \tag{2.2}$$

Suppose that b_0 and b_1 are the solutions of the above system, we can describe the relationship between x and y by the regression line $\hat{y} = b_0 + b_1 x$ which is called the fitted regression line by convention. It is more convenient to solve for b_0 and b_1 using the centralized linear model:

$$y_i = \beta_0^* + \beta_1 (x_i - \bar{x}) + \varepsilon_i,$$

where $\beta_0 = \beta_0^* - \beta_1 \bar{x}$. We need to solve for

$$\frac{\partial}{\partial \beta_0^*} \sum_{i=1}^{n} [y_i - (\beta_0^* + \beta_1 (x_i - \bar{x}))]^2 = 0$$

$$\frac{\partial}{\partial \beta_1} \sum_{i=1}^{n} [y_i - (\beta_0^* + \beta_1 (x_i - \bar{x}))]^2 = 0$$

Taking the partial derivatives with respect to β_0 and β_1 we have

$$\sum_{i=1}^{n} [y_i - (\beta_0^* + \beta_1 (x_i - \bar{x}))] = 0$$

$$\sum_{i=1}^{n} [y_i - (\beta_0^* + \beta_1 (x_i - \bar{x}))](x_i - \bar{x}) = 0$$

Note that

$$\sum_{i=1}^{n} y_i = n\beta_0^* + \sum_{i=1}^{n} \beta_1 (x_i - \bar{x}) = n\beta_0^* \tag{2.3}$$

Therefore, we have $\beta_0^* = \frac{1}{n} \sum_{i=1}^{n} y_i = \bar{y}$. Substituting β_0^* by $\bar{y}$ in (2.3) we obtain

$$\sum_{i=1}^{n} [y_i - (\bar{y} + \beta_1 (x_i - \bar{x}))](x_i - \bar{x}) = 0.$$

Denote b_0 and b_1 be the solutions of the system (2.1) and (2.2). Now it is easy to see

$$b_1 = \frac{\sum_{i=1}^n (y_i - \bar{y})(x_i - \bar{x})}{\sum_{i=1}^n (x_i - \bar{x})^2} = \frac{S_{xy}}{S_{xx}} \tag{2.4}$$

and

$$b_0 = b_0^* - b_1\bar{x} = \bar{y} - b_1\bar{x} \tag{2.5}$$

The fitted value of the simple linear regression is defined as $\hat{y}_i = b_0 + b_1 x_i$. The difference between y_i and the fitted value $\hat{y}_i$, $e_i = y_i - \hat{y}_i$, is referred to as the regression residual. Regression residuals play an important role in the regression diagnosis on which we will have extensive discussions later. Regression residuals can be computed from the observed responses y_i's and the fitted values $\hat{y}_i$'s, therefore, residuals are observable. It should be noted that the error term ε_i in the regression model is unobservable. Thus, regression error is unobservable and regression residual is observable. Regression error is the amount by which an observation differs from its expected value; the latter is based on the whole population from which the statistical unit was chosen randomly. The expected value, the average of the entire population, is typically unobservable.

Example 2.1. If the average height of 21-year-old male is 5 feet 9 inches, and one randomly chosen male is 5 feet 11 inches tall, then the "error" is 2 inches; if the randomly chosen man is 5 feet 7 inches tall, then the "error" is -2 inches. It is as if the measurement of man's height was an attempt to measure the population average, so that any difference between man's height and average would be a measurement error.

A residual, on the other hand, is an observable estimate of unobservable error. The simplest case involves a random sample of n men whose heights are measured. The sample average is used as an estimate of the population average. Then the difference between the height of each man in the sample and the unobservable population average is an error, and the difference between the height of each man in the sample and the observable sample average is a residual. Since residuals are observable we can use residual to estimate the unobservable model error. The detailed discussion will be provided later.

2.3 Statistical Properties of the Least Squares Estimation

In this section we discuss the statistical properties of the least squares estimates for the simple linear regression. We first discuss statistical properties without the distributional assumption on the error term, but we shall assume that $E(\epsilon_i) = 0$, $\text{Var}(\epsilon_i) = \sigma^2$, and ϵ_i's for $i = 1, 2, \cdots, n$ are independent.

Theorem 2.1. *The least squares estimator b_0 is an unbiased estimate of β_0.*

Proof.

$$Eb_0 = E(\bar{y} - b_1\bar{x}) = E\Big(\frac{1}{n}\sum_{i=1}^{n} y_i\Big) - Eb_1\bar{x} = \frac{1}{n}\sum_{i=1}^{n} Ey_i - \bar{x}Eb_1$$

$$= \frac{1}{n}\sum_{i=1}^{n}(\beta_0 + \beta_1 x_i) - \beta_1\bar{x} = \frac{1}{n}\sum_{i=1}^{n}\beta_0 + \beta_1\frac{1}{n}\sum_{i=1}^{n} x_i - \beta_1\bar{x} = \beta_0.$$

□

Theorem 2.2. *The least squares estimator b_1 is an unbiased estimate of β_1.*

Proof.

$$\begin{aligned}
E(b_1) &= E\Big(\frac{S_{xy}}{S_{xx}}\Big) \\
&= \frac{1}{S_{xx}} E\frac{1}{n}\sum_{i=1}^{n}(y_i - \bar{y})(x_i - \bar{x}) \\
&= \frac{1}{S_{xx}}\frac{1}{n}\sum_{i=1}^{n}(x_i - \bar{x})Ey_i \\
&= \frac{1}{S_{xx}}\frac{1}{n}\sum_{i=1}^{n}(x_i - \bar{x})(\beta_0 + \beta_1 x_i) \\
&= \frac{1}{S_{xx}}\frac{1}{n}\sum_{i=1}^{n}(x_i - \bar{x})\beta_1 x_i \\
&= \frac{1}{S_{xx}}\frac{1}{n}\sum_{i=1}^{n}(x_i - \bar{x})\beta_1(x_i - \bar{x}) \\
&= \frac{1}{S_{xx}}\frac{1}{n}\sum_{i=1}^{n}(x_i - \bar{x})^2\beta_1 = \frac{S_{xx}}{S_{xx}}\beta_1 = \beta_1
\end{aligned}$$

□

Theorem 2.3. $Var(b_1) = \dfrac{\sigma^2}{nS_{xx}}$.

Proof.

$$
\begin{aligned}
\text{Var}(b_1) &= \text{Var}\Big(\frac{S_{xy}}{S_{xx}}\Big)\\
&= \frac{1}{S_{xx}^2}\text{Var}\Big(\frac{1}{n}\sum_{i=1}^{n}(y_i-\bar{y})(x_i-\bar{x})\Big)\\
&= \frac{1}{S_{xx}^2}\text{Var}\Big(\frac{1}{n}\sum_{i=1}^{n}y_i(x_i-\bar{x})\Big)\\
&= \frac{1}{S_{xx}^2}\frac{1}{n^2}\sum_{i=1}^{n}(x_i-\bar{x})^2\text{Var}(y_i)\\
&= \frac{1}{S_{xx}^2}\frac{1}{n^2}\sum_{i=1}^{n}(x_i-\bar{x})^2\sigma^2 = \frac{\sigma^2}{nS_{xx}}
\end{aligned}
$$

□

Theorem 2.4. *The least squares estimator b_1 and $\bar{y}$ are uncorrelated. Under the normality assumption of y_i for $i = 1, 2, \cdots, n$, b_1 and $\bar{y}$ are normally distributed and independent.*

Proof.

$$
\begin{aligned}
\text{Cov}(b_1,\bar{y}) &= \text{Cov}(\frac{S_{xy}}{S_{xx}},\ \bar{y})\\
&= \frac{1}{S_{xx}}\ \text{Cov}(S_{xy},\ \bar{y})\\
&= \frac{1}{nS_{xx}}\ \text{Cov}\Big(\sum_{i=1}^{n}(x_i-\bar{x})(y_i-\bar{y}),\ \bar{y}\Big)\\
&= \frac{1}{nS_{xx}}\ \text{Cov}\Big(\sum_{i=1}^{n}(x_i-\bar{x})y_i,\ \bar{y}\Big)\\
&= \frac{1}{n^2S_{xx}}\ \text{Cov}\Big(\sum_{i=1}^{n}(x_i-\bar{x})y_i,\ \sum_{i=1}^{n}y_i\Big)\\
&= \frac{1}{n^2S_{xx}}\ \sum_{i,j=1}^{n}(x_i-\bar{x})\ \text{Cov}(y_i,y_j)
\end{aligned}
$$

Note that $E\varepsilon_i = 0$ and ε_i's are independent we can write

$$
\text{Cov}(y_i,y_j) = E[\ (y_i-Ey_i)(y_j-Ey_j)\ ] = E(\varepsilon_i,\ \varepsilon_j) = \begin{cases}\sigma^2, & \text{if } i=j\\ 0, & \text{if } i\neq j\end{cases}
$$

Thus, we conclude that

$$\text{Cov}(b_1, \bar{y}) = \frac{1}{n^2 S_{xx}} \sum_{i=1}^{n} (x_i - \bar{x})\sigma^2 = 0.$$

Recall that zero correlation is equivalent to the independence between two normal variables. Thus, we conclude that b_0 and $\bar{y}$ are independent. □

Theorem 2.5. *Var*$(b_0) = \Big(\frac{1}{n} + \frac{\bar{x}^2}{nS_{xx}}\Big)\sigma^2$.

Proof.

$$\begin{aligned}\text{Var}(b_0) &= \text{Var}(\bar{y} - b_1\bar{x})\\ &= \text{Var}(\bar{y}) + (\bar{x})^2 \text{Var}(b_1)\\ &= \frac{\sigma^2}{n} + \bar{x}^2 \frac{\sigma^2}{nS_{xx}}\\ &= \Big(\frac{1}{n} + \frac{\bar{x}^2}{nS_{xx}}\Big)\sigma^2\end{aligned}$$

□

The properties $1 - 5$, especially the variances of b_0 and b_1, are important when we would like to draw statistical inference on the intercept and slope of the simple linear regression.

Since the variances of least squares estimators b_0 and b_1 involve the variance of the error term in the simple regression model. This error variance is unknown to us. Therefore, we need to estimate it. Now we discuss how to estimate the variance of the error term in the simple linear regression model. Let y_i be the observed response variable, and $\hat{y}_i = b_0 + b_1 x_i$, the fitted value of the response. Both y_i and $\hat{y}_i$ are available to us. The true error σ_i in the model is not observable and we would like to estimate it. The quantity $y_i - \hat{y}_i$ is the empirical version of the error ε_i. This difference is regression residual which plays an important role in regression model diagnosis. We propose the following estimation of the error variance based on e_i:

$$s^2 = \frac{1}{n-2} \sum_{i=1}^{n} (y_i - \hat{y}_i)^2$$

Note that in the denominator is $n-2$. This makes s^2 an unbiased estimator of the error variance σ^2. The simple linear model has two parameters, therefore, $n - 2$ can be viewed as $n-$ number of parameters in simple linear regression model. We will see in later chapters that it is true for all general linear models. In particular, in a multiple linear regression model with p parameters the denominator should be $n-p$ in order to construct an unbiased estimator of the error variance σ^2. Detailed discussion can be found in later chapters. The unbiasness of estimator s^2 for the simple linear regression can be shown in the following derivations.

$$y_i - \hat{y}_i = y_i - b_0 - b_1 x_i = y_i - (\bar{y} - b_1\bar{x}) - b_1 x_i = (y_i - \bar{y}) - b_1(x_i - \bar{x})$$

It follows that

$$\sum_{i=1}^{n}(y_i - \hat{y}_i) = \sum_{i=1}^{n}(y_i - \bar{y}) - b_1\sum_{i=1}^{n}(x_i - \bar{x}) = 0.$$

Note that $(y_i - \hat{y}_i)x_i = [(y_i - \bar{y}) - b_1(x_i - \bar{x})]x_i$, hence we have

$$\sum_{i=1}^{n}(y_i - \hat{y}_i)x_i = \sum_{i=1}^{n}[(y_i - \bar{y}) - b_1(x_i - \bar{x})]x_i$$

$$= \sum_{i=1}^{n}[(y_i - \bar{y}) - b_1(x_i - \bar{x})](x_i - \bar{x})$$

$$= \sum_{i=1}^{n}(y_i - \bar{y})(x_i - \bar{x}) - b_1\sum_{i=1}^{n}(x_i - \bar{x})^2$$

$$= n(S_{xy} - b_1 S_{xx}) = n\Big(S_{xy} - \frac{S_{xy}}{S_{xx}}S_{xx}\Big) = 0$$

To show that s^2 is an unbiased estimate of the error variance, first we note that

$$(y_i - \hat{y}_i)^2 = [(y_i - \bar{y}) - b_1(x_i - \bar{x})]^2,$$

therefore,

$$\sum_{i=1}^{n}(y_i - \hat{y}_i)^2 = \sum_{i=1}^{n}[(y_i - \bar{y}) - b_1(x_i - \bar{x})]^2$$

$$= \sum_{i=1}^{n}(y_i - \bar{y})^2 - 2b_1\sum_{i=1}^{n}(x_i - \bar{x})(y_i - \bar{y}_i) + b_1^2\sum_{i=1}^{n}(x_i - \bar{x})^2$$

$$= \sum_{i=1}^{n}(y_i - \bar{y})^2 - 2nb_1 S_{xy} + nb_1^2 S_{xx}$$

$$= \sum_{i=1}^{n}(y_i - \bar{y})^2 - 2n\frac{S_{xy}}{S_{xx}}S_{xy} + n\frac{S_{xy}^2}{S_{xx}^2}S_{xx}$$

$$= \sum_{i=1}^{n}(y_i - \bar{y})^2 - n\frac{S_{xy}^2}{S_{xx}}$$

Since

$$(y_i - \bar{y})^2 = [\beta_1(x_i - \bar{x}) + (\varepsilon_i - \bar{\varepsilon})]^2$$

and

$$(y_i - \bar{y})^2 = \beta_1^2(x_i - \bar{x})^2 + (\varepsilon_i - \bar{\varepsilon})^2 + 2\beta_1(x_i - \bar{x})(\varepsilon_i - \bar{\varepsilon}),$$

therefore,

$$E(y_i - \bar{y})^2 = \beta_1^2(x_i - \bar{x})^2 + E(\varepsilon_i - \bar{\varepsilon})^2 = \beta_1^2(x_i - \bar{x})^2 + \frac{n-1}{n}\sigma^2,$$

and

$$\sum_{i=1}^{n} E(y_i - \bar{y})^2 = n\beta_1^2 S_{xx} + \sum_{i=1}^{n} \frac{n-1}{n}\sigma^2 = n\beta_1^2 S_{xx} + (n-1)\sigma^2.$$

Furthermore, we have

$$\begin{aligned} E(S_{xy}) &= E\Big(\frac{1}{n}\sum_{i=1}^{n}(x_i - \bar{x})(y_i - \bar{y})\Big) \\ &= \frac{1}{n}E\sum_{i=1}^{n}(x_i - \bar{x})y_i \\ &= \frac{1}{n}\sum_{i=1}^{n}(x_i - \bar{x})Ey_i \\ &= \frac{1}{n}\sum_{i=1}^{n}(x_i - \bar{x})(\beta_0 + \beta_1 x_i) \\ &= \frac{1}{n}\beta_1\sum_{i=1}^{n}(x_i - \bar{x})x_i \\ &= \frac{1}{n}\beta_1\sum_{i=1}^{n}(x_i - \bar{x})^2 = \beta_1 S_{xx} \end{aligned}$$

and

$$\mathrm{Var}\Big(S_{xy}\Big) = \mathrm{Var}\Big(\frac{1}{n}\sum_{i=1}^{n}(x_i - \bar{x})y_i\Big) = \frac{1}{n^2}\sum_{i=1}^{n}(x_i - \bar{x})^2\mathrm{Var}(y_i) = \frac{1}{n}S_{xx}\sigma^2$$

Thus, we can write

$$E(S_{xy}^2) = \mathrm{Var}(S_{xy}) + [E(S_{xy})]^2 = \frac{1}{n}S_{xx}\sigma^2 + \beta_1^2 S_{xx}^2$$

and

$$E\Big(\frac{nS_{xy}^2}{S_{xx}}\Big) = \sigma^2 + n\beta_1^2 S_{xx}.$$

Finally, $E(\hat{\sigma}^2)$ is given by:

$$E\sum_{i=1}^{n}(y_i - \hat{y})^2 = n\beta_1^2 S_{xx} + (n-1)\sigma^2 - n\beta_1^2 S_{xx} - \sigma^2 = (n-2)\sigma^2.$$

In other words, we prove that

$$E(s^2) = E\left(\frac{1}{n-2}\sum_{i=1}^{n}(y_i - \hat{y})^2\right) = \sigma^2.$$

Thus, s^2, the estimation of the error variance, is an unbiased estimator of the error variance σ^2 in the simple linear regression. Another view of choosing $n-2$ is that in the simple linear regression model there are n observations and two restrictions on these observations:

(1) $\sum_{i=1}^{n}(y_i - \hat{y}) = 0,$

(2) $\sum_{i=1}^{n}(y_i - \hat{y})x_i = 0.$

Hence the error variance estimation has $n-2$ degrees of freedom which is also the number of total observations − total number of the parameters in the model. We will see similar feature in the multiple linear regression.

2.4 Maximum Likelihood Estimation

The maximum likelihood estimates of the simple linear regression can be developed if we assume that the dependent variable y_i has a normal distribution: $y_i \sim N(\beta_0 + \beta_1 x_i, \sigma^2)$. The likelihood function for $(y_1, y_2, \cdots, y_n)$ is given by

$$\mathrm{L} = \prod_{i=1}^{n} f(y_i) = \frac{1}{(2\pi)^{n/2}\sigma^n} e^{(-1/2\sigma^2)\sum_{i=1}^{n}(y_i - \beta_0 - \beta_1 x_i)^2}.$$

The estimators of β_0 and β_1 that maximize the likelihood function L are equivalent to the estimators that minimize the exponential part of the likelihood function, which yields the same estimators as the least squares estimators of the linear regression. Thus, under the normality assumption of the error term the MLEs of β_0 and β_1 and the least squares estimators of β_0 and β_1 are exactly the same.

After we obtain b_1 and b_0, the MLEs of the parameters β_0 and b_1, we can compute the fitted value $\hat{y}_i$, and the likelihood function in terms of the fitted values.

$$\mathrm{L} = \prod_{i=1}^{n} f(y_i) = \frac{1}{(2\pi)^{n/2}\sigma^n} e^{(-1/2\sigma^2)\sum_{i=1}^{n}(y_i-\hat{y}_i)^2}$$

We then take the partial derivative with respect to σ^2 in the log likelihood function $\log(L)$ and set it to zero:

$$\frac{\partial \log(L)}{\partial \sigma^2} = -\frac{n}{2\sigma^2} + \frac{1}{2\sigma^4}\sum_{i=1}^{n}(y_i - \hat{y}_i)^2 = 0$$

The MLE of σ^2 is $\hat{\sigma}^2 = \frac{1}{n}\sum_{i=1}^{n}(y_i - \hat{y}_i)^2$. Note that it is a biased estimate of σ^2, since we know that $s^2 = \frac{1}{n-2}\sum_{i=1}^{n}(y_i - \hat{y}_i)^2$ is an unbiased estimate of the error variance σ^2. $\frac{n}{n-2}\hat{\sigma}^2$ is an unbiased estimate of σ^2. Note also that the $\hat{\sigma}^2$ is an asymptotically unbiased estimate of σ^2, which coincides with the classical theory of MLE.

2.5 Confidence Interval on Regression Mean and Regression Prediction

Regression models are often constructed based on certain conditions that must be verified for the model to fit the data well, and to be able to predict the response for a given regressor as accurate as possible. One of the main objectives of regression analysis is to use the fitted regression model to make prediction. Regression prediction is the calculated response value from the fitted regression model at data point which is not used in the model fitting. Confidence interval of the regression prediction provides a way of assessing the quality of prediction. Often the following regression prediction confidence intervals are of interest:

- A confidence interval for a single pint on the regression line.
- A confidence interval for a single future value of y corresponding to a chosen value of x.
- A confidence region for the regression line as a whole.

If a particular value of predictor variable is of special importance, a confidence interval for the corresponding response variable y at particular regressor x may be of interest.

A confidence interval of interest can be used to evaluate the accuracy of a single future value of y at a chosen value of regressor x. Confidence interval estimator for a future value of y provides confidence interval for an estimated value y at x with a desirable confidence level $1 - \alpha$.

It is of interest to compare the above two different kinds of confidence interval. The second kind has larger confidence interval which reflects the less accuracy resulting from the estimation of a single future value of y rather than the mean value computed for the first kind confidence interval.

When the entire regression line is of interest, a confidence region can provide simultaneous statements about estimates of y for a number of values of the predictor variable x. i.e., for a set of values of the regressor the $100(1 - \alpha)$ percent of the corresponding response values will be in this interval.

To discuss the confidence interval for regression line we consider the fitted value of the regression line at $x = x_0$, which is $\hat{y}(x_0) = b_0 + b_1 x_0$ and the mean value at $x = x_0$ is $E(\hat{y}|x_0) = \beta_0 + \beta_1 x_0$. Note that b_1 is independent of $\bar{y}$ we have

$$\begin{aligned}
\text{Var}(\hat{y}(x_0)) &= \text{Var}(b_0 + b_1 x_0) \\
&= \text{Var}(\bar{y} - b_1(x_0 - \bar{x})) \\
&= \text{Var}(\bar{y}) + (x_0 - \bar{x})^2 \text{Var}(b_1) \\
&= \frac{1}{n}\sigma^2 + (x_0 - \bar{x})^2 \frac{1}{S_{xx}} \sigma^2 \\
&= \sigma^2 \Big[\frac{1}{n} + \frac{(x_0 - \bar{x})^2}{S_{xx}} \Big]
\end{aligned}$$

Replacing σ by s, the standard error of the regression prediction at x_0 is given by

$$s_{\hat{y}(x_0)} = s\sqrt{\frac{1}{n} + \frac{(x_0 - \bar{x})^2}{S_{xx}}}$$

If $\varepsilon \sim N(0, \sigma^2)$ the $(1 - \alpha)100\%$ of confidence interval on $E(\hat{y}|x_0) = \beta_0 + \beta_1 x_0$ can be written as

$$\hat{y}(x_0) \pm t_{\alpha/2,n-2}\, s\sqrt{\frac{1}{n}+\frac{(x_0-\bar{x})^2}{S_{xx}}}.$$

We now discuss confidence interval on the regression prediction. Denoting the regression prediction at x_0 by y_0 and assuming that y_0 is independent of $\hat{y}(x_0)$, where $y(x_0) = b_0 + b_1 x_0$, and $E(y - \hat{y}(x_0)) = 0$, we have

$$\mathrm{Var}\big(y_0 - \hat{y}(x_0)\big) = \sigma^2 + \sigma^2\Big[\frac{1}{n}+\frac{(x_0-\bar{x})^2}{S_{xx}}\Big] = \sigma^2\Big[1+\frac{1}{n}+\frac{(x_0-\bar{x})^2}{S_{xx}}\Big].$$

Under the normality assumption of the error term

$$\frac{y_0 - \hat{y}(x_0)}{\sigma\sqrt{1+\frac{1}{n}+\frac{(x_0-\bar{x})^2}{S_{xx}}}} \sim N(0,1).$$

Substituting σ with s we have

$$\frac{y_0 - \hat{y}(x_0)}{s\sqrt{1+\frac{1}{n}+\frac{(x_0-\bar{x})^2}{S_{xx}}}} \sim t_{n-2}.$$

Thus the $(1-\alpha)100\%$ confidence interval on regression prediction y_0 can be expressed as

$$\hat{y}(x_0) \pm t_{\alpha/2,n-2}\, s\sqrt{1+\frac{1}{n}+\frac{(x_0-\bar{x})^2}{S_{xx}}}.$$

2.6 Statistical Inference on Regression Parameters

We start with the discussions on the total variance of regression model which plays an important role in the regression analysis. In order to partition the total variance $\sum_{i=1}^{n}(y_i-\bar{y})^2$, we consider the fitted regression equation $\hat{y}_i = b_0 + b_1 x_i$, where $b_0 = \bar{y} - b_1\bar{x}$ and $b_1 = S_{xy}/S_{xx}$. We can write

$$\bar{\hat{y}} = \frac{1}{n}\sum_{i=1}^{n}\hat{y}_i = \frac{1}{n}\sum_{i=1}^{n}[(\bar{y}-b_1\bar{x})+b_1x_i] = \frac{1}{n}\sum_{i=1}^{n}[\bar{y}+b_1(x_i-\bar{x})] = \bar{y}.$$

For the regression response y_i, the total variance is $\frac{1}{n}\sum_{i=1}^{n}(y_i - \bar{y})^2$. Note that the product term is zero and the total variance can be partitioned into two parts:

$$\frac{1}{n}\sum_{i=1}^{n}(y_i - \bar{y})^2 = \frac{1}{n}\sum_{i=1}^{n}[(y_i - \hat{y})^2 + (\hat{y}_i - \bar{y})]^2$$
$$= \frac{1}{n}\sum_{i=1}^{n}(\hat{y}_i - \bar{y})^2 + \frac{1}{n}\sum_{i=1}^{n}(y_i - \hat{y})^2 = \text{SS}_{Reg} + \text{SS}_{Res}$$
$$= \text{Variance explained by regression} + \text{Variance unexplained}$$

It can be shown that the product term in the partition of variance is zero:

$$\sum_{i=1}^{n}(\hat{y}_i - \bar{y})(y_i - \hat{y}_i) \qquad (\text{use the fact that } \sum_{i=1}^{n}(y_i - \hat{y}_i) = 0)$$
$$= \sum_{i=1}^{n}\hat{y}_i(y_i - \hat{y}_i) = \sum_{i=1}^{n}\big[b_0 + b_1(x_i - \bar{x})\big](y_i - \hat{y})$$
$$= b_1\sum_{i=1}^{n}x_i(y_i - \hat{y}_i) = b_1\sum_{i=1}^{n}x_i[y_i - b_0 - b_1(x_i - \bar{x})]$$
$$= b_1\sum_{i=1}^{n}x_i\big[(y_i - \bar{y}) - b_1(x_i - \bar{x})\big]$$
$$= b_1\Big[\sum_{i=1}^{n}(x_i - \bar{x})(y_i - \bar{y}) - b_1\sum_{i=1}^{n}(x_i - \bar{x})^2\Big]$$
$$= b_1[S_{xy} - b_1 S_{xx}] = b_1[S_{xy} - (S_{xy}/S_{xx})S_{xx}] = 0$$

The degrees of freedom for SS_{Reg} and SS_{Res} are displayed in Table 2.2.

Table 2.2 Degrees of Freedom in Partition of Total Variance

SS_{Total}	=	SS_{Reg}	+	SS_{Res}
n-1	=	1	+	n-2

To test the hypothesis $H_0 : \beta_1 = 0$ versus $H_1 : \beta_1 \neq 0$ it is needed to assume that $\varepsilon_i \sim N(0, \sigma^2)$. Table 2.3 lists the distributions of SS_{Reg}, SS_{Res} and SS_{Total} under the hypothesis H_0. The test statistic is given by

$$F = \frac{SS_{Reg}}{SS_{Res}/n-2} \sim F_{1,\, n-2},$$

which is a one-sided, upper-tailed F test. Table 2.4 is a typical regression Analysis of Variance (ANOVA) table.

Table 2.3 Distributions of Partition of Total Variance

SS	df	Distribution
SS_{Reg}	1	$\sigma^2\chi_1^2$
SS_{Res}	n-2	$\sigma^2\chi_{n-2}^2$
SS_{Total}	n-1	$\sigma^2\chi_{n-1}^2$

Table 2.4 ANOVA Table 1

Source	SS	df	MS	F
Regression	SS_{Reg}	1	$SS_{Reg}/1$	$F = \frac{MS_{Reg}}{s^2}$
Residual	SS_{Res}	n-2	s^2	
Total	SS_{Total}	n-1		

To test for regression slope β_1, it is noted that b_1 follows the normal distribution

$$b_1 \sim N\Big(\beta_1, \frac{\sigma^2}{SS_{xx}}\Big)$$

and

$$\Big(\frac{b_1-\beta_1}{s}\Big)\sqrt{S_{xx}} \sim t_{n-2},$$

which can be used to test $H_0 : \beta_1 = \beta_{10}$ versus $H_1 : \beta_1 \neq \beta_{10}$. Similar approach can be used to test for the regression intercept. Under the normality assumption of the error term

$$b_0 \sim N\Big[\beta_0, \sigma^2(\frac{1}{n} + \frac{\bar{x}^2}{S_{xx}})\Big].$$

Therefore, we can use the following t test statistic to test $H_0 : \beta_0 = \beta_{00}$ versus $H_1 : \beta_0 \neq \beta_{00}$.

$$t = \frac{b_0 - \beta_0}{s\sqrt{1/n + (\bar{x}^2/S_{xx})}} \sim t_{n-2}$$

It is straightforward to use the distributions of b_0 and b_1 to obtain the $(1 - \alpha)100\%$ confidence intervals of β_0 and β_1:

$$b_0 \pm t_{\alpha/2,n-2}\, s\sqrt{\frac{1}{n} + \frac{\bar{x}^2}{S_{xx}}},$$

and

$$b_1 \pm t_{\alpha/2,n-2}\, s\sqrt{\frac{1}{S_{xx}}}.$$

Suppose that the regression line pass through (0, β_0). i.e., the y intercept is a known constant β_0. The model is given by $y_i = \beta_0 + \beta_1 x_i + \varepsilon_i$ with known constant β_0. Using the least squares principle we can estimate β_1:

$$b_1 = \frac{\sum x_i y_i}{\sum x_i^2}.$$

Correspondingly, the following test statistic can be used to test for $H_0 : \beta_1 = \beta_{10}$ versus $H_1 : \beta_1 \neq \beta_{10}$. Under the normality assumption on ε_i

$$t = \frac{b_1 - \beta_{10}}{s}\sqrt{\sum_{i=1}^{n} x_i^2} \sim t_{n-1}$$

Note that we only have one parameter for the fixed y-intercept regression model and the t test statistic has $n-1$ degrees of freedom, which is different from the simple linear model with 2 parameters.

The quantity R^2, defined as below, is a measurement of regression fit:

$$R^2 = \frac{SS_{Reg}}{SS_{Total}} = \frac{\sum_{i=1}^{n}(\hat{y}_i - \bar{y})^2}{\sum_{i=1}^{n}(y_i - \bar{y})^2} = 1 - \frac{SS_{Res}}{SS_{Total}}$$

Note that $0 \leq \mathrm{R}^2 \leq 1$ and it represents the proportion of total variation explained by regression model.

Quantity $\text{CV} = \frac{s}{\bar{y}} \times 100$ is called the coefficient of variation, which is also a measurement of quality of fit and represents the spread of noise around the regression line. The values of R^2 and CV can be found from Table 2.7, an ANOVA table generated by SAS procedure REG.

We now discuss simultaneous inference on the simple linear regression. Note that so far we have discussed statistical inference on β_0 and β_1 individually. The individual test means that when we test $H_0 : \beta_0 = \beta_{00}$ we only test this H_0 regardless of the values of β_1. Likewise, when we test $H_0 : \beta_1 = \beta_{10}$ we only test H_0 regardless of the values of β_0. If we would like to test whether or not a regression line falls into certain region we need to test the multiple hypothesis: $H_0 : \beta_0 = \beta_{00}, \beta_1 = \beta_{10}$ simultaneously. This falls into the scope of multiple inference. For the multiple inference on β_0 and β_1 we notice that

$$\begin{pmatrix} b_0 - \beta_0, b_1 - \beta_1 \end{pmatrix} \begin{pmatrix} n & \sum_{i=1}^n x_i \\ \sum_{i=1}^n x_i & \sum_{i=1}^n x_i^2 \end{pmatrix} \begin{pmatrix} b_0 - \beta_0 \\ b_1 - \beta_1 \end{pmatrix}$$

$$\sim 2s^2 F_{2,n-2}.$$

Thus, the $(1 - \alpha)100\%$ confidence region of the β_0 and β_1 is given by

$$\begin{pmatrix} b_0 - \beta_0, b_1 - \beta_1 \end{pmatrix} \begin{pmatrix} n & \sum_{i=1}^n x_i \\ \sum_{i=1}^n x_i & \sum_{i=1}^n x_i^2 \end{pmatrix} \begin{pmatrix} b_0 - \beta_0 \\ b_1 - \beta_1 \end{pmatrix}$$

$$\leq 2s^2 F_{\alpha,2,n-2},$$

where $F_{\alpha,2,n-2}$ is the upper tail of the αth percentage point of the F-distribution. Note that this confidence region is an ellipse.

2.7 Residual Analysis and Model Diagnosis

One way to check performance of a regression model is through regression residual, i.e., $e_i = y_i - \hat{y}_i$. For the simple linear regression a scatter plot of e_i against x_i provides a good graphic diagnosis for the regression model. An evenly distributed residuals around mean zero is an indication of a good regression model fit.

We now discuss the characteristics of regression residuals if a regression model is misspecified. Suppose that the correct model should take the quadratic form:

$$y_i = \beta_0 + \beta_1(x_i - \bar{x}) + \beta_2 x_i^2 + \varepsilon_i$$

with $E(\varepsilon_i) = 0$. Assume that the incorrectly specified linear regression model takes the following form:

$$y_i = \beta_0 + \beta_1(x_i - \bar{x}) + \varepsilon_i^*.$$

Then $\varepsilon_i^* = \beta_2 x_i^2 + \varepsilon_i^*$ which is unknown to the analyst. Now, the mean of the error for the simple linear regression is not zero at all and it is a function of x_i. From the quadratic model we have

$$b_0 = \bar{y} = \beta_0 + \beta_2 \bar{x}^2 + \bar{\varepsilon}$$

and

$$b_1 = \frac{S_{xy}}{S_{xx}} = \frac{\sum_{i=1}^n (x_i - \bar{x})(\beta_0 + \beta_1(x_i - \bar{x}) + \beta_2 x_i^2 + \varepsilon_i)}{S_{xx}}$$

$$b_1 = \beta_1 + \beta_2 \frac{\sum_{i=1}^n (x_i - \bar{x}) x_i^2}{S_{xx}} + \frac{\sum_{i=1}^n (x_i - \bar{x}) \varepsilon_i}{S_{xx}}.$$

It is easy to know that

$$E(b_0) = \beta_0 + \beta_2 \bar{x}^2$$

and

$$E(b_1) = \beta_1 + \beta_2 \frac{\sum_{i=1}^n (x_i - \bar{x}) x_i^2}{S_{xx}}.$$

Therefore, the estimators b_0 and b_1 are biased estimates of β_0 and β_1. Suppose that we fit the linear regression model and the fitted values are given by $\hat{y}_i = b_0 + b_1(x_i - \bar{x})$, the expected regression residual is given by

$$\begin{aligned}
E(e_i) = E(y_i - \hat{y}_i) &= \big[\beta_0 + \beta_1(x_i - \bar{x}) + \beta_2 x_i^2\big] - \big[E(b_0) + E(b_1)(x_i - \bar{x})\big] \\
&= \big[\beta_0 + \beta_1(x_i - \bar{x}) + \beta_2 x_i^2\big] - \big[\beta_0 + \beta_2 \bar{x}^2\big] \\
&\quad - \Big[\beta_1 + \beta_2 \frac{\sum_{i=1}^n (x_i - \bar{x}) x_i^2}{S_{xx}}\Big](x_i - \bar{x}) \\
&= \beta_2 \Big[(x_i^2 - \bar{x}^2) - \frac{\sum_{i=1}^n (x_i - \bar{x}) x_i^2}{S_{xx}}\Big]
\end{aligned}$$

If $\beta_2 = 0$ then the fitted model is correct and $E(y_i - \hat{y}_i) = 0$. Otherwise, the expected value of residual takes the quadratic form of x_i's. As a result, the plot of residuals against x_i's should have a curvature of quadratic appearance.

Statistical inference on regression model is based on the normality assumption of the error term. The least squares estimators and the MLEs of the regression parameters are exactly identical only under the normality assumption of the error term. Now, question is how to check the normality of the error term? Consider the residual $y_i - \hat{y}_i$: we have $E(y_i - \hat{y}_i) = 0$ and

$$\text{Var}(y_i - \hat{y}_i) = Var(y_i) + \text{Var}(\hat{y}_i) - 2\text{Cov}(y_i, \hat{y}_i)$$
$$= \sigma^2 + \sigma^2\Big[\frac{1}{n} + \frac{(x_i - \bar{x})^2}{S_{xx}}\Big] - 2\text{Cov}(y_i, \bar{y} + b_1(x_i - \bar{x}))$$

We calculate the last term

$$\text{Cov}(y_i,\ \bar{y} + b_1(x_i - \bar{x})) = \text{Cov}(y_i,\ \bar{y}) + (x_i - \bar{x})\text{Cov}(y_i, b_1)$$
$$= \frac{\sigma^2}{n} + (x_i - \bar{x})\text{Cov}(y_i,\ S_{xy}/S_{xx})$$
$$= \frac{\sigma^2}{n} + (x_i - \bar{x})\frac{1}{S_{xx}}\text{Cov}\Big(y_i,\ \sum_{i=1}^{n}(x_i - \bar{x})(y_i - \bar{y})\Big)$$
$$= \frac{\sigma^2}{n} + (x_i - \bar{x})\frac{1}{S_{xx}}\text{Cov}\Big(y_i,\ \sum_{i=1}^{n}(x_i - \bar{x})y_i\Big) = \frac{\sigma^2}{n} + \frac{(x_i - \bar{x})^2}{S_{xx}}\sigma^2$$

Thus, the variance of the residual is given by

$$\text{Var}(e_i) = Var(y_i - \hat{y}_i) = \sigma^2\Big[1 - \Big(\frac{1}{n} + \frac{(x_i - \bar{x})^2}{S_{xx}}\Big)\Big],$$

which can be estimated by

$$s_{e_i} = s\Big[1 - \Big(\frac{1}{n} + \frac{(x_i - \bar{x})^2}{S_{xx}}\Big)\Big].$$

If the error term in the simple linear regression is correctly specified, i.e., error is normally distributed, the standardized residuals should behave like the standard normal random variable. Therefore, the quantile of the standardized residuals in the simple linear regression will be similar to the quantile of the standardized normal random variable. Thus, the plot of the quantile of the standardized residuals versus the normal quantile should follow a straight line in the first quadrant if the normality assumption on the error term is correct. It is usually called the normal plot and has been used as a useful tool for checking the normality of the error term in simple linear regression. Specifically, we can

(1) Plot ordered residual $\dfrac{y_i - \hat{y}_i}{s}$ against the normal quantile $Z\left(\frac{i-0.375}{n+0.25}\right)$

(2) Plot ordered standardized residual $\dfrac{y_i - \hat{y}_i}{s_{e_i}}$ against the normal quantile $Z\left(\frac{i-0.375}{n+0.25}\right)$

2.8 Example

The SAS procedure REG can be used to perform regression analysis. It is convenient and efficient. The REG procedure provides the most popular parameter estimation, residual analysis, regression diagnosis. We present the example of regression analysis of the density and stiffness data using SAS.

```
data example1;
input density  stiffness @@;
datalines;
 9.5  14814   8.4    17502   9.8  14007  11.0  19443   8.3   7573
 9.9  14191   8.6     9714   6.4   8076   7.0   5304   8.2  10728
17.4  43243  15.0    25319  15.2  28028  16.4  41792  16.7  49499
15.4  25312  15.0    26222  14.5  22148  14.8  26751  13.6  18036
25.6  96305  23.4   104170  24.4  72594  23.3  49512  19.5  32207
21.2  48218  22.8    70453  21.7  47661  19.8  38138  21.3  53045
;
proc reg data=example1 outest=out1 tableout;
model stiffness=density/all;
run;

ods rtf file="C:\Example1_out1.rtf";
proc print data=out1;
title "Parameter Estimates and CIs";
run;
ods rtf close;
```

```
*Trace ODS to find out the names of the output data sets;
ods trace on;
ods show;

ods rtf file="C:\Example1_out2.rtf";
proc reg data=Example1 alpha=0.05;
model stiffness=density;
ods select Reg.MODEL1.Fit.stiffness.ANOVA;
ods select Reg.MODEL1.Fit.stiffness.FitStatistics;
ods select Reg.MODEL1.Fit.stiffness.ParameterEstimates;
ods rtf close;

proc reg data=Example1;
model stiffness=density;
output out=out3 p=yhat r=yresid student=sresid;
run;

ods rtf file="C:\Example1_out3.rtf";
proc print data=out3;
title "Predicted Values and Residuals";
run;
ods rtf close;
```

The above SAS code generate the following output tables 2.5, 2.6, 2.7, 2.8, and 2.9.

Table 2.5 Confidence Intervals on Parameter Estimates

Obs	MODEL	TYPE	DEPVAR	RMSE	Intercept	density
1	Model1	Parms	stiffness	11622.44	-25433.74	3884.98
2	Model1	Stderr	stiffness	11622.44	6104.70	370.01
3	Model1	T	stiffness	11622.44	-4.17	10.50
4	Model1	P-value	stiffness	11622.44	0.00	0.00
5	Model1	L95B	stiffness	11622.44	-37938.66	3127.05
6	Model1	U95B	stiffness	11622.44	-12928.82	4642.91

Data Source: density and stiffness data

The following is an example of SAS program for computing the confidence band of regression mean, the confidence band for regression predic-

Table 2.6 ANOVA Table 2

Source	DF	Sum of Squares	Mean Square	F Value	Pr >F
Model	1	14891739363	14891739363	110.24	<.0001
Error	28	3782270481	135081089		
Corrected Total	29	18674009844			

Data Source: density and stiffness data

Table 2.7 Regression Table

Root MSE	11622.00	R-Square	0.7975
Dependent Mean	34667.00	Adj R-Sq	0.7902
Coeff Var	33.53		

Data Source: density and stiffness data

Table 2.8 Parameter Estimates of Simple Linear Regression

Variable	DF	Parameter Estimate	Standard Error	t Value	Pr > $\|t\|$
Intercept	1	-25434.00	6104.70	-4.17	0.0003
density	1	3884.98	370.01	10.50	< .0001

Data Source: density and stiffness data

tion, and probability plot (QQ-plot and PP-plot).

```
data Example2;
input density  stiffness @@;
datalines;
9.5   14814  8.4  17502   9.8  14007  11     19443   8.3  7573
9.9   14191  8.6   9714   6.4   8076  7       5304   8.2  10728
17.4  43243  15   25319  15.2  28028  16.4  41792  16.7  49499
15.4  25312  15   26222  14.5  22148  14.8  26751  13.6  18036
25.6  96305  23.4 104170 24.4  72594  23.3  49512  19.5  32207
21.2  48218  22.8  70453 21.7  47661  19.8  38138  21.3  53045
;
across=1 cborder=red offset=(0,0)
shape=symbol(3,1) label=none value=(height=1);
symbol1 c=black value=- h=1;
symbol2 c=red;
```

Table 2.9 Table for Fitted Values and Residuals

Obs	density	stiffness	yhat	yresid
1	9.5	14814	11473.53	3340.47
2	8.4	17502	7200.06	10301.94
3	9.8	14007	12639.02	1367.98
4	11	19443	17300.99	2142.01
5	8.3	7573	6811.56	761.44
6	9.9	14191	13027.52	1163.48
7	8.6	9714	7977.05	1736.95
8	6.4	8076	-569.90	8645.90
9	7.0	5304	1761.09	3542.91
10	8.2	10728	6423.06	4304.94
11	17.4	43243	42164.84	1078.16
12	15.0	25319	32840.89	-7521.89
13	15.2	28028	33617.89	-5589.89
14	16.4	41792	38279.86	3512.14
15	16.7	49499	39445.35	10053.65
16	15.4	25312	34394.89	-9082.89
17	15.0	26222	32840.89	-6618.89
18	14.5	22148	30898.41	-8750.41
19	14.8	26751	32063.90	-5312.90
20	13.6	18036	27401.93	-9365.93
21	25.6	96305	74021.64	22283.36
22	23.4	104170	65474.69	38695.31
23	24.4	72594	69359.67	3234.33
24	23.3	49512	65086.19	-15574.19
25	19.5	32207	50323.28	-18116.28
26	21.2	48218	56927.74	-8709.74
27	22.8	70453	63143.70	7309.30
28	21.7	47661	58870.23	-11209.23
29	19.8	38138	51488.78	-13350.78
30	21.3	53045	57316.24	-4271.24

Data Source: density and stiffness data

```
symbol3 c=blue;
symbol4 c=blue;
proc reg data=Example2;
     model density=stiffness /noprint p r;
     output out=out p=pred r=resid LCL=lowpred
            UCL=uppred LCLM=lowreg UCLM=upreg;
run;
ods rtf file="C:\Example2.rtf";
ods graphics on;
title "PP Plot";
```

```
plot npp.*r./caxis=red ctext=blue nostat cframe=ligr;
run;
title "QQ Plot";
plot r.*nqq. /noline  mse
     caxis=red ctext=blue cframe=ligr;
run;

*Compute confidence band of regression mean;
plot  density*stiffness/conf caxis=red ctext=blue
      cframe=ligr legend=legend1;
      run;

*Compute confidence band of regression prediction;
plot density*stiffness/pred caxis=red ctext=blue
     cframe=ligr legend=legend1;
run;
ods graphics off;
ods rtf close;
quit;
```

The regression scatterplot, residual plot, 95% confidence bands for regression mean and prediction are presented in Fig. 2.1.

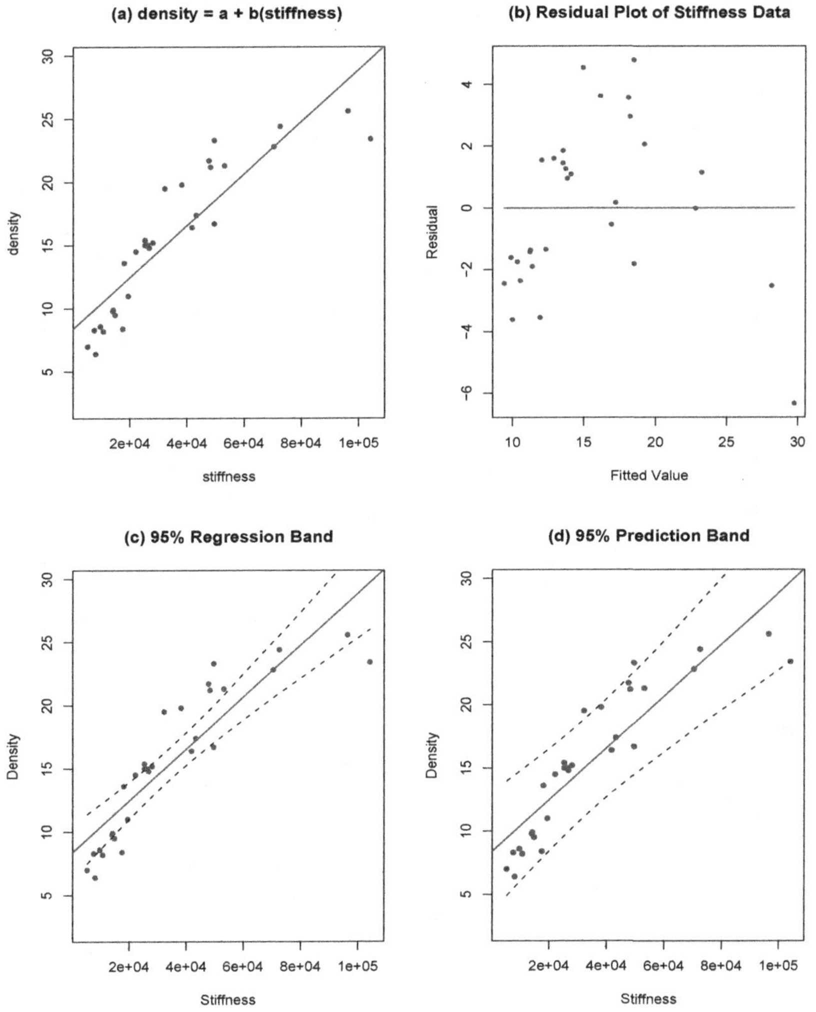

Fig. 2.1 (a) Regression Line and Scatter Plot. (b) Residual Plot, (c) 95% Confidence Band for Regression Mean. (d) 95% Confidence Band for Regression Prediction.

The Q-Q plot for regression model density$=\beta_0 + \beta_1$ stiffness is presented in Fig. 2.2.

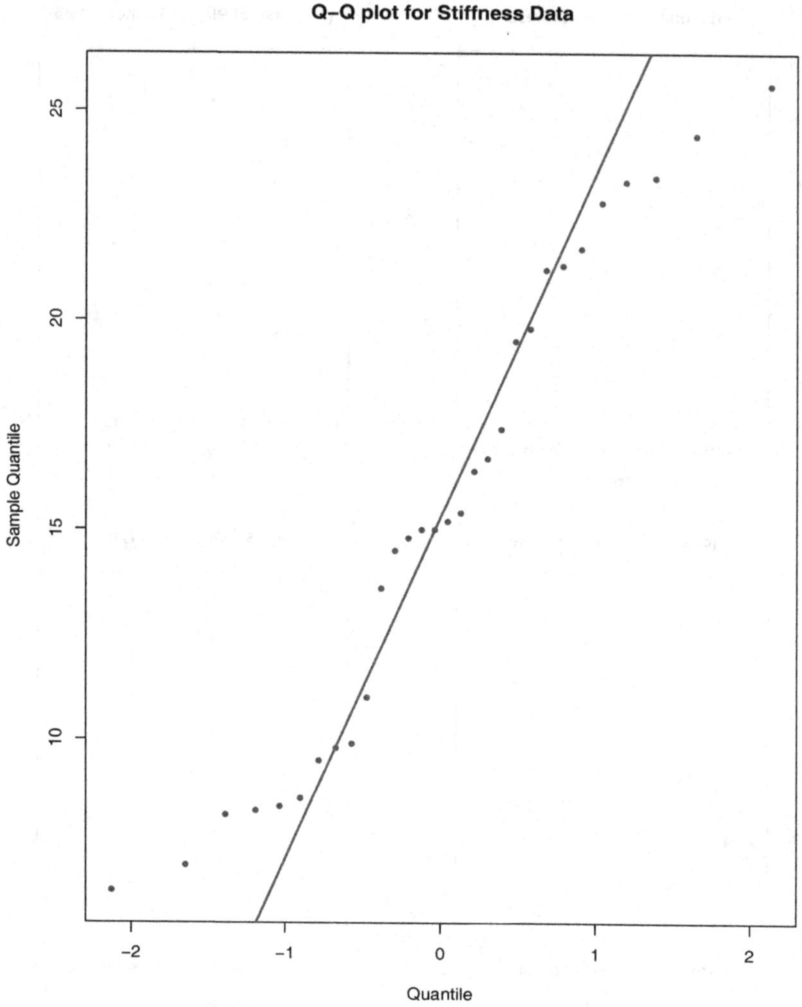

Fig. 2.2 Q-Q Plot for Regression Model density=$\beta_0 + \beta_1$ stiffness + ε.

Problems

1. Consider a set of data $(x_i,\ y_i)$, $i = 1, 2, \cdots, n$, and the following two regression models:

$$y_i = \beta_0 + \beta_1 x_i + \varepsilon, \qquad (i = 1, 2, \cdots, n), \quad \text{Model A}$$
$$y_i = \gamma_0 + \gamma_1 x_i + \gamma_2 x_i^2 + \varepsilon, \quad (i = 1, 2, \cdots, n), \quad \text{Model B}$$

Suppose both models are fitted to the same data. Show that

$$SS_{Res,\ A} \geq SS_{Res,\ B}$$

If more higher order terms are added into the above Model B, i.e.,

$$y_i = \gamma_0 + \gamma_1 x_i + \gamma_2 x_i^2 + \gamma_3 x_i^3 + \cdots + \gamma_k x_i^k + \varepsilon, \quad (i = 1, 2, \cdots, n),$$

show that the inequality $SS_{Res,\ A} \geq SS_{Res,\ B}$ still holds.

2. Consider the zero intercept model given by

$$y_i = \beta_1 x_i + \varepsilon_i, \quad (i = 1, 2, \cdots, n)$$

where the ε_i's are independent normal variables with constant variance σ^2. Show that the $100(1-\alpha)\%$ confidence interval on $E(y|x_0)$ is given by

$$b_1 x_0 + t_{\alpha/2,\ n-1} s \sqrt{\frac{x_0^2}{\sum_{i=1}^n x_i^2}}$$

where $s = \sqrt{\sum_{i=1}^{n} (y_i - b_1 x_i)/(n-1)}$ and $b_1 = \dfrac{\sum_{i=1}^n y_i x_i}{\sum_{i=1}^n x_i^2}$.

3. Derive and discuss the $(1-\alpha)100\%$ confidence interval on the slope β_1 for the simple linear model with zero intercept.

4. Consider the fixed zero intercept regression model

$$y_i = \beta_1 x_i + \varepsilon_i, \quad (i = 1, 2, \cdots, n)$$

The appropriate estimator of σ^2 is given by

$$s^2 = \sum_{i=1}^{n} \frac{(y_i - \hat{y}_i)^2}{n-1}$$

Show that s^2 is an unbiased estimator of σ^2.

Table 2.10 Data for Two Parallel Regression Lines

x	y
x_1	y_1
$\vdots$	$\vdots$
x_{n_1}	y_{n_1}
x_{n_1+1}	y_{n_1+1}
$\vdots$	$\vdots$
$x_{n_1+n_2}$	$y_{n_1+n_2}$

5. Consider a situation in which the regression data set is divided into two parts as shown in Table 2.10.
The regression model is given by
$$y_i = \begin{cases} \beta_0^{(1)} + \beta_1 x_i + \varepsilon_i, & i = 1, 2, \cdots, n_1; \\ \\ \beta_0^{(2)} + \beta_1 x_i + \varepsilon_i, & i = n_1 + 1, \cdots, n_1 + n_2. \end{cases}$$
In other words, there are two regression lines with common slope. Using the centered regression model
$$y_i = \begin{cases} \beta_0^{(1*)} + \beta_1 (x_i - \bar{x}_1) + \varepsilon_i, & i = 1, 2, \cdots, n_1; \\ \\ \beta_0^{(2*)} + \beta_1 (x_i - \bar{x}_2) + \varepsilon_i, & i = n_1 + 1, \cdots, n_1 + n_2, \end{cases}$$
where $\bar{x}_1 = \sum_{i=1}^{n_1} x_i/n_1$ and $\bar{x}_2 = \sum_{i=n_1+1}^{n_1+n_2} x_i/n_2$. Show that the least squares estimate of β_1 is given by
$$b_1 = \frac{\sum_{i=1}^{n_1}(x_i - \bar{x}_1)y_i + \sum_{i=n_1+1}^{n_1+n_2}(x_i - \bar{x}_2)y_i}{\sum_{i=1}^{n_1}(x_i - \bar{x}_1)^2 + \sum_{i=n_1+1}^{n_1+n_2}(x_i - \bar{x}_2)^2}$$
6. Consider two simple linear models
$$Y_{1j} = \alpha_1 + \beta_1 x_{1j} + \varepsilon_{1j}, \ j = 1, 2, \cdots, n_1$$
and
$$Y_{2j} = \alpha_2 + \beta_2 x_{2j} + \varepsilon_{2j}, \ j = 1, 2, \cdots, n_2$$
Assume that $\beta_1 \neq \beta_2$ the above two simple linear models intersect. Let x_0 be the point on the x-axis at which the two linear models intersect. Also assume that ε_{ij} are independent normal variable with a variance σ^2. Show that

(a). $x_0 = \dfrac{\alpha_1 - \alpha_2}{\beta_1 - \beta_2}$

(b). Find the maximum likelihood estimates (MLE) of x_0 using the least squares estimators $\hat{\alpha}_1$, $\hat{\alpha}_2$, $\hat{\beta}_1$, and $\hat{\beta}_2$.

(c). Show that the distribution of Z, where

$$Z = (\hat{\alpha}_1 - \hat{\alpha}_2) + x_0(\hat{\beta}_1 - \hat{\beta}_2),$$

is the normal distribution with mean 0 and variance $A^2\sigma^2$, where

$$A^2 = \frac{\sum x_{1j}^2 - 2x_0 \sum x_{1j} + x_0^2 n_1}{n_1 \sum (x_{1j} - \bar{x}_1)^2} + \frac{\sum x_{2j}^2 - 2x_0 \sum x_{2j} + x_0^2 n_2}{n_2 \sum (x_{2j} - \bar{x}_2)^2}.$$

(d). Show that $U = N\hat{\sigma}^2/\sigma^2$ is distributed as $\chi^2(N)$, where $N = n_1 + n_2 - 4$.

(e). Show that U and Z are independent.

(f). Show that $W = Z^2/A^2\hat{\sigma}^2$ has the F distribution with degrees of freedom 1 and N.

(g). Let $S_1^2 = \sum (x_{1j} - \bar{x}_1)^2$ and $S_2^2 = \sum (x_{2j} - \bar{x}_2)^2$, show that the solution of the following quadratic equation about x_0, $q(x_0) = ax_0^2 + 2bx_0 + c = 0$,

$$\Big[(\hat{\beta}_1 - \hat{\beta}_2)^2 - \Big(\frac{1}{S_1^2} + \frac{1}{S_2^2}\Big)\hat{\sigma}^2 F_{\alpha,1,N}\Big] x_0^2$$
$$+ 2\Big[(\hat{\alpha}_1 - \hat{\alpha}_2)(\hat{\beta}_1 - \hat{\beta}_2) + \Big(\frac{\bar{x}_1}{S_1^2} + \frac{\bar{x}_2}{S_2^2}\Big)\hat{\sigma}^2 F_{\alpha,1,N}\Big] x_0$$
$$+ \Big[(\hat{\alpha}_1 - \hat{\alpha}_2)^2 - \Big(\frac{\sum x_{1j}^2}{n_1 S_1^2} + \frac{\sum x_{2j}^2}{n_2 S_2^2}\Big)\hat{\sigma}^2 F_{\alpha,1,N}\Big] = 0.$$

Show that if $a \geq 0$ and $b^2 - ac \geq 0$, then $1 - \alpha$ confidence interval on x_0 is

$$\frac{-b - \sqrt{b^2 - ac}}{a} \leq x_0 \leq \frac{-b + \sqrt{b^2 - ac}}{a}.$$

7. Observations on the yield of a chemical reaction taken at various temperatures were recorded in Table 2.11:

   (a). Fit a simple linear regression and estimate β_0 and β_1 using the least squares method.

   (b). Compute 95% confidence intervals on $E(y|x)$ at 4 levels of temperatures in the data. Plot the upper and lower confidence intervals around the regression line.

Table 2.11 Chemical Reaction Data

temperature (C^0)	yield of chemical reaction (%)
150	77.4
150	77.4
150	77.4
150	77.4
150	77.4
150	77.4
150	77.4
150	77.4
150	77.4
150	77.4
150	77.4
150	77.4

Data Source: Raymond H. Myers, Classical and Modern Regression Analysis With Applications, P77.

(c). Plot a 95% confidence band on the regression line. Plot on the same graph for part (b) and comment on it.

8. The study "Development of LIFETEST, a Dynamic Technique to Assess Individual Capability to Lift Material" was conducted in Virginia Polytechnic Institute and State University in 1982 to determine if certain static arm strength measures have influence on the "dynamic lift" characteristics of individual. 25 individuals were subjected to strength tests and then were asked to perform a weight-lifting test in which weight was dynamically lifted overhead. The data are in Table 2.12:

   (a). Find the linear regression line using the least squares method.
   (b). Define the joint hypothesis $H_0 : \beta_0 = 0, \ \beta_1 = 2.2$. Test this hypothesis problem using a 95% joint confidence region and β_0 and β_1 to draw your conclusion.
   (c). Calculate the studentized residuals for the regression model. Plot the studentized residuals against x and comment on the plot.

Table 2.12 Weight-lifting Test Data

Individual	Arm Strength (x)	Dynamic Lift (y)
1	17.3	71.4
2	19.5	48.3
3	19.5	88.3
4	19.7	75.0
5	22.9	91.7
6	23.1	100.0
7	26.4	73.3
8	26.8	65.0
9	27.6	75.0
10	28.1	88.3
11	28.1	68.3
12	28.7	96.7
13	29.0	76.7
14	29.6	78.3
15	29.9	60.0
16	29.9	71.7
17	30.3	85.0
18	31.3	85.0
19	36.0	88.3
20	39.5	100.0
21	40.4	100.0
22	44.3	100.0
23	44.6	91.7
24	50.4	100.0
25	55.9	71.7

Data Source: Raymond H. Myers, Classical and Modern Regression Analysis With Applications, P76.

Chapter 3

Multiple Linear Regression

The general purpose of multiple linear regression is to seek for the linear relationship between a dependent variable and several independent variables. Multiple regression allows researchers to examine the effect of more than one independent variables on response at the same time. For some research questions, regression can be used to examine how much a particular set of independent variables can explain sufficiently the outcome. In other cases, multiple regression is used to examine the effect of outcome while accounting for more than one factor that could influence the outcome. In this chapter we discuss multiple linear regression. To facilitate the discussion of the theory of the multiple regression model we start with a brief introduction of the linear space and the projection in the linear space. Then we will introduce multiple linear model in matrix form. All subsequent discussions of multiple regression will be based on its matrix form.

3.1 Vector Space and Projection

First we briefly discuss the vector space, subspace, projection, and quadratic form of multivariate normal variable, which are useful in the discussions of the subsequent sections of this chapter.

3.1.1 *Vector Space*

A vector is a geometric object which has both magnitude and direction. A vector is frequently represented by a line segment connecting the initial point A with the terminal point B and denoted by $\overrightarrow{AB}$. The magnitude of the vector $\overrightarrow{AB}$ is the length of the segment and the direction of this vector characterizes the displacement of the point B relative to the point

A. Vectors can be added, subtracted, multiplied by a number, and flipped around (multiplying by number -1) so that the direction is reversed. These operations obey the familiar algebraic laws: commutativity, associativity, and distributivity. The sum of two vectors with the same initial point can be found geometrically using the parallelogram law. Multiplication by a positive number, commonly called a scalar in this context, amounts to changing the magnitude of vector, that is, stretching or compressing it while keeping its direction; multiplication by -1 preserves the magnitude of the vector but reverses its direction. Cartesian coordinates provide a systematic way of describing vectors and operations on them.

A vector space is a set of vectors that is closed under finite vector addition and scalar multiplication. The basic example is n-dimensional Euclidean space, where every element is represented by a list of real numbers, such as

$$\boldsymbol{x}^{'} = (x_1, x_2, \cdots, x_n).$$

Scalars are real numbers, addition is componentwise, and scalar multiplication is multiplication on each term separately. Suppose V is closed under vector addition on $\mathbb{R}^n$: if $u, v \in V$, then $u + v \in V$. V is also closed under scalar multiplication: if $a \in \mathbb{R}^1$, $v \in V$, then $av \in V$. Then V is a vector space (on $\mathbb{R}^n$). We will focus our discussion only on vector space on n-dimensional Euclidean space. For example, for any positive integer n, the space of all n-tuples of elements of real line $\mathbb{R}^1$ forms an n-dimensional real vector space sometimes called real coordinate space and denoted by $\mathbb{R}^n$. An element in $\mathbb{R}^n$ can be written as

$$\boldsymbol{x}^{'} = (x_1, x_2, \cdots, x_n),$$

where each x_i is an element of $\mathbb{R}^1$. The addition on $\mathbb{R}^n$ is defined by

$$\boldsymbol{x} + \boldsymbol{y} = (x_1 + y_1,\ x_2 + y_2,\ \cdots,\ x_n + y_n),$$

and the scalar multiplication on $\mathbb{R}^n$ is defined by

$$a\,\boldsymbol{x} = (ax_1,\ ax_2,\ \cdots,\ ax_n).$$

When $a = -1$ the vector $a\boldsymbol{x}$ has the same length as $\boldsymbol{x}$ but with a geometrically reversed direction.

It was F. Hausdorff who first proved that every vector space has a basis. A basis makes it possible to express every vector of the space as a unique tuple of the field elements, although caution must be exercised when a vector space does not have a finite basis. In linear algebra, a basis is a set of vectors that, in a linear combination, can represent every vector in a given vector space, and such that no element of the set can be represented as a linear combination of the others. In other words, a basis is a linearly independent spanning set. The following is an example of basis of $\mathbb{R}^n$:

$$\begin{aligned} \boldsymbol{e}_1' &= (1,\ 0,\ 0, \cdots,\ 0)_{1\times n} \\ \boldsymbol{e}_2' &= (0,\ 1,\ 0, \cdots,\ 0)_{1\times n} \\ \boldsymbol{e}_3' &= (0,\ 0,\ 1, \cdots,\ 0)_{1\times n} \\ &\vdots \\ \boldsymbol{e}_n' &= (0,\ 0,\ 0, \cdots,\ 1)_{1\times n}. \end{aligned}$$

Actually, the above vectors consist of the standard orthogonal basis of the vector space $\mathbb{R}^n$. Any vector $\boldsymbol{x}' = (x_1, x_2, \cdots, x_n)$ in the $\mathbb{R}^n$ can be a linear combination of $\boldsymbol{e}_1, \boldsymbol{e}_2, \cdots, \boldsymbol{e}_n$. In fact,

$$\boldsymbol{x} = x_1\boldsymbol{e}_1 + x_2\boldsymbol{e}_2 + x_3\boldsymbol{e}_3 + \cdots + x_n\boldsymbol{e}_n.$$

This representation is unique. i.e., if there is another representation such that

$$\boldsymbol{x} = x_1^*\boldsymbol{e}_1 + x_2^*\boldsymbol{e}_2 + x_3^*\boldsymbol{e}_3 + \cdots + x_n^*\boldsymbol{e}_n,$$

then

$$\begin{aligned} &(x_1 - x_1^*)\boldsymbol{e}_1 + (x_2 - x_2^*)\boldsymbol{e}_2 + \cdots + (x_n - x_n^*)\boldsymbol{e}_n \\ &= (x_1 - x_1^*,\ x_2 - x_2^*,\ \cdots, x_2 - x_2^*) = (0,\ 0, \cdots,\ 0). \end{aligned}$$

Therefore, we have $x_i = x_i^*$ for all $i = 1, 2, \cdots, n$.

Given a vector space V, a nonempty subset W of V that is closed under addition and scalar multiplication is called a subspace of V. The intersection of all subspaces containing a given set of vectors is called its span. If no vector can be removed without changing the span, the vectors in this set is said to be linearly independent. A linearly independent set whose span is V is called a basis for V. A vector span by two vectors v and w can be defined as: $\boldsymbol{x} : \boldsymbol{x} = av + bw$, for all $(a, b) \in \mathbb{R}^2$. Note that v and w may not be necessarily independent. If a vector space S is spanned by a set of independent vectors $v_1, v_2, \cdots, v_p$, i.e., S is the set of vectors

$$\{\boldsymbol{x} : \boldsymbol{x} = a_1 + v_1 + a_2v_2 + \cdots + a_pv_p, \text{ for all } (a_1, a_2, \cdots, a_p) \in \mathbb{R}^p\},$$

then the dimension of S is p. Vectors $v_1, v_2, \cdots, v_p$ are the basis of the vector space S. The dimension of a vector space S is the largest number of a set of independent vectors in S. If the dimension of a linear space S is p we write $\text{Dim}(S) = p$.

3.1.2 *Linearly Independent Vectors*

If there exist a finite number of distinct vectors $v_1, v_2, \cdots, v_n$ in vector space V and scalars $a_1, a_2, \cdots, a_n$, not all zero, such that

$$a_1 v_1 + a_2 v_2 + a_3 v_3 + \cdots + a_n v_n = 0,$$

then the vectors $v_1, v_2, \cdots, v_n$ are said to be linearly dependent. If $v_1, v_2, \cdots, v_n$ are dependent then out of these n vectors there is at least one vector that can be expressed as a linear combination of other vectors. Note that the zero on the right is the zero vector, not the number zero. If no such scalars exist, then the vectors $v_1, v_2, \cdots, v_n$ are said to be linearly independent. This condition can be reformulated as follows: whenever $a_1, a_2, \cdots, a_n$ are scalars such that

$$a_1 v_1 + a_2 v_2 + a_3 v_3 + \cdots + a_n v_n = 0,$$

we have $a_i = 0$ for $i = 1, 2, \cdots, n$, then $v_1, v_2, \cdots, v_n$ are linearly independent.

A basis of a vector space V is defined as a subset of vectors in V that are linearly independent and these vectors span space V. Consequently, if $(v_1, v_2, \cdots, v_n)$ is a list of vectors in V, then these vectors form a basis if and only if every vector $\boldsymbol{x} \in V$ can be uniquely expressed by a linear combination of $v_1, v_2, \cdots, v_p$. i.e.,

$$\boldsymbol{x} = a_1 v_1 + a_2 v_2 + \cdots + a_n v_n, \text{ for any } \boldsymbol{x} \in \text{V}.$$

The number of basis vectors in V is called the dimension of linear space V. Note that a vector space can have more than one basis, but the number of vectors which form a basis of the vector space V is always fixed. i.e., the dimension of vector space V is fixed but there will be more than one basis. In fact, if the dimension of vector space V is n, then any n linearly independent vectors in V form its basis.

3.1.3 *Dot Product and Projection*

If $\boldsymbol{x} = (x_1, x_2, \cdots, x_n)$ and $\boldsymbol{y} = (y_1, y_2, \cdots, y_n)$ are two vectors in a vector Euclidean space $\mathbb{R}^n$. The dot product of two vectors $\boldsymbol{x}$ and $\boldsymbol{y}$ is defined as

$$\boldsymbol{x} \cdot \boldsymbol{y} = x_1 y_1 + x_2 y_2 + \cdots + x_n y_n.$$

Two vectors are said to be orthogonal if their dot product is 0. If θ is the angle between two vectors $(x_1, x_2, \cdots, x_n)$ and $(y_1, y_2, \cdots, y_n)$, the cosine of the angle between the two vectors is defined as

$$\cos(\theta) = \frac{\boldsymbol{x} \cdot \boldsymbol{y}}{|\boldsymbol{x}||\boldsymbol{y}|} = \frac{x_1 y_1 + x_2 y_2 + \cdots + x_n y_n}{\sqrt{x_1^2 + x_2^2 + \cdots + x_n^2}\sqrt{y_1^2 + y_2^2 + \cdots + y_n^2}} \tag{3.1}$$

Two orthogonal vectors meet at 90°; i.e., they are perpendicular. One important application of the dot product is projection. The projection of a vector $\boldsymbol{y}$ onto another vector $\boldsymbol{x}$ forms a new vector that has the same direction as the vector $\boldsymbol{x}$ and the length $|\boldsymbol{y}|\cos(\theta)$, where $|\boldsymbol{y}|$ denotes the length of vector $\boldsymbol{y}$ and θ is the angle between two vectors $\boldsymbol{x}$ and $\boldsymbol{y}$. We write this projection as $P_{\boldsymbol{x}}\boldsymbol{y}$. The projection vector can be expressed as

$$\begin{aligned} P_{\boldsymbol{x}}\boldsymbol{y} &= |\boldsymbol{y}|\cos(\theta)\frac{\boldsymbol{x}}{|\boldsymbol{x}|} = |\boldsymbol{y}|\frac{\boldsymbol{x}\cdot\boldsymbol{y}}{|\boldsymbol{x}||\boldsymbol{y}|}\frac{\boldsymbol{x}}{|\boldsymbol{x}|} \\ &= \frac{x_1y_1 + x_2y_2 + \cdots + x_ny_n}{x_1^2 + x_2^2 + \cdots + x_n^2}\boldsymbol{x} = \lambda\boldsymbol{x}, \end{aligned} \tag{3.2}$$

where λ is a scalar and

$$\lambda = \frac{x_1y_1 + x_2y_2 + \cdots + x_ny_n}{x_1^2 + x_2^2 + \cdots + x_n^2} = \frac{\boldsymbol{x}\cdot\boldsymbol{y}}{\boldsymbol{x}\boldsymbol{x}}.$$

Thus, the projection of $\boldsymbol{y}$ onto vector $\boldsymbol{x}$ is a vector $\boldsymbol{x}$ multiplying a scalar λ where λ is the $\cos(\theta)$ and θ is the angle between two vectors $\boldsymbol{x}$ and $\boldsymbol{y}$.

If $\boldsymbol{x}$ and $\boldsymbol{y}$ are two vectors in $\mathbb{R}^n$. Consider the difference vector between the vector $\boldsymbol{e}$, $\boldsymbol{e} = \lambda\boldsymbol{x} - \boldsymbol{y}$, and $\lambda = \boldsymbol{x}\cdot\boldsymbol{y}/\boldsymbol{x}\cdot\boldsymbol{x}$. The vector $\boldsymbol{e}$ is perpendicular to the vector $\boldsymbol{x}$ when $\lambda = (\boldsymbol{x}\cdot\boldsymbol{y})/(\boldsymbol{x}\cdot\boldsymbol{x})$. To see this we simply calculate the dot product of $\boldsymbol{e}$ and $\boldsymbol{x}$:

$$\boldsymbol{e}\cdot\boldsymbol{x} = (\lambda\boldsymbol{x} - \boldsymbol{y})\cdot\boldsymbol{x} = \lambda\boldsymbol{x}\cdot\boldsymbol{x} - \boldsymbol{x}\cdot\boldsymbol{y} = \left(\frac{\boldsymbol{x}\cdot\boldsymbol{y}}{\boldsymbol{x}\cdot\boldsymbol{x}}\right)\boldsymbol{x}\cdot\boldsymbol{x} - \boldsymbol{x}\cdot\boldsymbol{y} = 0$$

Thus, the angle between $\boldsymbol{e}$ and $\boldsymbol{x}$ is indeed 90°, i.e., they are perpendicular to each other. In addition, since $\boldsymbol{e}$ is perpendicular to $\boldsymbol{x}$, it is the vector with the shortest distance among all the vectors starting from the end of $\boldsymbol{y}$ and ending at any point on $\boldsymbol{x}$.

If a vector space has a basis and the length of the basis vectors is a unity then this basis is an orthonormal basis. Any basis divided by its length forms an orthonormal basis. If S is a p-dimensional subspace of a vector space V, then it is possible to project vectors in V onto S. If the subspace S has an orthonormal basis $(w_1, w_2, \cdots, w_p)$, for any vector $\boldsymbol{y}$ in V, the projection of $\boldsymbol{y}$ onto the subspace S is

$$P_S\boldsymbol{y} = \sum_{i=1}^{p}(\boldsymbol{y}\cdot w_i)w_i. \tag{3.3}$$

Let vector spaces S and T be the two subspaces of a vector space V and union $S \cup T = V$. If for any vector $\boldsymbol{x} \in S$ and any vector $\boldsymbol{y} \in T$, the dot product $\boldsymbol{x}\cdot\boldsymbol{y} = 0$, then the two vector spaces S and T are said to be orthogonal. Or we can say that T is the orthogonal space of S, denoted by

$T = S^{\perp}$. Thus, for a vector space V, if S is a vector subspace in V, then $V = S \cup S^{\perp}$. Any vector $\boldsymbol{y}$ in V can be written uniquely as $\boldsymbol{y}_S + \boldsymbol{y}_S^{\perp}$, where $\boldsymbol{y}_S \in S$ and $\boldsymbol{y}_S^{\perp}$ is in $S^{\perp}$, the orthogonal subspace of S.

A projection of a vector onto a linear space S is actually a linear transformation of the vector and can be represented by a projection matrix times the vector. A projection matrix P is an $n \times n$ square matrix that gives the projection from $\mathbb{R}^n$ onto subspace S. The columns of P are the projections of the standard basis vectors, and S is the image of P. For the projection matrix we have the following theorems.

Theorem 3.1. *A square matrix P is a projection matrix if and only if it is idempotent, i.e., $P^2 = P$.*

Theorem 3.2. *Let $U = (u_1, u_2, \cdots, u_k)$ be an orthonormal basis for a subspace W of linear space V. The matrix UU' is a projection matrix of V onto W. i.e., for any vector $v \in V$ the projection of v onto W is $Proj_W v = UU'v$.*

The matrix UU' is called the projection matrix for the subspace W. It does not depend on the choice of orthonormal basis. If we do not start with an orthonormal basis of W, we can still construct the projection matrix. This can be summarized in the following theorem.

Theorem 3.3. *Let $A = (a_1, a_2, \cdots, a_k)$ be any basis for a subspace W of V. The matrix $A(A'A)^{-1}A'$ is a projection matrix of V onto W. i.e., for any vector $v \in V$ the projection of v onto W is*

$$Proj_W v = A(A'A)^{-1}A'v. \tag{3.4}$$

To understand the above three theorems the following lemma is important.

Lemma 3.1. *Suppose that A is an $n \times k$ matrix whose columns are linearly independent. Then AA' is invertible.*

Proof. Consider the transformation A: $\mathbb{R}^k \to \mathbb{R}^k$ determined by A. Since the columns of A are linearly independent, this transformation is one-to-one. In addition, the null space of A' is orthogonal to the column space of A. Thus, A' is one-to-one on the column space of A, and as a result, $A'A$ is one-to-one transformation $\mathbb{R}^k \to \mathbb{R}^k$. By invertible matrix theorem, $A'A$ is invertible. □

Let's now derive the projection matrix for the column space of A. Note that any element of the column space of A is a linear combination of the columns of A, i.e., $x_1a_1 + x_2a_2 + \cdots + x_ka_k$. If we write

$$\boldsymbol{x} = \begin{pmatrix} x_1 \\ x_2 \\ \vdots \\ x_k \end{pmatrix},$$

then we have

$$x_1a_1 + x_2a_2 + \cdots + x_ka_k = A\boldsymbol{x}.$$

Now, for any vector $v \in \mathbb{R}^n$, we denote the projection of v onto W by x_p.

$$Proj_W v = Ax_p.$$

The projection matrix can be found by calculating x_p. The projection of vector v onto W is characterized by the fact that $v - Proj_W v$ is orthogonal to any vector w in W. Thus we have

$$w \cdot (v - Proj_W v) = 0$$

for all w in W. Since $w = A\boldsymbol{x}$ for some $\boldsymbol{x}$, we have

$$A\boldsymbol{x} \cdot (v - Ax_p) = 0$$

for all $\boldsymbol{x}$ in $\mathbb{R}^n$. Write this dot product in terms of matrices yields

$$(A\boldsymbol{x})'(v - A\boldsymbol{x}_p) = 0$$

which is equivalent to

$$(x'A')(v - A\boldsymbol{x}_p) = 0$$

Converting back to dot products we have

$$\boldsymbol{x} \cdot A'(v - A\boldsymbol{x}_p) = 0$$

We get

$$A'v = A'A\boldsymbol{x}_p$$

Since $A'A$ is invertible we have

$$(A'A)^{-1}A'v = \boldsymbol{x}_p$$

Since $A\boldsymbol{x}_p$ is the desired projection, we have

$$A(A'A)^{-1}A'v = A\boldsymbol{x}_p = Proj_W v$$

Therefore, we conclude that the projection matrix for W is $A(A^{'}A)^{-1}A^{'}$.

Projection matrix is very useful in the subsequent discussions of linear regression model $Y = \boldsymbol{X\beta} + \varepsilon$. A squared matrix, $P = \boldsymbol{X}(\boldsymbol{XX}^{'})^{-1}\boldsymbol{X}^{'}$, is constructed using the design matrix. It can be easily verified that P is an idempotent matrix:

$$P^2 = \boldsymbol{X}(\boldsymbol{XX}^{'})^{-1}\boldsymbol{X}^{'}\boldsymbol{X}(\boldsymbol{XX}^{'})^{-1}\boldsymbol{X}^{'} = P.$$

Thus, $P = \boldsymbol{X}(\boldsymbol{XX}^{'})^{-1}\boldsymbol{X}^{'}$ is a projection matrix. In addition, if we define a matrix as $I - P = I - \boldsymbol{X}(\boldsymbol{XX}^{'})^{-1}\boldsymbol{X}^{'}$. It is easy to see that $I - P$ is also idempotent. In fact,

$$(I - P)^2 = I - 2P + P^2 = I - 2P + P = I - P.$$

Therefore, $I - P = I - \boldsymbol{X}(\boldsymbol{XX}^{'})^{-1}\boldsymbol{X}^{'}$ is a projection matrix. In the subsequent sections we will see how these projection matrices are used to obtain the best linear unbiased estimator (BLUE) for the linear regression model and how they are used in regression model diagnosis.

3.2 Matrix Form of Multiple Linear Regression

In many scientific research it is often needed to determine the relationship between a response (or dependent) variable (y) and more than one regressors (or independent variables) ($x_1, x_2, \cdots, x_k$). A general form of a multiple linear regression model is given by

$$y = \beta_0 + \beta_1 x_1 + \beta_2 x_2 + \cdots + \beta_k x_k + \varepsilon \tag{3.5}$$

where ε is the random error. Here, regressors $x_1, x_2, \cdots, x_k$ may contain regressors and their higher order terms. In the classical setting, it is assumed that the error term ε has the normal distribution with a mean 0 and a constant variance σ^2.

The first impression of the multiple regression may be a response plane. However, some regressors may be higher order terms of other regressors, or may even be functions of regressors as long as these functions do not contain unknown parameters. Thus, multiple regression model can be a response surface of versatile shapes. Readers may already realize the difference between a linear model and a nonlinear model.

Definition 3.1. A linear model is defined as a model that is linear in regression parameters, i.e., linear in β_i's.

The following are examples of linear regression models in which the response variable y is a linear function of regression parameters:

$$y = \beta_0 + \beta_1 x_1 + \beta_2 x_2 + \beta_3 x_3 + \varepsilon,$$
$$y = \beta_0 + \beta_1 x_1 + \beta_2 x_1^2 + \beta_3 x_2 + \varepsilon,$$
$$y = \beta_0 + \beta_1 x_1 + \beta_2 x_1^2 + \beta_3 x_2 + \beta_4 x_1 x_2 + \varepsilon,$$
$$y = \beta_0 + \beta_1 x_1 + \beta_2 ln(x_1) + \beta_2 ln(x_2) + \varepsilon,$$
$$y = \beta_0 + \beta_2 1_{(x_1>5)} + \beta_2 1_{(x_2>10)} + \beta_3 x_3 + \varepsilon.$$

In the last model $1_{(x_1>5)}$ is an indicator function taking value 1 if $x_1 > 5$ and 0 otherwise. Examples of non-linear regression model may be given by

$$y = \beta_0 + \beta_1 x_1 + \beta_2 x_2^{\gamma} + \varepsilon,$$
$$y = \frac{1}{\lambda + \exp(\beta_0 + \beta_1 x_1 + \beta_2 x_2 + \cdots + \beta_k x_k)} + \varepsilon,$$

where the response variable cannot be expressed as a linear function of regression parameters.

3.3 Quadratic Form of Random Variables

Definition 3.2. Let $y' = (y_1, y_2, \cdots, y_n)$ be n real variables and a_{ij} be $n \times n$ real numbers, where $i, j = 1, 2, \cdots, n$. A quadratic form of $y_1, y_2, \cdots, y_n$ is defined as

$$f(y_1, y_2, \cdots, y_n) = \sum_{i,j=1}^{n} a_{ij} y_i y_j.$$

This quadratic form can be written in the matrix form: $y'Ay$, where A is an $n \times n$ matrix $A = (a_{ij})_{n\times n}$. Quadratic form plays an important role in the discussions of linear regression model. In the classical setting the parameters of a linear regression model are estimated via minimizing the sum of squared residuals:

$$\begin{aligned} \mathbf{b} &= (b_0, b_1, \cdots, b_k) \\ &= \mathbf{arg\,min}_{(\beta_0, \beta_1, \cdots, \beta_k)} \sum_{i=1}^{n} \Big[y_i - (\beta_0 + \beta_1 x_{1i} + \beta_2 x_{2i} + \cdots + \beta_k x_{ki}) \Big]^2. \end{aligned}$$

This squared residual is actually a quadratic form. Thus, it is important to discuss some general properties of this quadratic form that will be used in the subsequent discussions.

3.4 Idempotent Matrices

In this section we discuss properties of the idempotent matrix and its applications in the linear regression. First we define the idempotent matrix.

Definition 3.3. An $n \times n$ symmetric matrix A is idempotent if $A^2 = A$.

Let $\alpha = (\alpha_1, \alpha_2, \cdots, \alpha_k)$ be a k-dimensional vector and A is a $k \times k$ matrix. $\alpha' A\alpha$ is a quadratic form of $\alpha_1, \alpha_2, \cdots, \alpha_k$. When A is an idempotent matrix, the corresponding quadratic form has its particular properties. The quadratic form with idempotent matrices are used extensively in linear regression analysis. We now discuss the properties of idempotent matrix.

Theorem 3.4. *Let $A_{n\times n}$ be an idempotent matrix of rank p, then the eigenvalues of A are either 1 or 0.*

Proof. Let λ_i and v_i be the eigenvalue and the corresponding normalized eigenvector of the matrix A, respectively. We then have $Av_i = \lambda_i v_i$, and $v_i' A v_i = \lambda_i v_i' v_i = \lambda_i$. On the other hand, since $A^2 = A$, we can write

$$\lambda_i = v_i' A v_i = v_i' A^2 v_i = v_i' A' A v_i = (Av_i)' A v_i = (\lambda_i v_i)' (\lambda_i v_i) = \lambda_i^2.$$

Hence, we have $\lambda_i(\lambda_i - 1) = 0$, which yields either $\lambda_i = 1$ or $\lambda_i = 0$. This completes the proof. □

It is easy to know that p eigenvalues of A are 1 and $n - p$ eigenvalues of A are zero. Therefore, the rank of an idempotent matrix A is the sum of its non-zero eigenvalues.

Definition 3.4. Let $A = (a_{i,j})_{n\times n}$ be an $n \times n$ matrix, trace of A is defined as the sum of the orthogonal elements. i.e.,

$$tr(A) = a_{11} + a_{22} + \cdots + a_{nn}.$$

If A is a symmetric matrix then the sum of all squared elements of A can be expressed by $tr(A^2)$. i.e., $\sum_{i,j} a_{ij}^2 = tr(A^2)$. It is easy to verify that $tr(AB) = tr(BA)$ for any two $n \times n$ matrices A and B. The following theorem gives the relationship between the rank of matrix A and and trace of A when A is an idempotent matrix.

Theorem 3.5. *If A is an idempotent matrix then $tr(A) = rank(A) = p$.*

Proof. If the rank of an $n \times n$ idempotent matrix A is p then A has p eigenvalues of 1 and $n - p$ eigenvalues of 0. Thus, we can write $rank(A) =$

$\sum_{i=1}^{n} \lambda_i = p$. Since $A^2 = A$, the eigenvalues of the idempotent matrix A is either 1 or 0. From matrix theory there is an orthogonal matrix V such that

$$V'AV = \begin{pmatrix} I_p & 0 \\ 0 & 0 \end{pmatrix}.$$

Therefore, we have

$$tr(V'AV) = tr(VV'A) = tr(A) = tr\begin{pmatrix} I_p & 0 \\ 0 & 0 \end{pmatrix} = p = rank(A).$$

Here we use the simple fact: $tr(AB) = tr(BA)$ for any matrices $A_{n\times n}$ and $B_{n\times n}$. □

A quadratic form of a random vector $\boldsymbol{y}' = (y_1, y_2, \cdots, y_n)$ can be written in a matrix form $\boldsymbol{y}'A\boldsymbol{y}$, where A is an $n \times n$ matrix. It is of interest to find the expectation and variance of $\boldsymbol{y}'A\boldsymbol{y}$. The following theorem gives the expected value of $\boldsymbol{y}'A\boldsymbol{y}$ when the components of $\boldsymbol{y}$ are independent.

Theorem 3.6. *Let $\boldsymbol{y}' = (y_1, y_2, \cdots, y_n)$ be an $n \times 1$ random vector with mean $\mu' = (\mu_1, \mu_2, \cdots, \mu_n)$ and variance σ^2 for each component. Further, it is assumed that $y_1, y_2, \cdots, y_n$ are independent. Let A be an $n \times n$ matrix, $\boldsymbol{y}'A\boldsymbol{y}$ is a quadratic form of random variables. The expectation of this quadratic form is given by*

$$E(\boldsymbol{y}'A\boldsymbol{y}) = \sigma^2 tr(A) + \mu'A\mu. \tag{3.6}$$

Proof. First we observe that

$$\boldsymbol{y}'A\boldsymbol{y} = (\boldsymbol{y} - \mu)'A(\boldsymbol{y} - \mu) + 2\mu'A(\boldsymbol{y} - \mu) + \mu'A\mu.$$

We can write

$$E(\boldsymbol{y}'A\boldsymbol{y}) = E[(\boldsymbol{y} - \mu)'A(\boldsymbol{y} - \mu)] + 2E[\mu'A(\boldsymbol{y} - \mu)] + \mu'A\mu$$

$$= E\Big[\sum_{i,j=1}^{n} a_{ij}(y_i - \mu_i)(y_j - \mu_j)\Big] + 2\mu'AE(\boldsymbol{y} - \mu) + \mu'A\mu$$

$$= \sum_{i=1}^{n} a_{ii}E(y_i - \mu_i)^2 + \mu'A\mu = \sigma^2 tr(A) + \mu'A\mu.$$

□

We now discuss the variance of the quadratic form $\boldsymbol{y}'A\boldsymbol{y}$.

Theorem 3.7. *Let $\boldsymbol{y}$ be an $n \times 1$ random vector with mean $\mu' = (\mu_1, \mu_2, \cdots, \mu_n)$ and variance σ^2 for each component. It is assumed that $y_1, y_2, \cdots, y_n$ are independent. Let A be an $n \times n$ symmetric matrix, $E(y_i - \mu_i)^4 = \mu_i^{(4)}$, $E(y_i - \mu_i)^3 = \mu_i^{(3)}$, and $a' = (a_{11}, a_{22}, \cdots, a_{nn})$. The variance of the quadratic form $Y'AY$ is given by*

$$\begin{aligned} Var(\boldsymbol{y}'A\boldsymbol{y}) &= (\mu^{(4)} - 3\sigma^2)a'a + \sigma^4(2tr(A^2) + [tr(A)]^2) \\ &\quad + 4\sigma^2\mu'A^2\mu + 4\mu^{(3)}a'A\mu. \end{aligned} \tag{3.7}$$

Proof. Let $Z = \boldsymbol{y} - \mu$, $A = (A_1, A_2, \cdots, A_n)$, and $b = (b_1, b_2, \cdots, b_n) = \mu'(A_1, A_2, \cdots, A_n) = \mu'A$ we can write

$$\begin{aligned} \boldsymbol{y}'A\boldsymbol{y} &= (\boldsymbol{y}' - \mu)A(\boldsymbol{y} - \mu) + 2\mu'A(\boldsymbol{y} - \mu) + \mu'A\mu \\ &= Z'AZ + 2bZ + \mu'A\mu. \end{aligned}$$

Thus

$$\mathrm{Var}(\boldsymbol{y}'A\boldsymbol{y}) = \mathrm{Var}(Z'AZ) + 4Var(bZ) + 4\mathrm{Cov}(Z'AZ,\ bZ).$$

We then calculate each term separately:

$$\begin{aligned} (Z'AZ)^2 &= \sum_{ij} a_{ij}a_{lm}Z_iZ_jZ_lZ_m \\ E(Z'AZ)^2 &= \sum_{i\,j\,l\,m} a_{ij}a_{lm}E(Z_iZ_jZ_lZ_m) \end{aligned}$$

Note that

$$E(Z_iZ_jZ_lZ_m) = \begin{cases} \mu^{(4)}, & \text{if } i = j = k = l; \\ \sigma^4, & \text{if } i = j,\ l = k \text{ or } i = l,\ j = k, \text{ or } i = k,\ j = l\ ; \\ 0, & \text{else.} \end{cases}$$

We have

$$\begin{aligned} E(Z'AZ)^2 &= \sum_{i\,j\,l\,m} a_{ij}a_{lm}E(Z_iZ_jZ_lZ_m) \\ &= \mu^{(4)}\sum_{i=1}^{n} a_{ii}^2 + \sigma^4\Big(\sum_{i \neq k} a_{ii}a_{kk} + \sum_{i \neq j} a_{ij}^2 + \sum_{i \neq j} a_{ij}a_{ji}\Big) \end{aligned}$$

Since A is symmetric, $a_{ij} = a_{ji}$, we have

$$\begin{aligned}
&\sum_{i\neq j} a_{ij}^2 + \sum_{i\neq j} a_{ij}a_{ji} \\
&= 2\sum_{i\neq j} a_{ij}^2 = 2\sum_{i,j} a_{ij}^2 - 2\sum_{i=j} a_{ij}^2 \\
&= 2tr(A^2) - 2\sum_{i=1}^{n} a_{ii}^2 \\
&= 2tr(A^2) - 2a^{'}a
\end{aligned}$$

and

$$\begin{aligned}
&\sum_{i\neq k} a_{ii}a_{kk} = \sum_{i,k} a_{ii}a_{kk} - \sum_{i=k} a_{ii}a_{kk} \\
&= [tr(A)]^2 - \sum_{i=1}^{n} a_{ii}^2 = [tr(A)]^2 - a^{'}a.
\end{aligned}$$

So we can write

$$E(Z^{'}AZ)^2 = (\mu^{(4)} - 3\sigma^4)a^{'}a + \sigma^4(2tr(A^2) + [tr(A)]^2). \tag{3.8}$$

For Var(bZ) we have

$$\text{Var}(bZ) = b\text{Var}(Z)b^{'} = bb^{'}\sigma^2 = (\mu^{'}A)(\mu^{'}A)^{'}\sigma^2 = \mu^{'}A^2\mu\sigma^2. \tag{3.9}$$

To calculate Cov$(Z^{'}AZ,\ bZ)$, note that $EZ = 0$, we have

$$\begin{aligned}
&\text{Cov}(Z^{'}AZ,\ bZ) \\
&= \text{Cov}\Big(\sum_{i,j} a_{ij}Z_iZ_j, \sum_{k} b_kZ_k\Big) \\
&= \sum_{i,j,k} a_{ij}b_k\text{Cov}(Z_iZ_j,\ Z_k) \\
&= \sum_{i,j,k} a_{ij}b_kE[(Z_iZ_j - E(Z_iZ_j))Z_k] \\
&= \sum_{i,j,k} a_{ij}b_k[E(Z_iZ_jZ_k) - E(Z_iZ_j)EZ_k] \\
&= \sum_{i,j,k} a_{ij}b_k[E(Z_iZ_jZ_k)] \quad (\text{since } EZ_k = 0).
\end{aligned}$$

It is easy to know that

$$E(Z_iZ_jZ_k) = \begin{cases} \mu^{(3)}, & \text{if } i = j = k; \\ 0, & \text{else.} \end{cases}$$

Thus,

$$\text{Cov}(Z^{'}AZ,\ bZ) = \sum_{i=1}^{n} a_{ii}b_i\mu^{(3)}$$

$$= \sum_{i=1}^{n} a_{ii}\mu^{'}A_i\mu^{(3)} = \sum_{i=1}^{n} a_{ii}A_i^{'}\mu\ \mu^{(3)} = a^{'}A\mu\ \mu^{(3)}. \tag{3.10}$$

Combining the results above completes the proof. □

3.5 Multivariate Normal Distribution

A random variable Y is said to follow the normal distribution $N(\mu, \sigma^2)$ if and only if the probability density function of Y is

$$f(y) = \frac{1}{\sqrt{2\pi}\sigma}\exp\Big\{-\frac{(y-\mu)^2}{\sigma^2}\Big\} \text{ for } -\infty < y < \infty. \tag{3.11}$$

The cumulative distribution of Y is defined as

$$F(y) = P(Y \le y) = \frac{1}{\sqrt{2\pi}\sigma}\int_{-\infty}^{y}\exp\Big\{-\frac{(y-\mu)^2}{\sigma^2}\Big\}dy. \tag{3.12}$$

The moment generating function for the normal random variable $Y \sim N(\mu,\ \sigma)$ is

$$M(t) = E(e^{tY}) = \exp(t\mu + \frac{1}{2}t^2\sigma^2). \tag{3.13}$$

The multivariate normal distribution is an extension of the univariate normal distribution. A random vector $\boldsymbol{y}^{'} = (y_1, y_2, \cdots, y_p)$ is said to follow the multivariate normal distribution if and only if its probability density function has the following form

$$f(y_1, y_2, \cdots, y_p) \tag{3.14}$$
$$= \frac{1}{(2\pi)^{p/2}|\Sigma|^{1/2}}\exp\Big\{-\frac{1}{2}(\boldsymbol{y}-\boldsymbol{\mu})^{'}\Sigma^{-1}(\boldsymbol{y}-\boldsymbol{\mu})\Big\},$$

where $\Sigma = (\sigma_{ij})_{p\times p}$ is the covariance matrix of $\boldsymbol{y}$ and the inverse matrix Σ^{-1} exists. $\boldsymbol{\mu}^{'} = (\mu_1, \mu_2, \cdots, \mu_p)$ is the mean vector of $\boldsymbol{y}$.

When Σ is a diagonal matrix $\Sigma = diag(\sigma_1^2, \sigma_2^2, \cdots, \sigma_p^2)$, or $\sigma_{ij} = 0$ for all $i \neq j$, then $y_1, y_2, \cdots, y_p$ are not correlated since it is easy to know that the density function of $\boldsymbol{y}$ can be written as a product of p univariate normal density function:

$$\frac{1}{(2\pi)^{p/2}|\Sigma|^{1/2}}\exp\Big\{-\frac{1}{2}(\boldsymbol{y}-\boldsymbol{\mu})^{'}\Sigma^{-1}(\boldsymbol{y}-\boldsymbol{\mu})\Big\} = \prod_{i=1}^{p}\frac{1}{\sqrt{2\pi}\sigma_i}\exp\{-\frac{(y_i-\mu_i)^2}{\sigma_i^2}\}$$

Since density function of multivariate normal vector $\boldsymbol{y}$ is a product of density functions of $y_1, y_2, \cdots, y_p$, they are jointly independent. For multivariate normal variables, the uncorrelated normal random variables are jointly independent. We summarize this into the following theorem:

Theorem 3.8. *If random vector $\boldsymbol{y}^{'} = (y_1, y_2, \cdots, y_p)$ follows a multivariate normal distribution $N(\boldsymbol{\mu}, \Sigma)$ and the covariance matrix $\Sigma = (\sigma_{ij})_{p\times p}$ is a diagonal matrix $diag(\sigma_{11}, \sigma_{22}, \cdots, \sigma_{pp})$, then $y_1, y_2, \cdots, y_p$ are jointly independent.*

We now introduce the central χ^2 distribution. Let $y_1, y_2, \cdots, y_p$ be p independent standard normal random variables, i.e., $E(y_i) = 0$ and $\text{Var}(y_i) = 1$. The special quadratic form $Z = \sum_{i=1}^{p} y_i^2$ has the chi-square distribution with p degrees of freedom and non-centrality parameter $\lambda = 0$. In addition, the random variable Z has the density function

$$f(z) = \frac{1}{\Gamma(p/2)2^{p/2}}z^{(p-2)/2}e^{-z/2} \quad \text{for } 0 < z < \infty. \tag{3.15}$$

The moment generating function for Z is given by

$$M(t) = E(e^{tZ}) = (1-2t)^{-n/2} \quad \text{for } t < \frac{1}{2}. \tag{3.16}$$

Using this moment generating function it is easy to find $E(Z) = p$ and $\text{Var}(Z) = 2p$. In addition, the following results are obtained through direct calculations:

$$E(Z^2) = p\,(p+2),$$

$$E(\sqrt{Z}\,) = \frac{\sqrt{2}\,\Gamma[(p+1)/2]}{\Gamma(p/2)},$$

$$E\Big(\frac{1}{Z}\Big) = \frac{1}{p-2},$$

$$E\Big(\frac{1}{Z^2}\Big) = \frac{1}{(n-2)(n-4)},$$

$$E\Big(\frac{1}{\sqrt{Z}}\Big) = \frac{\Gamma[(p-1/2)]}{\sqrt{2}\,\Gamma(p/2)}.$$

3.6 Quadratic Form of the Multivariate Normal Variables

The distribution of the quadratic form $\boldsymbol{y}'A\boldsymbol{y}$ when $\boldsymbol{y}$ follows the multivariate normal distribution plays a significant role in the discussion of linear regression methods. We should further discuss some theorems about the distribution of the quadratic form based upon the mean and covariance matrix of a normal vector $\boldsymbol{y}$, as well as the matrix A.

Theorem 3.9. *Let $\boldsymbol{y}$ be an $n \times 1$ normal vector and $\boldsymbol{y} \sim N(0, I)$. Let A be an idempotent matrix of rank p. i.e., $A^2 = A$. The quadratic form $\boldsymbol{y}'A\boldsymbol{y}$ has the chi-square distribution with p degrees of freedom.*

Proof. Since A is an idempotent matrix of rank p. The eigenvalues of A are 1's and 0's. Moreover, there is an orthogonal matrix V such that

$$VAV' = \begin{pmatrix} I_p & 0 \\ 0 & 0 \end{pmatrix}.$$

Now, define a new vector $\boldsymbol{z} = V\boldsymbol{y}$ and $\boldsymbol{z}$ is a multivariate normal vector. $E(\boldsymbol{z}) = VE(\boldsymbol{y}) = 0$ and $\text{Cov}(\boldsymbol{z}) = \text{Cov}(V\boldsymbol{y}) = V\text{Cov}(\boldsymbol{y})V' = VI_pV' = I_p$. Thus, $\boldsymbol{z} \sim N(0, I_p)$. Notice that V is an orthogonal matrix and

$$\boldsymbol{y}'A\boldsymbol{y} = (V'\boldsymbol{z})'AV'\boldsymbol{z} = \boldsymbol{z}'VAV'\boldsymbol{z} = \boldsymbol{z}'I_p\boldsymbol{z} = \sum_{i=1}^{p} z_i^2.$$

By the definition of the chi-square random variable, $\sum_{i=1}^{p} z_i^2$ has the chi-square distribution with p degrees of freedom. □

The above theorem is for the quadratic form of a normal vector $\boldsymbol{y}$ when $E\boldsymbol{y} = 0$. This condition is not completely necessary. However, if this condition is removed, i.e., if $E(\boldsymbol{y}) = \mu \neq 0$ the quadratic form of $\boldsymbol{y}'A\boldsymbol{y}$ still follows the chi-square distribution but with a non-centrality parameter $\lambda = \frac{1}{2}\boldsymbol{\mu}'A\boldsymbol{\mu}$. We state the theorem and the proofs of the theorem should follow the same lines as the proofs of the theorem for the case of $\mu = 0$.

Theorem 3.10. *Let $\boldsymbol{y}$ be an $n \times 1$ normal vector and $y \sim N(\mu, I)$. Let A be an idempotent matrix of rank p. The quadratic form $\boldsymbol{y}'A\boldsymbol{y}$ has the chi-square distribution with degrees of freedom p and the non-centrality parameter $\lambda = \frac{1}{2}\mu'A\mu$.*

We now discuss more general situation where the normal vector $\boldsymbol{y}$ follows a multivariate normal distribution with mean $\boldsymbol{\mu}$ and covariance matrix Σ.

Theorem 3.11. *Let $\boldsymbol{y}$ be a multivariate normal vector with mean $\boldsymbol{\mu}$ and covariance matrix Σ. If $A\Sigma$ is an idempotent matrix of rank p, The quadratic form of $\boldsymbol{y}'A\boldsymbol{y}$ follows a chi-square distribution with degrees of freedom p and non-centrality parameter $\lambda = \frac{1}{2}\mu'A\mu$.*

Proof. First, for covariance matrix Σ there exists an orthogonal matrix Γ such that $\Sigma = \Gamma\Gamma'$. Define $Z = \Gamma^{-1}(\boldsymbol{y} - \mu)$ and Z is a normal vector with $E(Z) = 0$ and

$$\begin{aligned}\text{Cov}(Z) &= \text{Cov}(\Gamma^{-1}(\boldsymbol{y} - \mu)) = \Gamma^{-1}\text{Cov}(\boldsymbol{y})\Gamma'^{-1} = \Gamma^{-1}\Sigma\Gamma'^{-1} \\ &= \Gamma^{-1}(\Gamma\Gamma')\Gamma'^{-1} = I_p.\end{aligned}$$

i.e., $Z \sim N(0, I)$. Moreover, since $\boldsymbol{y} = \Gamma Z + \mu$ we have

$$\boldsymbol{y}'A\boldsymbol{y} = [\Gamma Z + \mu)]'A(\Gamma Z + \mu) = (Z' + \Gamma'^{-1}\mu)'(\Gamma' A\Gamma)(Z + \Gamma'^{-1}\mu) = V'BV,$$

where $V = Z' + \Gamma'^{-1}\mu \sim N(\Gamma'^{-1}\mu,\ I_p)$ and $B = \Gamma' A\Gamma$. We now need to show that B is an idempotent matrix. In fact,

$$B^2 = (\Gamma' A\Gamma)(\Gamma' A\Gamma) = \Gamma'(A\Gamma\Gamma' A)\Gamma$$

Since $A\Sigma$ is idempotent we can write

$$A\Sigma = A\Gamma\Gamma' = A\Sigma A\Sigma = (A\Gamma\Gamma' A)\Gamma\Gamma' = (A\Gamma\Gamma' A)\Sigma.$$

Note that Σ is non-singular we have

$$A = A\Gamma\Gamma' A.$$

Thus,

$$B^2 = \Gamma'(A\Gamma\Gamma' A)\Gamma = \Gamma' A\Gamma = B.$$

i.e., B is an idempotent matrix. This concludes that $V' BV$ is a chi-square random variable with degrees of freedom p. To find the non-centrality parameter we have

$$\begin{aligned}\lambda &= \frac{1}{2}(\Gamma'^{-1}\mu)' B(\Gamma'^{-1}\mu) \\ &= \frac{1}{2}\mu'\Gamma'^{-1}(\Gamma' A\Gamma)\Gamma'^{-1}\mu = \frac{1}{2}\mu' A\mu.\end{aligned}$$

This completes the proof. □

3.7 Least Squares Estimates of the Multiple Regression Parameters

The multiple linear regression model is typically stated in the following form

$$y_i = \beta_0 + \beta_1 x_{1i} + \beta_2 x_{2i} + \cdots + \beta_k x_{ki} + \varepsilon_i,$$

where y_i is the dependent variable, $\beta_0, \beta_1, \beta_2, \cdots, \beta_k$ are the regression coefficients, and ε_i's are the random errors assuming $E(\varepsilon_i) = 0$ and $\text{Var}(\varepsilon_i) = \sigma^2$ for $i = 1, 2, \cdots, n$. In the classical regression setting the error term is assumed to be normally distributed with a constant variance σ^2. The regression coefficients are estimated using the least squares principle. It should be noted that it is not necessary to assume that the regression error term follows the normal distribution in order to find the least squares estimation of the regression coefficients. It is rather easy to show that under the assumption of normality of the error term, the least squares estimation of the regression coefficients are exactly the same as the maximum likelihood estimations (MLE) of the regression coefficients.

The multiple linear model can also be expressed in the matrix format

$$\boldsymbol{y} = \boldsymbol{X\beta} + \boldsymbol{\varepsilon},$$

where

$$\boldsymbol{X} = \begin{pmatrix} x_{11} & x_{12} & \cdots & x_{1k} \\ x_{21} & x_{22} & \cdots & x_{2k} \\ \cdots \\ x_{n1} & x_{n2} & \cdots & x_{nk} \end{pmatrix} \quad \boldsymbol{\beta} = \begin{pmatrix} \beta_0 \\ \beta_1 \\ \beta_2 \\ \cdots \\ \beta_{k-1} \end{pmatrix} \quad \boldsymbol{\varepsilon} = \begin{pmatrix} \varepsilon_1 \\ \varepsilon_2 \\ \varepsilon_3 \\ \cdots \\ \varepsilon_n \end{pmatrix} \tag{3.17}$$

The matrix form of the multiple regression model allows us to discuss and present many properties of the regression model more conveniently and efficiently. As we will see later the simple linear regression is a special case of the multiple linear regression and can be expressed in a matrix format. The least squares estimation of $\boldsymbol{\beta}$ can be solved through the least squares principle:

$$\boldsymbol{b} = \mathbf{arg\,min}_{\boldsymbol{\beta}}[(\boldsymbol{y} - \boldsymbol{X\beta})'(\boldsymbol{y} - \boldsymbol{X\beta})],$$

where $\boldsymbol{b}' = (b_0, b_1, \cdots b_{k-1})'$, a k-dimensional vector of the estimations of the regression coefficients.

Theorem 3.12. *The least squares estimation of $\boldsymbol{\beta}$ for the multiple linear regression model $\boldsymbol{y} = \boldsymbol{X\beta} + \boldsymbol{\varepsilon}$ is $\boldsymbol{b} = (\boldsymbol{X}'\boldsymbol{X})^{-1}\boldsymbol{X}'\boldsymbol{y}$, assuming $(\boldsymbol{X}'\boldsymbol{X})$ is a non-singular matrix. Note that this is equivalent to assuming that the column vectors of $\boldsymbol{X}$ are independent.*

Proof. To obtain the least squares estimation of $\boldsymbol{\beta}$ we need to minimize the residual of sum squares by solving the following equation:

$$\frac{\partial}{\partial \boldsymbol{b}}[(\boldsymbol{y} - \boldsymbol{Xb})'(\boldsymbol{y} - \boldsymbol{Xb})] = 0,$$

or equivalently,

$$\frac{\partial}{\partial \boldsymbol{b}}[(\boldsymbol{y}'\boldsymbol{y} - 2\boldsymbol{y}'\boldsymbol{Xb} + \boldsymbol{b}'\boldsymbol{X}'\boldsymbol{Xb})] = 0.$$

By taking partial derivative with respect to each component of $\boldsymbol{\beta}$ we obtain the following normal equation of the multiple linear regression model:

$$\boldsymbol{X}'\boldsymbol{Xb} = \boldsymbol{X}'\boldsymbol{y}.$$

Since $\boldsymbol{X}'\boldsymbol{X}$ is non-singular it follows that $\boldsymbol{b} = (\boldsymbol{X}'\boldsymbol{X})^{-1}\boldsymbol{X}'\boldsymbol{y}$. This completes the proof. □

We now discuss statistical properties of the least squares estimation of the regression coefficients. We first discuss the unbiasness of the least squares estimation $\boldsymbol{b}$.

Theorem 3.13. *The estimator* $\boldsymbol{b} = (\boldsymbol{X}'\boldsymbol{X})^{-1}\boldsymbol{X}'\boldsymbol{y}$ *is an unbiased estimator of* $\boldsymbol{\beta}$. *In addition,*

$$Var(\boldsymbol{b}) = (\boldsymbol{X}'\boldsymbol{X})^{-1}\sigma^2. \tag{3.18}$$

Proof. We notice that

$$E\boldsymbol{b} = E((\boldsymbol{X}'\boldsymbol{X})^{-1}\boldsymbol{X}'\boldsymbol{y}) = (\boldsymbol{X}'\boldsymbol{X})^{-1}\boldsymbol{X}'E(\boldsymbol{y}) = (\boldsymbol{X}'\boldsymbol{X})^{-1}\boldsymbol{X}'\boldsymbol{X}\boldsymbol{\beta} = \boldsymbol{\beta}.$$

This completes the proof of the unbiasness of $\boldsymbol{b}$. Now we further discuss how to calculate the variance of $\boldsymbol{b}$. The variance of the $\boldsymbol{b}$ can be computed directly:

$$\begin{aligned}
&\mathrm{Var}(\boldsymbol{b}) = \mathrm{Var}((\boldsymbol{X}'\boldsymbol{X})^{-1}\boldsymbol{X}'\boldsymbol{y})\\
&= (\boldsymbol{X}'\boldsymbol{X})^{-1}\boldsymbol{X}'\mathrm{Var}(\boldsymbol{b})((\boldsymbol{X}'\boldsymbol{X})^{-1}\boldsymbol{X}')'\\
&= (\boldsymbol{X}'\boldsymbol{X})^{-1}\boldsymbol{X}'\boldsymbol{X}(\boldsymbol{X}'\boldsymbol{X})^{-1}\sigma^2 = (\boldsymbol{X}'\boldsymbol{X})^{-1}\sigma^2.
\end{aligned}$$

□

Another parameter in the classical linear regression is the variance σ^2, a quantity that is unobservable. Statistical inference on regression coefficients and regression model diagnosis highly depend on the estimation of error variance σ^2. In order to estimate σ^2, consider the residual sum of squares:

$$\boldsymbol{e}^t\boldsymbol{e} = (\boldsymbol{y} - \boldsymbol{X}\boldsymbol{b})'(\boldsymbol{y} - \boldsymbol{X}\boldsymbol{b}) = \boldsymbol{y}'[I - \boldsymbol{X}(\boldsymbol{X}'\boldsymbol{X})^{-1}\boldsymbol{X}']\boldsymbol{y} = \boldsymbol{y}'P\boldsymbol{y}.$$

This is actually a distance measure between observed $\boldsymbol{y}$ and fitted regression value $\hat{\boldsymbol{y}}$. Note that it is easy to verify that $P = [I - \boldsymbol{X}(\boldsymbol{X}'\boldsymbol{X})^{-1}\boldsymbol{X}']$ is idempotent. i.e.,

$$P^2 = [I - \boldsymbol{X}(\boldsymbol{X}'\boldsymbol{X})^{-1}\boldsymbol{X}'][I - \boldsymbol{X}(\boldsymbol{X}'\boldsymbol{X})^{-1}\boldsymbol{X}'] = [I - \boldsymbol{X}(\boldsymbol{X}'\boldsymbol{X})^{-1}\boldsymbol{X}'] = P.$$

Therefore, the eigenvalues of P are either 1 or 0. Note that the matrix $\boldsymbol{X}(\boldsymbol{X}'\boldsymbol{X})^{-1}\boldsymbol{X}'$ is also idempotent. Thus, we have

$$\begin{aligned}
&rank(\boldsymbol{X}(\boldsymbol{X}'\boldsymbol{X})^{-1}\boldsymbol{X}') = tr(\boldsymbol{X}(\boldsymbol{X}'\boldsymbol{X})^{-1}\boldsymbol{X}')\\
&= tr(\boldsymbol{X}'\boldsymbol{X}(\boldsymbol{X}'\boldsymbol{X})^{-1}) = tr(I_p) = p.
\end{aligned}$$

Since $tr(A - B) = tr(A) - tr(B)$ we have

$$
\begin{aligned}
& rank(I - \boldsymbol{X}(\boldsymbol{X}'\boldsymbol{X})^{-1}\boldsymbol{X}') = tr(I - \boldsymbol{X}(\boldsymbol{X}'\boldsymbol{X})^{-1}\boldsymbol{X}') \\
&= tr(I_n) - tr(\boldsymbol{X}'\boldsymbol{X}(\boldsymbol{X}'\boldsymbol{X})^{-1}) = n - p
\end{aligned}
$$

The residual of sum squares in the multiple linear regression is $\boldsymbol{e}'\boldsymbol{e}$ which can be written as a quadratic form of the response vector $\boldsymbol{y}$.

$$
\boldsymbol{e}'\boldsymbol{e} = (\boldsymbol{y} - \boldsymbol{X}\boldsymbol{b})'(\boldsymbol{y} - \boldsymbol{X}\boldsymbol{b}) = \boldsymbol{y}'(I - \boldsymbol{X}(\boldsymbol{X}'\boldsymbol{X})^{-1}\boldsymbol{X}')\boldsymbol{y}.
$$

Using the result of the mathematical expectation of the quadratic form we have

$$
\begin{aligned}
& E(\boldsymbol{e}'\boldsymbol{e}) = E\Big[\boldsymbol{y}'(I - \boldsymbol{X}(\boldsymbol{X}'\boldsymbol{X})^{-1}\boldsymbol{X}')\boldsymbol{y}\Big] \\
&= (\boldsymbol{X}\beta)'(I - \boldsymbol{X}(\boldsymbol{X}'\boldsymbol{X})^{-1}\boldsymbol{X}')(\boldsymbol{X}\beta) + \sigma^2(n - p) \\
&= (\boldsymbol{X}\beta)'(\boldsymbol{X}\beta - \boldsymbol{X}(\boldsymbol{X}'\boldsymbol{X})^{-1}\boldsymbol{X}'\boldsymbol{X}\beta) + \sigma^2(n - p) = \sigma^2(n - p)
\end{aligned}
$$

We summarize the discussions above into the following theorem:

Theorem 3.14. *The unbiased estimator of the variance in the multiple linear regression is given by*

$$
s^2 = \frac{\boldsymbol{e}'\boldsymbol{e}}{n - p} = \frac{\boldsymbol{y}'(I - \boldsymbol{X}(\boldsymbol{X}'\boldsymbol{X})^{-1}\boldsymbol{X}')\boldsymbol{y}}{n - p} = \frac{1}{n - p}\sum_{i=1}^{n}(y_i - \hat{y}_i)^2. \quad (3.19)
$$

Let $P = \boldsymbol{X}(\boldsymbol{X}'\boldsymbol{X})^{-1}\boldsymbol{X}'$. The vector $\boldsymbol{y}$ can be partitioned into two vectors $(I - P)\boldsymbol{y} = (I - \boldsymbol{X}(\boldsymbol{X}'\boldsymbol{X})^{-1}\boldsymbol{X}')\boldsymbol{y}$ and $P\boldsymbol{y} = \boldsymbol{X}(\boldsymbol{X}'\boldsymbol{X})^{-1}\boldsymbol{X}')\boldsymbol{y}$. Assuming the normality of regression error term $(I - P)\boldsymbol{y}$ is independent of $P\boldsymbol{y}$. To see this we simply calculate the covariance of $(I - P)\boldsymbol{y}$ and $P\boldsymbol{y}$:

$$
\begin{aligned}
& \text{Cov}\Big((I - P)\boldsymbol{y}, P\boldsymbol{y}\Big) \\
&= (I - P)\text{Cov}(\boldsymbol{y})P = (I - P)P\sigma^2 \\
&= (I - \boldsymbol{X}(\boldsymbol{X}'\boldsymbol{X})^{-1}\boldsymbol{X}')\boldsymbol{X}(\boldsymbol{X}'\boldsymbol{X})^{-1}\boldsymbol{X}'\sigma^2 \\
&= [\boldsymbol{X}(\boldsymbol{X}'\boldsymbol{X})^{-1}\boldsymbol{X}' - (\boldsymbol{X}(\boldsymbol{X}'\boldsymbol{X})^{-1}\boldsymbol{X}')\boldsymbol{X}(\boldsymbol{X}'\boldsymbol{X})^{-1}\boldsymbol{X}']\sigma^2 \\
&= (\boldsymbol{X}(\boldsymbol{X}'\boldsymbol{X})^{-1}\boldsymbol{X}' - \boldsymbol{X}(\boldsymbol{X}'\boldsymbol{X})^{-1}\boldsymbol{X}')\sigma^2 = 0
\end{aligned}
$$

Since $(I - P)\boldsymbol{y}$ and $P\boldsymbol{y}$ are normal vectors, the zero covariance implies that they are independent of each other. Thus, the quadratic functions

$\boldsymbol{y}'(I-P)\boldsymbol{y}$ and $\boldsymbol{y}'P\boldsymbol{y}$ are independent as well. When P is idempotent, the quadratic function of a normal vector $\boldsymbol{y}'P\boldsymbol{y}$ follows the chi-square distribution with degrees of freedom p, where $p = rank(P)$. This property can be used to construct the F test statistic that is commonly used in the hypothesis testing problem for multiple linear regression.

The above calculations can be simplified if we introduce the following theorem for the two linear transformations of a multivariate normal variable $\boldsymbol{y}$.

Theorem 3.15. *Let $\boldsymbol{y} \sim N(\mu, I)$ and A and B be two matrices. Two normal vectors $A\boldsymbol{y}$ and $B\boldsymbol{y}$ are independent if and only if $AB' = 0$.*

Proof. Recall that the independence of two normal vectors is equivalent to zero covariance between them. We calculate the covariance of $A\boldsymbol{y}$ and $B\boldsymbol{y}$.

$$\text{Cov}(A\boldsymbol{y}, B\boldsymbol{y}) = A\text{Cov}(\boldsymbol{y})B' = AB'$$

Thus, the independence of two normal vectors $A\boldsymbol{y}$ and $B\boldsymbol{y}$ is equivalent to $AB' = 0$. □

By using this theorem we can easily show that $(I-P)\boldsymbol{y}$ and $P\boldsymbol{y}$ are independent. In fact, because P is idempotent, therefore, $(I-P)P = P - P^2 = P - P = 0$. The result follows immediately.

3.8 Matrix Form of the Simple Linear Regression

The simple linear regression model is a special case of the multiple linear regression and can be expressed in the matrix format. In particular,

$$\boldsymbol{X} = \begin{pmatrix} 1 & x_1 \\ 1 & x_2 \\ \cdots & \\ 1 & x_n \end{pmatrix}, \quad \boldsymbol{\beta} = \begin{pmatrix} \beta_0 \\ \beta_1 \end{pmatrix}, \quad \boldsymbol{\varepsilon} = \begin{pmatrix} \varepsilon_1 \\ \varepsilon_2 \\ \cdots \\ \varepsilon_n \end{pmatrix}.$$

The formula for calculating $\boldsymbol{b}$ in matrix format can be applied to the simple linear regression.

$$\boldsymbol{X}'\boldsymbol{X} = \begin{pmatrix} n & \sum x_i \\ \sum x_i & \sum x_i^2 \end{pmatrix}.$$

It is not difficult to solve for $(\boldsymbol{X}'\boldsymbol{X})^{-1}$ analytically. In fact, the inverse matrix of $\boldsymbol{X}'\boldsymbol{X}$ is given by

$$(\boldsymbol{X}'\boldsymbol{X})^{-1} = \frac{1}{n\sum_{i=1}^{n} x_i^2 - (\sum_{i=1}^{n} x_i)^2} \begin{pmatrix} \sum x_i^2 & -\sum x_i \\ -\sum x_i & n \end{pmatrix}$$

$$= \frac{1}{n\sum_{i=1}^{n}(x_i - \bar{x})^2} \begin{pmatrix} \sum x_i^2 & -\sum x_i \\ -\sum x_i & n \end{pmatrix}.$$

The least squares estimation of the simple linear regression can be calculated based on its matrix form:

$$\boldsymbol{b} = (\boldsymbol{X}'\boldsymbol{X})^{-1}\boldsymbol{X}'\boldsymbol{y} == \frac{1}{n\sum_{i=1}^{n}(x_i - \bar{x})^2} \begin{pmatrix} \sum x_i^2 & -\sum x_i \\ -\sum x_i & n \end{pmatrix} \begin{pmatrix} \sum y_i \\ \sum x_i y_i \end{pmatrix}$$

$$= \frac{1}{n\sum_{i=1}^{n}(x_i - \bar{x})^2} \begin{pmatrix} \sum x_i^2 \sum y_i - \sum x_i \sum x_i y_i \\ n\sum x_i y_i - \sum x_i \sum y_i \end{pmatrix}$$

$$= \begin{pmatrix} \dfrac{\sum x_i^2 \sum y_i - \sum x_i \sum x_i y_i}{\sum_{i=1}^{n}(x_i - \bar{x})^2} \\ \\ \dfrac{\sum (x_i - \bar{x})(y_i - \bar{y})}{\sum_{i=1}^{n}(x_i - \bar{x})^2} \end{pmatrix}.$$

Thus, we have

$$\boldsymbol{b}_0 = \frac{\sum x_i^2 \sum y_i - \sum x_i \sum x_i y_i}{\sum_{i=1}^{n}(x_i - \bar{x})^2} = \bar{y} - \boldsymbol{b}_1 \bar{x}$$

and

$$\boldsymbol{b}_1 = \frac{\sum (x_i - \bar{x})(y_i - \bar{y})}{\sum_{i=1}^{n}(x_i - \bar{x})^2}.$$

The results are exactly identical to the results derived in Chapter 2. The unbiasness of the $\boldsymbol{b}$ and the covariance of $\boldsymbol{b}$ can be shown for the simple linear regression using its matrix form as well. We left this to the readers.

3.9 Test for Full Model and Reduced Model

Before an appropriate linear regression model is chosen it is often unknown how many variables should be included in the regression model. A linear regression model with more variables may not always perform better than the regression model with less variables when both models are compared in terms of residual of sum squares. To compare two regression models in terms of the independent variables included in the models we need to test if the regression model with more independent variables performs statistically better than the regression model with less independent variables. To this end, we define the full regression model as:

$$\boldsymbol{y} = \boldsymbol{X}_1\boldsymbol{\beta}_1 + \boldsymbol{X}_2\boldsymbol{\beta}_2 + \boldsymbol{\varepsilon}, \tag{3.20}$$

and the reduced regression model as:

$$\boldsymbol{y} = \boldsymbol{X}_2\boldsymbol{\beta}_2 + \boldsymbol{\varepsilon}. \tag{3.21}$$

A full linear regression model is the model with more independent variables and a reduced model is the model with a subset of the independent variables in the full model. In other words, the reduced regression model is the model nested in the full regression model. We would like to test the following hypothesis

$$H_0 : \boldsymbol{\beta}_1 = 0 \text{ versus } H_1 : \boldsymbol{\beta}_1 \neq 0.$$

Under the null hypothesis H_0, the error term of the regression model $\boldsymbol{\varepsilon} \sim N(0, \sigma^2 I_n)$. Denote $\boldsymbol{X} = (\boldsymbol{X}_1, \boldsymbol{X}_2)$, where $\boldsymbol{X}_1$ is an $n \times p_1$ matrix, $\boldsymbol{X}_2$ is an $n \times p_2$ matrix, and n is the total number of observations. A test statistic needs to be constructed in order to compare the full regression model with the reduced regression regression model. Consider the difference between the SSE of the full model and the SSE of the reduced model:

$$SSE_{reduced} = \boldsymbol{y}'(I - \boldsymbol{X}_2(\boldsymbol{X}_2'\boldsymbol{X}_2)^{-1}\boldsymbol{X}_2')\boldsymbol{y}$$

and

$$SSE_{full} = \boldsymbol{y}'(I - \boldsymbol{X}(\boldsymbol{X}'\boldsymbol{X})^{-1}\boldsymbol{X}')\boldsymbol{y},$$

$$SSE_{reduced} - SSE_{full} = \boldsymbol{y}'\Big(\boldsymbol{X}(\boldsymbol{X}'\boldsymbol{X})^{-1}\boldsymbol{X}' - \boldsymbol{X}_2(\boldsymbol{X}_2'\boldsymbol{X}_2)^{-1}\boldsymbol{X}_2'\Big)\boldsymbol{y}.$$

The matrices $\boldsymbol{X}(\boldsymbol{X}'\boldsymbol{X})^{-1}\boldsymbol{X}'$ and $\boldsymbol{X}_2(\boldsymbol{X}_2'\boldsymbol{X}_2)^{-1}\boldsymbol{X}_2'$ are idempotent. In addition, it can be shown that the matrix $\Big(\boldsymbol{X}(\boldsymbol{X}'\boldsymbol{X})^{-1}\boldsymbol{X}' -$

$\boldsymbol{X}_2(\boldsymbol{X}_2'\boldsymbol{X}_2)^{-1}\boldsymbol{X}_2')$ is also idempotent and the rank of this matrix is p_1 which is the dimension of $\boldsymbol{\beta}_1$:

$$\begin{aligned}
&\text{Rank of } \left(\boldsymbol{X}(\boldsymbol{X}'\boldsymbol{X})^{-1}\boldsymbol{X}' - \boldsymbol{X}_2(\boldsymbol{X}_2'\boldsymbol{X}_2)^{-1}\boldsymbol{X}_2'\right) \\
&= tr\left(\boldsymbol{X}(\boldsymbol{X}'\boldsymbol{X})^{-1}\boldsymbol{X}' - \boldsymbol{X}_2(\boldsymbol{X}_2'\boldsymbol{X}_2)^{-1}\boldsymbol{X}_2'\right) \\
&= tr\left(\boldsymbol{X}(\boldsymbol{X}'\boldsymbol{X})^{-1}\boldsymbol{X}'\right) - tr\left(\boldsymbol{X}_2(\boldsymbol{X}_2'\boldsymbol{X}_2)^{-1}\boldsymbol{X}_2'\right) \\
&= tr(\boldsymbol{X}'\boldsymbol{X}(\boldsymbol{X}'\boldsymbol{X})^{-1}) - tr(\boldsymbol{X}_2'\boldsymbol{X}_2(\boldsymbol{X}_2'\boldsymbol{X}_2)^{-1}) \\
&= (p_1 + p_2) - p_2 = p_1
\end{aligned}$$

The distribution of the following quadratic form is the chi-square distribution with degrees of freedom p_1:

$$\boldsymbol{y}'\left(\boldsymbol{X}(\boldsymbol{X}'\boldsymbol{X})^{-1}\boldsymbol{X}' - \boldsymbol{X}_2(\boldsymbol{X}_2'\boldsymbol{X}_2)^{-1}\boldsymbol{X}_2'\right)\boldsymbol{y} \sim \sigma^2\chi^2_{p_1}.$$

Note that the matrix $I - \boldsymbol{X}(\boldsymbol{X}'\boldsymbol{X})^{-1}\boldsymbol{X}'$ is idempotent and its rank is $n - p_1$. Applying the theorem of the distribution of the quadratic form, it can be shown that total sum of residuals

$$s^2 = \boldsymbol{y}'\left(I - \boldsymbol{X}(\boldsymbol{X}'\boldsymbol{X})^{-1}\boldsymbol{X}'\right)\boldsymbol{y} \sim \sigma^2\chi^2_{n-p},$$

where p is the total number of parameters. In addition, It can be shown that s^2 is independent of $SSE_{reduced} - SSE_{full}$. In fact, we only need to show

$$\left[\boldsymbol{X}(\boldsymbol{X}'\boldsymbol{X})^{-1}\boldsymbol{X}' - \boldsymbol{X}_2(\boldsymbol{X}_2'\boldsymbol{X}_2)^{-1}\boldsymbol{X}_2'\right]\left[I - \boldsymbol{X}(\boldsymbol{X}'\boldsymbol{X})^{-1}\boldsymbol{X}'\right] = 0.$$

It is easy to verify that

$$\left[I - \boldsymbol{X}(\boldsymbol{X}'\boldsymbol{X})^{-1}\boldsymbol{X}'\right]\left[\boldsymbol{X}(\boldsymbol{X}'\boldsymbol{X})^{-1}\boldsymbol{X}'\right] = 0.$$

It remains to show

$$\left[I - \boldsymbol{X}(\boldsymbol{X}'\boldsymbol{X})^{-1}\boldsymbol{X}'\right]\left[\boldsymbol{X}_2(\boldsymbol{X}_2'\boldsymbol{X}_2)^{-1}\boldsymbol{X}_2'\right] = 0.$$

It is straightforward that

$$\left[I - \boldsymbol{X}(\boldsymbol{X}'\boldsymbol{X})^{-1}\boldsymbol{X}'\right]\boldsymbol{X} = \left[I - \boldsymbol{X}(\boldsymbol{X}'\boldsymbol{X})^{-1}\boldsymbol{X}'\right](\boldsymbol{X}_1, \boldsymbol{X}_2) = 0.$$

Note that $\boldsymbol{X} = (\boldsymbol{X}_1, \boldsymbol{X}_2)$ we have

$$\Big[I - \boldsymbol{X}(\boldsymbol{X}'\boldsymbol{X})^{-1}\boldsymbol{X}'\Big]\boldsymbol{X}_2 = 0.$$

Therefore,

$$\Big[I - \boldsymbol{X}(\boldsymbol{X}'\boldsymbol{X})^{-1}\boldsymbol{X}'\Big]\boldsymbol{X}_2\Big[(\boldsymbol{X}_2'\boldsymbol{X}_2)^{-1}\boldsymbol{X}_2'\Big] = 0.$$

Thus, we can construct the following F test statistic:

$$F = \frac{\boldsymbol{y}'\Big(\boldsymbol{X}(\boldsymbol{X}'\boldsymbol{X})^{-1}\boldsymbol{X}' - \boldsymbol{X}_2(\boldsymbol{X}_2'\boldsymbol{X}_2)^{-1}\boldsymbol{X}_2'\Big)\boldsymbol{y}/p_1}{\boldsymbol{y}'\Big(I - \boldsymbol{X}(\boldsymbol{X}'\boldsymbol{X})^{-1}\boldsymbol{X}'\Big)\boldsymbol{y}/n-p} \sim F_{p_1, n-p}. \quad (3.22)$$

This test statistic can be used to test hypothesis $H_0 : \boldsymbol{\beta}_1 = 0$ versus $H_1 : \boldsymbol{\beta}_1 \neq 0$.

3.10 Test for General Linear Hypothesis

Consider the following multiple linear regression model

$$\boldsymbol{y} = \boldsymbol{\beta}\boldsymbol{X} + \boldsymbol{\varepsilon},$$

where $\boldsymbol{\varepsilon} \sim N(0, \sigma^2 I_n)$. It may be of interest to test the linear function of model parameters. This can be formulated into the following general linear hypothesis testing problem:

$$H_0 : C\boldsymbol{\beta} = d \text{ versus } H_1 : C\boldsymbol{\beta} \neq d.$$

Here, C is a $r \times p$ matrix of rank r and $r \leq p$, p is the number of parameters in the regression model, or the dimension of $\boldsymbol{\beta}$. Suppose that $\boldsymbol{b}$ is the least squares estimation of $\boldsymbol{\beta}$ then we have

$$\boldsymbol{b} \sim N(\boldsymbol{\beta}, \sigma^2(\boldsymbol{X}'\boldsymbol{X})^{-1})$$

and

$$C\boldsymbol{b} \sim N(C\boldsymbol{\beta}, \sigma^2 C(\boldsymbol{X}'\boldsymbol{X})^{-1}C').$$

Under the null hypothesis $H_0 : C\boldsymbol{\beta} = d$, we have

$$[Cb - d\,]'[C(X'X)^{-1}C']^{-1}[Cb - d\,] \sim \sigma^2\chi_r^2,$$

therefore, the statistic that can be used for testing $H_0 : C\beta = d$ versus $H_1 : C\beta \neq d$ is the F test statistic in the following form:

$$F = \frac{(Cb - d)'[C(X'X)^{-1}C']^{-1}(Cb - d)}{rs^2} \sim F_{r,n-p}. \qquad (3.23)$$

3.11 The Least Squares Estimates of Multiple Regression Parameters Under Linear Restrictions

Sometimes, we may have more knowledge about regression parameters, or we would like to see the effect of one or more independent variables in a regression model when the restrictions are imposed on other independent variables. This way, the parameters in such a regression model may be useful for answering a particular scientific problem of interest. Although restrictions on regression model parameters could be non-linear we only deal with the estimation of parameters under general linear restrictions. Consider a linear regression model

$$y = \beta X + \varepsilon.$$

Suppose that it is of interest to test the general linear hypothesis: $H_0 : C\beta = d$ versus $H_1 : C\beta \neq d$, where d is a known constant vector. We would like to explore the relationship of SSEs between the full model and the reduced model. Here, the full model is referred to as the regression model without restrictions on the parameters and the reduced model is the model with the linear restrictions on parameters. We would like to find the least squares estimation of β under the general linear restriction $C\beta = d$. Here C is a $r \times p$ matrix of rank r and $r \leq p$. With a simple linear transformation the general linear restriction $C\beta = d$ can be rewritten as $C\beta^* = 0$. So, without loss of generality, we consider homogeneous linear restriction: $C\beta = 0$. This will simplify the derivations. The estimator we are seeking for will minimize the least squares $(y - X\beta)'(y - X\beta)$ under the linear restriction $C\beta = 0$. This minimization problem under the linear restriction can be solved by using the method of the Lagrange multiplier. To this end, we construct the objective function $Q(\beta, \lambda)$ with Lagrange multiplier λ:

$$
\begin{aligned}
Q(\boldsymbol{\beta}, \lambda) &= (\boldsymbol{y} - \boldsymbol{X\beta})^{'}(\boldsymbol{y} - \boldsymbol{X\beta}) + 2\lambda C\boldsymbol{\beta} \\
&= \boldsymbol{y}^{'}\boldsymbol{y} + \boldsymbol{\beta}^{'}\boldsymbol{X}^{'}\boldsymbol{X\beta} - \boldsymbol{\beta}^{'}\boldsymbol{X}^{'}\boldsymbol{y} - \boldsymbol{y}^{'}\boldsymbol{X\beta} + 2\lambda C\boldsymbol{\beta}
\end{aligned}
$$

To minimize the objective function $Q(\boldsymbol{\beta}, \lambda)$, we take the partial derivatives with respect to each component of $\boldsymbol{\beta}$ and with respect to λ which yields the following normal equations:

$$
\begin{cases}
\boldsymbol{X}^{'}\boldsymbol{X\beta} + C\lambda = \boldsymbol{X}^{'}\boldsymbol{y} \\
C\boldsymbol{\beta} = 0
\end{cases}
$$

The solutions of the above normal equation are least squares estimators of the regression model parameters under the linear restriction $C\boldsymbol{\beta} = 0$. The normal equation can be written in the form of blocked matrix:

$$
\begin{pmatrix} \boldsymbol{X}^{'}\boldsymbol{X} & C^{'} \\ C & 0 \end{pmatrix} \begin{pmatrix} \boldsymbol{\beta} \\ \lambda \end{pmatrix} = \begin{pmatrix} \boldsymbol{X}^{'}\boldsymbol{y} \\ 0 \end{pmatrix} \tag{3.24}
$$

The normal equation can be easily solved if one can find the inverse matrix on the left of the above normal equation. Formula of inverse blocked matrix can be used to solve the solution of the system. To simplify the notations we denote $\boldsymbol{X}^{'}\boldsymbol{X} = A$, and the inverse matrix in blocked form is given by

$$
\begin{pmatrix} \boldsymbol{X}^{'}\boldsymbol{X} & C^{'} \\ C & 0 \end{pmatrix}^{-1} = \begin{pmatrix} A & C^{'} \\ C & 0 \end{pmatrix}^{-1}
$$

$$
= \begin{pmatrix} A^{-1} - A^{-1}C^{'}(CA^{-1}C^{'})^{-1}CA^{-1} & A^{-1}C(CA^{-1}C^{'})^{-1} \\ (CA^{-1}C^{'})^{-1}C^{'}A^{-1} & -(CA^{-1}C^{'})^{-1} \end{pmatrix}
$$

By multiplying the blocked inverse matrix on the both sides of the above normal equation the least squares estimator of $\boldsymbol{\beta}$ under the linear restriction is given by

$$
\boldsymbol{b}^{*} = (A^{-1} - A^{-1}C^{'}(CA^{-1}C^{'})^{-1}CA^{-1})\boldsymbol{X}^{'}\boldsymbol{y}. \tag{3.25}
$$

For the full model (the model without restriction)

$$
SSE_{full} = \boldsymbol{y}^{'}(I - \boldsymbol{X}A^{-1}\boldsymbol{X}^{'})\boldsymbol{y}.
$$

For the reduced model (the model with a linear restriction):

$$SSE_{red} = \boldsymbol{y}'(I - \boldsymbol{X}A^{-1}\boldsymbol{X}' + \boldsymbol{X}A^{-1}C'(CA^{-1}C')^{-1}CA^{-1}\boldsymbol{X}')\boldsymbol{y}$$

Note that $\boldsymbol{b} = (\boldsymbol{X}'\boldsymbol{X})^{-1}\boldsymbol{X}'\boldsymbol{y}$ and we have

$$SSE_{red} - SSE_{full} = \boldsymbol{y}'(\boldsymbol{X}A^{-1}C(CA^{-1}C')^{-1}CA^{-1}\boldsymbol{X}')\boldsymbol{y}$$
$$= \boldsymbol{y}'\boldsymbol{X}(\boldsymbol{X}'\boldsymbol{X})^{-1}C'(C(\boldsymbol{X}'\boldsymbol{X})^{-1}C')^{-1}C(\boldsymbol{X}'\boldsymbol{X})^{-1}\boldsymbol{X}'\boldsymbol{y}$$
$$= (C\boldsymbol{b})'(C(\boldsymbol{X}'\boldsymbol{X})^{-1}C')^{-1}C\boldsymbol{b}.$$

Under the normality assumption the above expression is a quadratic form of the normal variables. It can be shown that it has the chi-square distribution with degrees of freedom as the rank of the matrix $C(\boldsymbol{X}'\boldsymbol{X})^{-1}C'$, which is r, the number of parameters in the model. Thus, we can write

$$(C\boldsymbol{b})'[C(\boldsymbol{X}'\boldsymbol{X})^{-1}C']^{-1}(C\boldsymbol{b}) \sim \sigma^2\chi^2_r. \tag{3.26}$$

It can be shown that the s^2 is independent of the above χ^2 variable. Finally, we can construct the F test statistic:

$$F = \frac{(C\boldsymbol{b})'[C(\boldsymbol{X}'\boldsymbol{X})^{-1})C']^{-1}(C\boldsymbol{b})}{rs^2} \sim F_{r,\, n-p}, \tag{3.27}$$

which can be used to test the general linear hypothesis $H_0 : C\boldsymbol{\beta} = 0$ versus $H_1 : C\boldsymbol{\beta} \neq 0$.

3.12 Confidence Intervals of Mean and Prediction in Multiple Regression

We now discuss the confidence intervals on regression mean and regression prediction for multiple linear regression. For a given data point $\boldsymbol{x}_0'$ the fitted value is $\hat{y}|\boldsymbol{x}_0 = \boldsymbol{x}_0'\boldsymbol{b}$ and $Var(\hat{y}|\boldsymbol{x}_0) = \boldsymbol{x}_0'\text{Cov}(\boldsymbol{b})\boldsymbol{x}_0 = \boldsymbol{x}_0'(\boldsymbol{X}'\boldsymbol{X})^{-1}\boldsymbol{x}_0\sigma^2$. Note that under the normality assumption on the model error term $E(\hat{y}|\boldsymbol{x}_0) = E(x_0\boldsymbol{b}) = x_0'\boldsymbol{\beta}$ and

$$\frac{(\hat{y}|\boldsymbol{x}_0) - E(\hat{y}|\boldsymbol{x}_0)}{s\sqrt{\boldsymbol{x}_0'(\boldsymbol{X}'\boldsymbol{X})^{-1}\boldsymbol{x}_0}} \sim t_{n-p}$$

where n is the total number of observations and p is the number of the parameters in the regression model. Thus, the $(1-\alpha)100\%$ confidence interval for $E(\hat{y}|\boldsymbol{x}_0)$ is given by

$$(\hat{y}|\boldsymbol{x}_0) \pm t_{\alpha/2,n-p} s \sqrt{\boldsymbol{x}_0'(\boldsymbol{X}'\boldsymbol{X})^{-1}\boldsymbol{x}_0} \tag{3.28}$$

Using the arguments similar to that in Chapter 2 the confidence interval on regression prediction in multiple linear regression is given by:

$$(\hat{y}|\boldsymbol{x}_0) \pm t_{\alpha/2,n-p} s \sqrt{1 + \boldsymbol{x}_0'(\boldsymbol{X}'\boldsymbol{X})^{-1}\boldsymbol{x}_0} \tag{3.29}$$

3.13 Simultaneous Test for Regression Parameters

Instead of testing for regression parameters individually, we can simultaneously test for the model parameters. We describe this simultaneous hypothesis test for multiple regression parameters using the vector notation:

$$H_0 : \boldsymbol{\beta} = \boldsymbol{\beta}_0, \text{ versus } H_1 : \boldsymbol{\beta} \neq \boldsymbol{\beta}_0,$$

where $\boldsymbol{\beta}' = (\beta_0, \beta_1, \cdots, \beta_{p-1})$, a p-dimensional vector of regression parameters, and $\boldsymbol{\beta}_0' = (\beta_{00}, \beta_{10}, \cdots, \beta_{p-1,0})$, a p-dimensional constant vector. The above simultaneous hypothesis testing problem can be tested using the following F test statistic which has the F distribution with degrees of freedom p and $n-p$ under H_0:

$$F = \frac{(\boldsymbol{b} - \boldsymbol{\beta}_0)'(\boldsymbol{X}'\boldsymbol{X})^{-1}(\boldsymbol{b} - \boldsymbol{\beta}_0)}{ps^2} \sim F_{p,n-p}.$$

Here n is the total number of observations and p is the total number of regression parameters. To test simultaneously the regression parameters, for a given test level α, if the observed $\boldsymbol{b}$ satisfies the following inequality for a chosen cut-off $F_{\alpha,p,n-p}$,

$$Pr\Big(\frac{(\boldsymbol{b} - \boldsymbol{\beta}_0)'(\boldsymbol{X}'\boldsymbol{X})^{-1}(\boldsymbol{b} - \boldsymbol{\beta}_0)}{ps^2} \leq F_{\alpha,p,n-p}\Big) \geq 1 - \alpha,$$

then H_0 cannot be rejected. Otherwise, we accept H_1. The F test statistic can be used to construct the simultaneous confidence region for regression parameters $\boldsymbol{\beta}$:

$$\Big\{\boldsymbol{\beta} : \frac{(\boldsymbol{b} - \boldsymbol{\beta})'(\boldsymbol{X}'\boldsymbol{X})^{-1}(\boldsymbol{b} - \boldsymbol{\beta})}{ps^2} \leq F_{\alpha,p,n-p}\Big\}. \tag{3.30}$$

Note that this simultaneous confidence region of the regression parameters is an ellipsoid in $\mathbb{R}^p$.

3.14 Bonferroni Confidence Region for Regression Parameters

Instead of constructing an ellipsoid confidence region by a quadratic form of the regression coefficients we can set a higher confidence level for each parameter so that the joint confidence region for all regression coefficients has a confidence level $(1-\alpha)100\%$. This can be done using the Bonferroni approach. Suppose that we have p regression coefficients and would like to construct a $(1-\alpha)100\%$ joint confidence region for p regression parameters, instead of using $\alpha/2$ for each regression parameter we now use a higher level $\alpha/2p$ to construct the Bonferroni confidence interval for all regression parameters β_i, $i = 1, 2, \cdots, p$. i.e., we choose a cut-off $t_{\alpha/2p,\, n-p}$ and construct the following confidence interval for each regression parameter:

$$b_i \pm t_{\alpha/2p,\, n-p}(\text{standard error of } b_i).$$

Note that $\text{Cov}(\boldsymbol{b}) = (\boldsymbol{X}^{'}\boldsymbol{X})^{-1}\sigma^2$. The standard error of b_i can be estimated by the squared root of the diagonal elements in the matrix $(\boldsymbol{X}^{'}\boldsymbol{X})^{-1}s^2$. This confidence region is the p-dimensional rectangular in $\mathbb{R}^p$ and has a joint confidence level of not less than $1-\alpha$. Confidence region based on the Bonferroni approach is conservative but the calculation is simpler.

The Bonferroni method can also be used to construct the confidence bounds on regression mean. Suppose we have r data points $\boldsymbol{x}_1, \boldsymbol{x}_2, \cdots, \boldsymbol{x}_r$, and want to construct the Bonferroni simultaneous confidence intervals on regression means at points $\boldsymbol{x}_1, \boldsymbol{x}_2, \cdots, \boldsymbol{x}_r$. The following formula gives the simultaneous confidence intervals for regression means at the observations $\boldsymbol{x}_1, \boldsymbol{x}_2, \cdots, \boldsymbol{x}_r$:

$$\hat{y}(x_j) \pm t_{\alpha/2r,\, n-p}\, s\sqrt{x_j^{'}(\boldsymbol{X}^{'}\boldsymbol{X})^{-1}x_j} \tag{3.31}$$

The SAS code for calculating simultaneous confidence intervals on regression means and regression predictions are similar to those for the simple linear regression which was presented in the previous chapter for the simple linear regression. The only difference is to set a higher confidence level $(1-\alpha/2r)100\%$ for the simultaneous confidence intervals on regression means at $\boldsymbol{x}_1, \boldsymbol{x}_2, \cdots, \boldsymbol{x}_r$.

3.15 Interaction and Confounding

We have seen that linear regression is flexible enough to incorporate certain nonlinearity in independent variables via polynomial or other transformed terms of the independent variables. It is quite useful in many applications that the independent variables in the linear regression are categorical. In this section, we shall continue to demonstrate its great flexibility in handling and exploring interactions. We will also show how linear regression is used to evaluate the confounding effects among predictors.

A confounding variable (or factor) is an extraneous variable in a regression model that correlates (positively or negatively) with both the dependent variable and the independent variable. A confounding variable is associated with both the probable cause and the outcome. In clinical study, the common ways of experiment control of the confounding factor are case-control studies, cohort studies, and stratified analysis. One major problem is that confounding variables are not always known or measurable. An interaction in a regression model often refers to as the effect of two or more independent variables in the regression model is not simply additive. Such a term reflects that the effect of one independent variable depends on the values of one or more other independent variables in the regression model.

The concepts of both interaction and confounding are more methodological than analytic in statistical applications. A regression analysis is generally conducted for two goals: to predict the response Y and to quantify the relationship between Y and one or more predictors. These two goals are closely related to each other; yet one is more emphasized than the other depending on application contexts. For example, in spam detection, prediction accuracy is emphasized as determining whether or not an incoming email is a spam is of primary interest. In clinical trials, on the other hand, the experimenters are keenly interested to know if an investigational medicine is more effective than the *control* or *exposure*, for which the standard treatment or a placebo is commonly used, in treating some disease. The assessment of treatment effect is often desired in analysis of many clinical trials. Both interaction and confounding are more pertaining to the second objective.

Consider a regression analysis involving assessment of the association between the response and one (or more) predictor, which may be affected by other extraneous predictors. The predictor(s) of major interest can be either categorical or continuous. When it is categorical, it is often referred to as *treatment* in experimental designs. The difference it makes on the responses is cited as the *treatment effect.* The extraneous predictors that potentially influence the treatment effect are termed as *covariates* or *control variables.* Interaction and confounding can be viewed as different manners in which the covariates influence the treatment effect.

3.15.1 *Interaction*

By definition, *interaction* is referred to as the situation where the association of major concern or the treatment effect varies with the levels or values of the covariates. Consider, for example, the treatment-by-center interaction in a multi-center clinical trial, a common issue involved in a clinical trial that is conducted in different medical centers. If the treatment effect remains the same among different medical centers, then we say that no interaction exists between treatment and center; if the treatment is found effective, nevertheless, more or less across different medical centers, then we say interaction exists between treatment and center and interaction involved is referred to as *quantitative interaction*; if the new treatment is found effective than the control in some medical centers but harmful than the control in some other centers, then the interaction is referred to as *qualitative interaction.* There is a directional change in treatment effect across centers in the case of qualitative interaction while the treatment effect only differs in amount, not in direction of the comparison, with quantitative interactions. Quantitative interactions are quite common. But if qualitative interaction exists, it causes much more concerns. It is thus imperative in clinical trials to detect and, if exists, fully explore and test for qualitative interaction. In the following discussion, we shall treat these two types of interaction by the same token, while referring interested readers to Gail and Simon (1985) and Yan and Su (2005) for more discussion on their important differences.

In linear regression, interaction is commonly formulated by cross-product terms. Consider the regression setting of response Y and two continuous regressors X_1 and X_2. The interaction model can be stated as, ignoring the subscript i for observations,

$$y = \beta_0 + \beta_1 x_1 + \beta_2 x_2 + \beta_3 x_1 x_2 + \varepsilon. \tag{3.32}$$

Recall that in the additive or main effect model

$$y = \beta_0 + \beta_1 x_1 + \beta_2 x_2 + \varepsilon, \tag{3.33}$$

the association between Y and X_1 is mainly carried by its slope β_1, which corresponds to the amount of change in the mean response $E(Y)$ with one unit increase in X_1, holding X_2 fixed. Here the slope β_1, which does not depend on X_2, remains unchanged with different values of X_2 and hence can be interpreted as the *main effect* of X_1. Similar interpretation holds for the slope β_1 of X_2.

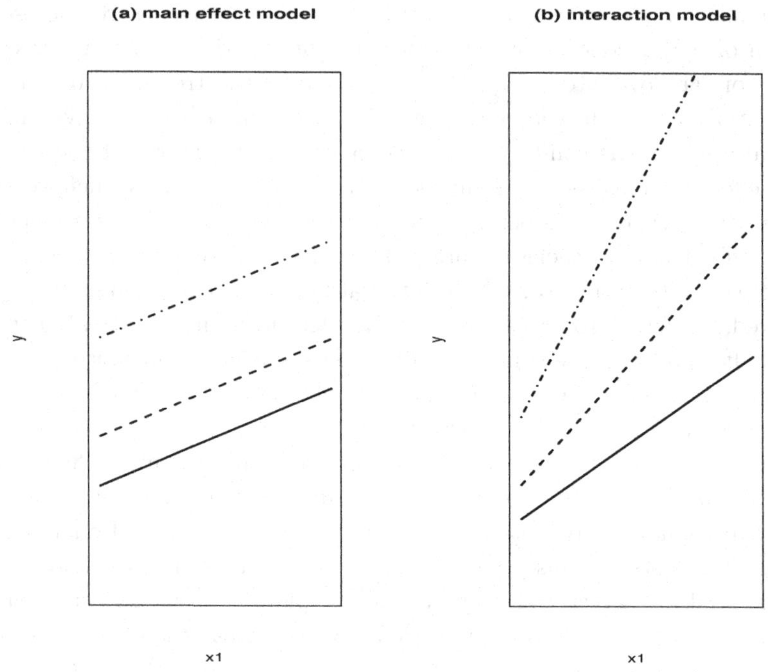

Fig. 3.1 Response Curves of Y Versus X_1 at Different Values of X_2 in Models (3.33) and (3.32).

In model (3.32), we can extract the 'slopes' for X_1 and X_2 by rewriting

$$\begin{aligned} E(y) &= (\beta_0 + \beta_2 x_2) + (\beta_1 + \beta_3 x_2) \cdot x_1 \\ &= (\beta_0 + \beta_1 x_1) + (\beta_2 + \beta_3 x_1) \cdot x_2 \end{aligned}$$

The slope for X_1 now becomes $(\beta_1 + \beta_3 x_2)$, which depends on what value X_2 is fixed at. For this reason, X_2 is said to be an effect-modifier of X_1.

It is instructive to plot the response curves for Y versus X_1 at different values of X_2 for models (3.33) and (3.32), as shown in Fig. 3.1. We can see that the response curves in the main effect model are parallel lines with the same slope and different intercepts while in the interaction model the lines are no longer parallel. This explains why no interaction is often viewed as synonymous to parallelism, a principle in interaction detection that is applicable to various settings such as two-way analysis of variance (ANOVA) and comparing two or more response curves. Analogously, the slope for X_2 is $(\beta_2 + \beta_3 x_1)$, which depends on what value X_1 is fixed at.

Interaction among predictors can be generally formulated as cross product terms. For instance, an interaction model for Y versus X_1, X_2, and X_3 can be written as

$$\begin{aligned} y &= \beta_0 + \beta_1 x_1 + \beta_2 x_2 + \beta_3 x_3 + \beta_4 x_1 x_2 + \beta_5 x_1 x_3 + \beta_6 x_2 x_3 + \beta_7 x_1 x_2 x_3 + \varepsilon \\ &= (\beta_0 + \beta_2 x_2 + \beta_3 x_3 + \beta_6 x_2 x_3) + (\beta_1 + \beta_4 x_2 + \beta_5 x_3 + \beta_7 x_2 x_3)\, x_1 + \varepsilon. \end{aligned}$$

The products involving two terms $x_i x_j$, $i \neq j$, are referred to as first-order interactions; the three-term cross-products such as $x_i x_j x_k$, $i \neq j \neq k$, are called second-order interactions; and so on for higher-order interactions in general. The higher order of the interaction, the more difficult it would be in model interpretation. As seen in the above model, the slope for X_1 is $(\beta_1 + \beta_4 x_2 + \beta_5 x_3 + \beta_7 x_2 x_3)$, which depends on both x_2 and x_3 values in a complicated manner. To retain meaningful and simple model interpretation, it is often advised to consider interactions only up to the second order. In reality, interaction can be of high order with a complicated form other than cross products, which renders interaction detection a dunting task sometimes.

3.15.2 *Confounding*

Confounding is generally related to the broad topic of variable controlling or adjustment. Variable controlling and adjustment, which plays an important role to help prevent bias and reduce variation in treatment effect assessment, can be incorporated into a study at two stages. The first stage is in the design of the study. Consider, for instance, a study where the objective is to compare the prices of soft drinks of different brands, say, (A, B, and C). In a completely randomized design, one randomly goes to a number of grocery stores, pick up a drink of Brand A from each store, and record its price; then another set of grocery stores are randomly selected for Brand B; and so on for Brand C. Data collected in this manner result in several independent random samples, one for each treatment or brand, and the analysis can be carried out using the one-way ANOVA technique. The potential problem with this design, however, is that the treatment effect, as measured by the differences in price among the three groups, would be contaminated due to heterogeneity in other factors. Imagine what would happen if it turns out that price data collected for Brand A are taken from stores, mostly located in Minnesota in winter times while data collected for Brand B are taken during summer times from stores mostly located in Florida. In this case, we will be unable to obtain a genuine evaluation of the price difference due to brands. A better approach in this study is to employ a randomized block design with grocery stores being blocks, which can be described as follows. One randomly selects a number of grocery stores first; at each store, pick up a Brand A drink, a Brand B drink, and a Brand C drink and record their prices. In this way, we are able to control for many other geographical and longitudinal factors. By controlling, it means to make sure that they have the same or similar values. In general, if we know which factors are potentially important, then we can make control for them beforehand by using block or stratified designs.

However, very often we are pretty much blind about which factors are important. Sometimes, even if we have a good idea about them according to previous studies or literatures, we nevertheless do not have the authority or convenience to perform the control beforehand in the design stage. This is the case in many *observational studies.* Or perhaps there are too many of them; it is impossible to control for all. In this case, the adjustment can still be made in a *post hoc* manner at the data analysis stage. This is exactly what the analysis of covariance (ANCOVA) is aimed for. The approach is to fit models by including the important covariates.

The conception of confounding is casted into the *post hoc* variable adjustment at the data analysis stage. In general, *confounding* occurs if interpretations of the treatment effect are statistically different when some covariates are excluded or included in the model. It is usually assessed through a comparison between a crude estimator of the treatment effect by ignoring the extraneous variables and an estimate after adjusting for the covariates. Consider a setting (Y vs. $Z, X_1, \ldots, X_p$), where variable Z denotes the treatment variable of major interest and $X_1, \ldots, X_p$ denote the associated covariates. The comparison can be carried out in terms of the following two models:

$$y = \beta_0 + \beta_1 z + \varepsilon \tag{3.34}$$

and

$$y = \beta_0 + \beta_1 z + \alpha_1 x_1 + \cdots + \alpha_p x_p + \varepsilon. \tag{3.35}$$

Let $\hat{\beta}_1^{(c)}$ denote the least squares estimator of β_1 in model (3.34), which gives a rudimentary assessment of the treatment effect. Let $\hat{\beta}_1^{(a)}$ denote the least squares estimator of β_1 in model (3.35), which evaluates the treatment effect after adjusting or controlling for covariates $(X_1, \ldots, X_p)$. We say confounding is present if these two estimates, combined with their standard errors, are statistically different from each other. In this case, $(X_1, \ldots, X_p)$ are called confounders (or confounding factors) of Z.

In the traditional assessment of confounding effects, a statistical test is not required, perhaps because the analytical properties of $(\hat{\beta}_1^{(a)} - \hat{\beta}_1^{(c)})$ are not easy to comprehend unless resampling techniques such as bootstrap is used. It is mainly up to field experts to decide on existence of confounders and hence can be subjective. Another important point about confounding is that its assessment would become irrelevant if the treatment is strongly interacted with covariates. Interaction should be assessed before looking for confounders as it no longer makes sense to purse the main or separate effect of the treatment when it really depends on the levels or values of the covariates.

3.16 Regression with Dummy Variables

In regression analysis, a dummy variable is one that takes the value 0 or 1 to indicate the absence or presence of some categorical effect that may be expected to shift outcome. The reason we say "dummy" because it is not a variable that carries value of actual magnitude. For example, in a clinical trial, it is often useful to define a dummy variable D and $D = 1$ represents treatment group and $D = 0$ indicates placebo group; or we can introduce dummy variable S and define $S = 1$ for male group and 0 for female group. The mean value of a dummy variable is the proportion of the cases in the category coded 1. The variance of a dummy variable is $\sum D_i^2/n - (\sum D_i/n)^2 = p - p^2 = p(1-p)$, where p is the proportion of the cases in the category coded 1. Example of a regression model with dummy variable gender is:

$$Y_i = \beta_0 + \beta_1 X_i + \beta_2 D_i + \varepsilon_i \tag{3.36}$$

where Y_i is the annual salary of a lawyer, $D_i = 1$ if the lawyer is male and $D_i = 0$ if the lawyer is female, and X_i is years of experience. This model assumes that there is a mean salary shift between male lawyers and female lawyers. Thus, the mean salary is $E(Y_i|D_i = 1) = \beta_0 + \beta_2 + \beta_1 X_i$ for a male lawyer and is $E(Y_i|D_i = 0) = \beta_0 + \beta_1 X_i$ for a female lawyer. A test of the hypothesis $H_0 : \beta_2 = 0$ is a test of the hypothesis that the wage is the same for male lawyers and female lawyers when they have the same years of experience.

Several dummy variables can be used together to deal with more complex situation where more than two categories are needed for regression analysis. For example, variable race usually refers to the concept of categorizing humans into populations on the basis of various sets of characteristics. A variable race can have more than two categories. Suppose that we wanted to include a race variable with three categories White/Asian/Black in a regression model. We need to create a whole new set of dummy variables as follows

$$\begin{cases} D_{1i} = 1, \text{ if the person is white} \\ D_{1i} = 0, \text{ otherwise} \\ D_{2i} = 1, \text{ if the person is asian} \\ D_{2i} = 0, \text{ otherwise} \end{cases}$$

Here, the 'black' person category is treated as the base category and there is no need to create a dummy variable for this base category. All salary comparisons between two races in the regression model will be relative to the base category. In general, if there are m categories that need to be considered in a regression model it is needed to create $m - 1$ dummy variables, since the inclusion of all categories will result in perfect collinearity. Suppose that we would like to model the relation between the salary of a lawyer in terms of years of experience (X_i) and his/her race determined jointly by two dummy variables D_{1i}, D_{2i}, we can use the following regression model with the two dummy variables :

$$Y_i = \beta_0 + \beta_1 X_i + \beta_2 D_{1i} + \beta_3 D_{2i} + \varepsilon_i, \tag{3.37}$$

where Y_i is the salary of the lawyer, X_i is years of his/her working experience, and D_{1i} and D_{2i} are dummy variables that determine the race of the lawyer. For example, based on the above regression model, the expected salary for a black lawyer with X_i years of working experience is

$$E(Y_i|D_{1i} = 0, D_{2i} = 0) = \beta_0 + \beta_1 X_i.$$

The expected salary for a white lawyer with X_i years of working experience is

$$E(Y_i|D_{1i} = 1, D_{2i} = 0) = \beta_0 + \beta_1 X_i + \beta_2.$$

The expected salary for an asian lawyer with X_i years of working experience is

$$E(Y_i|D_{1i} = 0, D_{2i} = 1) = \beta_0 + \beta_1 X_i + \beta_3.$$

In each case the coefficient of the dummy variable in the regression model represents the difference with the base race (the black lawyer's salary). Thus, the interpretation of β_2 is that a white lawyer earns β_2 more than a black lawyer, and the interpretation of β_3 is that an asian lawyer earns β_3 more than a black lawyer. The hypothesis test $H_0 : \beta_2 = 0$ is to test whether the wage is identical for a white lawyer and a black lawyer with same years of experience. And the hypothesis test $H_0 : \beta_3 = 0$ is to test that whether the wage is identical for an asian lawyer and a black lawyer with same years of experience.

Furthermore, if we would like to consider race effect and gender effect together, the following model with multiple dummy variables can be used:

$$Y_i = \beta_0 + \beta_1 X_i + \beta_2 D_{1i} + \beta_3 D_{2i} + \beta_4 D_i + \varepsilon_i \tag{3.38}$$

According to model (3.38), for example, the expected salary for a female black lawyer with X_i years of experience is

$$E(Y_i|D_{1i} = 0, D_{2i} = 0, D_i = 0) = \beta_0 + \beta_1 X_i.$$

For a black male lawyer with X_i years of experience, thc expected salary is

$$E(Y_i|D_{1i} = 0, D_{2i} = 0, D_i = 1) = \beta_0 + \beta_1 X_i + \beta_4.$$

For a white male lawyer with X_i years of experience, the expected salary is

$$E(Y_i|D_{1i} = 1, D_{2i} = 0, D_i = 1) = \beta_0 + \beta_1 X_i + \beta_2 + \beta_4.$$

The hypothesis test $H_0 : \beta_4 = 0$ is to test if the wage is the same for male lawyer and female lawyer with the same years of experience. The hypothesis test $H_0 : \beta_2 = \beta_4 = 0$ is to test if the wage is the same for male white lawyer and black lawyer with the same years of experience and if the gender has no impact on the salary.

If we have k categories then $k - 1$ dummy variables are needed. This is because in the classical regression it is required that none of exploratory variable should be a linear combination of remaining exploratory model variables to avoid collinearity. For example, we can use the dummy variable D and code $D = 1$ for male, if we also use another dummy variable $S = 0$ to indicate female, then there is a linear relation between D and S: $D = 1 - S$. Therefore, information become redundant. Thus, one dummy variable should be sufficient to represent information on gender. In general, $k - 1$ dummy variables are sufficient to represent k categories. Note that if D_i's, $i = 1, 2, \cdots, k - 1$, are $k - 1$ dummy variables then $D_i = 1$ represents a category out of the total k categories, and all $D_i = 0$ represents the base category out of the total k categories. Thus, $k - 1$ dummy variables are sufficient to represent k distinct categories.

If a regression model involves a nominal variable and the nominal variable has more than two levels, it is needed to create multiple dummy variables to replace the original nominal variable. For example, imagine that you wanted to predict depression level of a student according to status of freshman, sophomore, junior, or senior. Obviously, it has more than two levels. What you need to do is to recode "year in school" into a set of dummy variables, each of which has two levels. The first step in this process is to decide the number of dummy variables. This is simply $k - 1$, where k is the number of levels of the original nominal variable. In this instance, 3 dummy variables are needed to represent 4 categories of student status.

In order to create these variables, we are going to take 3 levels of "year in school", and create a variable corresponding to each level, which will have the value of yes or no (i.e., 1 or 0). In this example, we create three variables sophomore, junior, and senior. Each instance of "year in school" would then be recoded into a value for sophomore, junior, and senior. If a person is a junior, then variables sophomore and senior would be equal to 0, and variable junior would be equal to 1. A student with all variables sophomore, junior, and senior being all 0 is a freshman.

The decision as to which level is not coded is often arbitrary. The level which is not coded is the category to which all other categories will be compared. As such, often the biggest group will be the not-coded category. In a clinical trial often the placebo group or control group can be chosen as the not-coded group. In our example, freshman was not coded so that we could determine if being a sophomore, junior, or senior predicts a different depression level than being a freshman. Consequently, if the variable "junior" is significant in our regression, with a positive coefficient β, this would mean that juniors are significantly more depressive than freshmen. Alternatively, we could have decided to not code "senior", then the coefficients for freshman, sophomore and junior in the regression model would be interpreted as how much more depressive if being a freshman, sophomore, or junior predicts a different depressive level than being a senior.

For the purpose of illustration, the simple regression model with one dummy variable is shown in Fig. 3.2. In the figure, (a) represents regression model with only dummy variable and without regressor. The two groups are parallel. (b) represents the model with dummy variable and regressor x, but two groups are still parallel, (c) represents the model with dummy variable and regressor x. The two groups are not parallel but without crossover. (d) represents the model with dummy variable and regressor x. The two groups are not parallel and with crossover. In situations (c) and (d) we say that there is interaction which means that the response of one group is not always better/higher than the response of the other group by the same magnitude. Situation (c) is quantitative interaction and (d) is qualitative interaction or crossover interaction.

3.17 Collinearity in Multiple Linear Regression

3.17.1 *Collinearity*

What is the collinearity in multiple linear regression? The collinearity refers to the situation in which two or more independent variables in a multiple linear regression model are highly correlated. Let the regression model be $\boldsymbol{y} = \boldsymbol{X} + \boldsymbol{\varepsilon}$ with the design matrix $\boldsymbol{X} = (1, \boldsymbol{x}_1, \boldsymbol{x}_2, \cdots, \boldsymbol{x}_k)$. The collinearity occurs if the independent variable $\boldsymbol{x}_i$ is highly linearly correlated to another one or more independent variables $\boldsymbol{x}_{j1}, \boldsymbol{x}_{j2}, \cdots, \boldsymbol{x}_{jk}$. In other words, $\boldsymbol{x}_i$

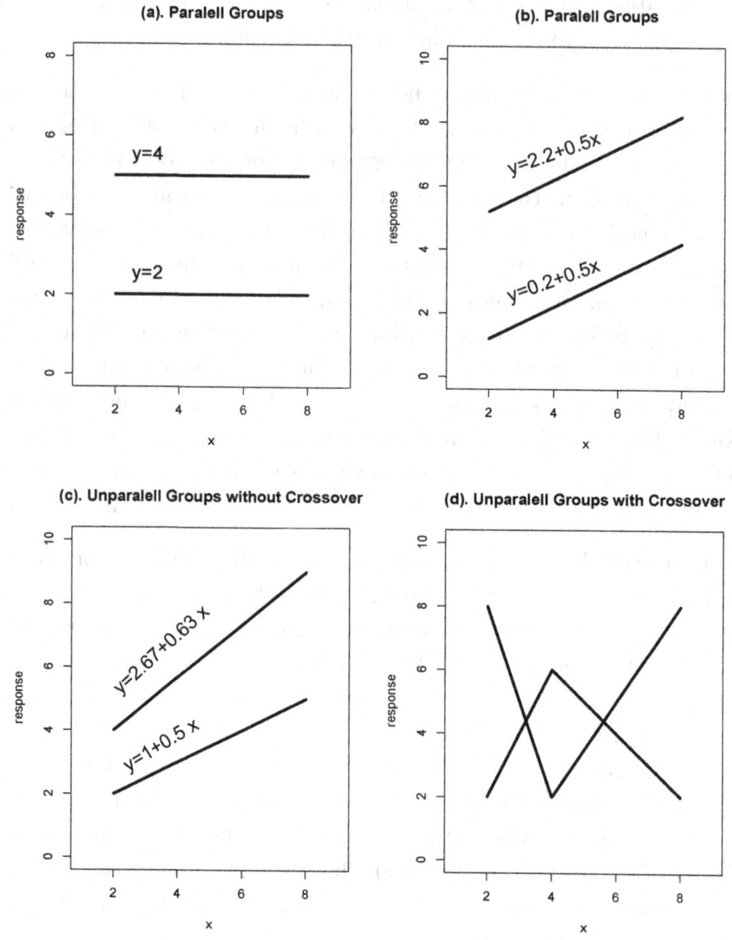

Fig. 3.2 Regression on Dummy Variables

can be almost linearly expressed by one or more other column vectors in $\boldsymbol{X}$. In this situation, the matrix $\boldsymbol{X}'\boldsymbol{X}$ is ill-conditioned or near singular. Although it is not completely singular, its eigenvalues may be close to zero and the eigenvalues of the inverse matrix $(\boldsymbol{X}'\boldsymbol{X})^{-1}$ tend to be very large which may cause instability of the least squares estimates of the regression parameters. If there is a perfect collinearity among column vectors of $\boldsymbol{X}$ then the matrix $\boldsymbol{X}'\boldsymbol{X}$ is not invertible. Therefore, it is problematic to solve for the unique least squares estimators of the regression coefficients from the normal equation. When the column vectors of the design matrix $\boldsymbol{X}$ is highly correlated, then the matrix $\boldsymbol{X}^t\boldsymbol{X}$ becomes ill-conditioned and the least squares estimator become less reliable even though we can find a unique solution of the normal equation. To see this let's look at the following example of two simple data sets (Tables 3.1 and 3.2).

Table 3.1 Two Independent Vectors

x_1	10	10	10	10	15	15	15	15
x_2	10	10	15	15	10	10	15	15

Table 3.2 Two Highly Correlated Vectors

x_1	10.0	11.0	11.9	12.7	13.3	14.2	14.7	15.0
x_2	10.0	11.4	12.2	12.5	13.2	13.9	14.4	15.0

The correlation matrix of the vectors in the first example data is a 2×2 identity matrix

$$\boldsymbol{X}'\boldsymbol{X} = \begin{pmatrix} 1 & 0 \\ 0 & 1 \end{pmatrix}.$$

Thus, its inverse matrix is also a 2×2 identity matrix. The correlation matrix of the two vectors in the second example data set is

$$\boldsymbol{X}'\boldsymbol{X} = \begin{pmatrix} 1.00000 & 0.99215 \\ 0.99215 & 1.00000 \end{pmatrix}$$

and its inverse matrix is given by

$$(\boldsymbol{X}'\boldsymbol{X})^{-1} = \begin{pmatrix} 63.94 & -63.44 \\ -64.44 & 63.94 \end{pmatrix}.$$

Note that for linear regression, $\text{Var}(\boldsymbol{b}) = (\boldsymbol{X}'\boldsymbol{X})^{-1}\sigma^2$. For the vectors in the first example data set we have

$$\frac{\text{Var}(b_1)}{\sigma^2} = \frac{\text{Var}(b_2)}{\sigma^2} = 1.$$

For the vectors in the second example data set we have

$$\frac{\text{Var}(b_1)}{\sigma^2} = \frac{\text{Var}(b_2)}{\sigma^2} = 63.94$$

The variances of the regression coefficients are inflated in the example of the second data set. This is because the collinearity of the two vectors in the second data set. The above example is the two extreme cases of the relationship between the two vectors. One is the case where two vectors are orthogonal to each other and the other is the case where two vectors are highly correlated.

Let us further examine the expected Euclidean distance between the least squares estimate $\boldsymbol{b}$ and the true parameter $\boldsymbol{\beta}$, $E(\boldsymbol{b}-\boldsymbol{\beta})'(\boldsymbol{b}-\boldsymbol{\beta})$ when collinearity exists among the column vectors of $\boldsymbol{X}$. First, it is easy to know that $E[(\boldsymbol{b}-\boldsymbol{\beta})'(\boldsymbol{b}-\boldsymbol{\beta})] = E(\boldsymbol{b}'\boldsymbol{b}) - \boldsymbol{\beta}'\boldsymbol{\beta}$. We then calculate $E(\boldsymbol{b}'\boldsymbol{b})$.

$$\begin{aligned}
& E(\boldsymbol{b}'\boldsymbol{b}) \\
&= E[(\boldsymbol{X}'\boldsymbol{X})^{-1}\boldsymbol{X}'\boldsymbol{y}'(\boldsymbol{X}'\boldsymbol{X})^{-1}\boldsymbol{X}'\boldsymbol{y}] \\
&= E[\boldsymbol{y}'\boldsymbol{X}(\boldsymbol{X}'\boldsymbol{X})^{-1}(\boldsymbol{X}'\boldsymbol{X})^{-1}\boldsymbol{X}'\boldsymbol{y}] \\
&= (\boldsymbol{X}\boldsymbol{\beta})'\boldsymbol{X}(\boldsymbol{X}'\boldsymbol{X})^{-1}(\boldsymbol{X}'\boldsymbol{X})^{-1}\boldsymbol{X}'\boldsymbol{X}\boldsymbol{\beta} + \sigma^2 tr[\boldsymbol{X}(\boldsymbol{X}'\boldsymbol{X})^{-1}(\boldsymbol{X}'\boldsymbol{X})^{-1}\boldsymbol{X}'] \\
&= \boldsymbol{\beta}'\boldsymbol{X}'\boldsymbol{X}(\boldsymbol{X}'\boldsymbol{X})^{-1}(\boldsymbol{X}'\boldsymbol{X})^{-1}\boldsymbol{X}'\boldsymbol{X}\boldsymbol{\beta} + \sigma^2 tr[\boldsymbol{X}'\boldsymbol{X}(\boldsymbol{X}'\boldsymbol{X})^{-1}(\boldsymbol{X}'\boldsymbol{X})^{-1}] \\
&= \boldsymbol{\beta}'\boldsymbol{\beta} + \sigma^2 tr[(\boldsymbol{X}'\boldsymbol{X})^{-1}]
\end{aligned}$$

Thus, we have

$$E[(\boldsymbol{b}-\boldsymbol{\beta})'(\boldsymbol{b}-\boldsymbol{\beta})] = \sigma^2 tr[(\boldsymbol{X}'\boldsymbol{X})^{-1}].$$

Note that $E[(\boldsymbol{b}-\boldsymbol{\beta})'(\boldsymbol{b}-\boldsymbol{\beta})]$ is the average Euclidean distance measure between the estimate $\boldsymbol{b}$ and the true parameter $\boldsymbol{\beta}$. Assuming that $(\boldsymbol{X}'\boldsymbol{X})$ has k distinct eigenvalues $\lambda_1, \lambda_2, \cdots, \lambda_k$, and the corresponding normalized eigenvectors $V = (v_1, v_2, \cdots, v_k)$, we can write

$$V'(\boldsymbol{X}'\boldsymbol{X})V = diag(\lambda_1, \lambda_2, \cdots, \lambda_k).$$

Moreover,

$$tr[V'(\boldsymbol{X}'\boldsymbol{X})V] = tr[VV'(\boldsymbol{X}'\boldsymbol{X})] = tr(\boldsymbol{X}'\boldsymbol{X}) = \sum_{i=1}^{k}\lambda_i.$$

Since the eigenvalues of $(X'X)^{-1}$ are $\frac{1}{\lambda_1}, \frac{1}{\lambda_2}, \cdots, \frac{1}{\lambda_k}$ we have

$$E(b'b) = \beta'\beta + \sigma^2 \sum_{i=1}^{k} \frac{1}{\lambda_i},$$

or it can be written as

$$E(\sum_{i=1}^{k} b_i^2) = \sum_{i=1}^{k} \beta_i^2 + \sigma^2 \sum_{i=1}^{k} \frac{1}{\lambda_i}. \tag{3.39}$$

Now it is easy to see that if one of λ is very small, say, $\lambda_i = 0.0001$, then roughly, $\sum_{i=1}^{k} b_i^2$ may over-estimate $\sum_{i=1}^{k} \beta_i^2$ by $1000\sigma^2$ times. The above discussions indicate that if some columns in X are highly correlated with other columns in X then the covariance matrix $(XX')^{-1}\sigma^2$ will have one or more large eigenvalues so that the mean Euclidean distance of $E[(b - \beta)'(b - \beta)]$ will be inflated. Consequently, this makes the estimation of the regression parameter β less reliable. Thus, the collinearity in column vectors of X will have negative impact on the least squares estimates of regression parameters and this need to be examined carefully when doing regression modeling.

How to deal with the collinearity in the regression modeling? One easy way to combat collinearity in multiple regression is to centralize the data. Centralizing the data is to subtract mean of the predictor observations from each observation. If we are not able to produce reliable parameter estimates from the original data set due to collinearity and it is very difficult to judge whether one or more independent variables can be deleted, one possible and quick remedy to combat collinearity in X is to fit the centralized data to the same regression model. This would possibly reduce the degree of collinearity and produce better estimates of regression parameters.

3.17.2 *Variance Inflation*

Collinearity can be checked by simply computing the correlation matrix of the original data X. As we have discussed, the variance inflation of the least squares estimator in multiple linear regression is caused by collinearity of the column vectors in X. When collinearity exists, the eigenvalues of the covariance matrix $(X'X)^{-1}\sigma^2$ become extremely large, which causes severe fluctuation in the estimates of regression parameters and makes these estimates less reliable. Variance inflation factor is the measure that can be used to quantify collinearity. The ith variance inflation factor is the scaled version of the multiple correlation coefficient between the ith independent variable and the rest of the independent variables. Specifically, the variance inflation factor for the ith regression coefficient is

$$\text{VIF}_i = \frac{1}{1 - R_i^2}, \tag{3.40}$$

where R_i^2 is the coefficient of multiple determination of regression produced by regressing the variable x_i against the other independent variables x_j, $j \neq i$. Measure of variance inflation is also given as the reciprocal of the above formula. In this case, they are referred to as *tolerances*.

If R_i equals zero (i.e., no correlation between x_i and the remaining independent variables), then VIF_i equals 1. This is the minimum value of variance inflation factor. For the multiple regression model it is recommended looking at the largest VIF value. A VIF value greater than 10 may be an indication of potential collinearity problems. The SAS procedure REG provides information on variance inflation factor and tolerance for each regression coefficient. The following example illustrates how to obtain this information using SAS procedure REG.

Example 3.1. SAS code for detection of collinearity and calculation the variance inflation factor.

```
Data example;
input x1 x2 x3 x4 x5 y;
datalines;
15.57    2463    472.92    18.0   4.45    566.52
44.02    2048   1339.75     9.5   6.92    696.82
20.42    3940    620.25    12.8   4.28   1033.15
18.74    6505    568.33    36.7   3.90   1603.62
49.20    5723   1497.60    35.7   5.50   1611.37
44.92   11520   1365.83    24.0   4.6    1613.27
55.48    5779   1687.00    43.3   5.62   1854.17
59.28    5969   1639.92    46.7   5.15   2160.55
94.39    8461   2872.33    78.7   6.18   2305.58
128.02  20106   3655.08   180.5   6.15   3503.93
96.00   13313   2912.00    60.9   5.88   3571.89
131.42  10771   3921.00   103.7   4.88   3741.40
127.21  15543   3865.67   126.8   5.50   4026.52
```

```
252.90  36194   7684.10   157.7    7.00  10343.81
409.20  34703  12446.33   169.4   10.78  11732.17
463.70  39204  14098.40   331.4    7.05  15414.94
510.22  86533  15524.00   371.6    6.35  18854.45
;
run;

proc reg  data=example corr alpha=0.05;
          model y=x1 x2 x3 x4 x5/tol vif collin;
run;

*Fit the regression model after deleting variable X1;
proc reg data=example corr alpha=0.05; ;
     model y=x2 x3 x4 x5/tol vif collin;
run;
```

The keyword TOL requests tolerance values for the estimates, VIF gives the variance inflation factors with the parameter estimates, and COLLIN requests a detailed analysis of collinearity among regressors. Variance inflation (VIF) is the reciprocal of tolerance (TOL). The above SAS procedures produce the following Table 3.3. The table shows that variables x_1 and x_3 are highly correlated. Due to this high correlation the variance inflation for both the variables x_1 and x_3 are rather significant and it can be found in Table 3.4.

We then delete variable x_1 and recalculate the correlation matrix. It can be seen that the variance inflations for all independent variables become much smaller after deleting x_1. The results of the correlation matrix and variance inflation are presented in Tables 3.5 and 3.6.

The least squares estimates in the regression model including the independent variables x_2, x_3, x_4 and x_5 behave much better than the model including all independent variables. The collinearity is eliminated by deleting one independent variable x_1 in this example.

3.18 Linear Model in Centered Form

The linear model can be rewritten in terms of centered x's as

$$\begin{aligned} y_i &= \beta_0 + \beta_1 x_{i1} + \beta_2 x_{i2} + \cdots + \beta_k x_{ik} + \varepsilon_i \\ &= \alpha + \beta_1(x_{i1} - \bar{x}_1) + \beta_2(x_{i2} - \bar{x}_2) + \cdots + \beta_k(x_{ik} - \bar{x}_k) + \varepsilon_i \quad (3.41) \end{aligned}$$

Table 3.3 Correlation Matrix for Variables $x_1, x_2, \cdots, x_5$

Variable	x_1	x_2	x_3	x_4	x_5	y
x1	1.0000	0.9074	0.9999	0.9357	0.6712	0.9856
x2	0.9074	1.0000	0.9071	0.9105	0.4466	0.9452
x3	0.9999	0.9071	1.0000	0.9332	0.6711	0.9860
x4	0.9357	0.9105	0.9332	1.0000	0.4629	0.9404
x5	0.6712	0.4466	0.6711	0.4629	1.0000	0.5786
y	0.9856	0.9452	0.9860	0.9404	0.5786	1.0000

Table 3.4 Parameter Estimates and Variance Inflation

Variable	Parameter	STD	t value	$P > \|t\|$	Tolerance	Inflation
Intercept	1962.95	1071.36	1.83	0.094		0
x1	-15.85	97.66	-0.16	0.874	0.0001042	9597.57
x2	0.06	0.02	2.63	0.023	0.12594	7.94
x3	1.59	3.09	0.51	0.617	0.000112	8933.09
x4	-4.23	7.18	-0.59	0.569	0.04293	23.29
x5	-394.31	209.64	-1.88	0.087	0.23365	4.28

Table 3.5 Correlation Matrix after Deleting Variable x_1

Variable	x_2	x_3	x_4	x_5	y
x2	1.0000	0.9071	0.9105	0.4466	0.9452
x3	0.9071	1.0000	0.9332	0.6711	0.9860
x4	0.9105	0.9332	1.0000	0.4629	0.9404
x5	0.4466	0.6711	0.4629	1.0000	0.5786
y	0.9452	0.9860	0.9404	0.5786	1.0000

Table 3.6 Variance Inflation after Deleting x_1

variable	parameter	std	t value	$P > \|t\|$	tolerance	inflation
intercept	2032.19	942.075	2.16	0.0520	0	
x2	0.056	0.020	2.75	0.0175	0.126	7.926
x3	1.088	0.153	7.10	$< .0001$	0.042	23.927
x4	-5.00	5.081	-0.98	0.3441	0.079	12.706
x5	-410.083	178.078	-2.30	0.0400	0.298	3.361

for $i = 1, \ldots, n$, where

$$\alpha = \beta_0 + \beta_1 \bar{x}_1 + \cdots + \beta_k \bar{x}_k$$

or

$$\beta_0 = \alpha - (\beta_1 \bar{x}_1 + \cdots + \beta_k \bar{x}_k) = \alpha - \bar{\mathbf{x}}' \boldsymbol{\beta}_1; \tag{3.42}$$

$\bar{\mathbf{x}} = (\bar{x}_1, \bar{x}_2, \ldots, \bar{x}_k)'$; $\boldsymbol{\beta}_1 = (\beta_1, \beta_2, \ldots, \beta_k)'$; and $\bar{x}_j$ denotes the sample average of x_{ij}'s for $j = 1, \ldots, k$. In the centered form, Y is regressed on centered X's, in which case the slope parameters in $\boldsymbol{\beta}_1$ remain the same. This centered form sometimes brings convenience in derivations of estimators of the linear models. Also, one can try the regression model in centered form when collinearity is observed among the independent variables and independent variables are difficult to be eliminated. Expressed in matrix form, model (3.41) becomes

$$\mathbf{y} = (\mathbf{j}, \mathbf{X}_c) \begin{pmatrix} \alpha \\ \boldsymbol{\beta}_1 \end{pmatrix} + \boldsymbol{\varepsilon}, \tag{3.43}$$

where

$$\mathbf{X}_c = \left(\mathbf{I} - \frac{1}{n} \mathbf{J} \right) \mathbf{X}_1 = (x_{ij} - \bar{x}_j)\,; \tag{3.44}$$

and $\mathbf{X}_1 = (x_{ij})$ for $i = 1, \ldots, n$ and $j = 1, \ldots, k$. Here matrix $\mathbf{X}_1$ is the sub-matrix of $\mathbf{X}$ after removing the first column of all 1's.

The matrix $\mathbf{C} = \mathbf{I} - 1/n \cdot \mathbf{J}$ is called the *centering matrix*, where $\mathbf{J} = \mathbf{j}\mathbf{j}'$ is an $n \times n$ matrix with all elements being 1. A geometric look at the centering matrix shows that

$$\begin{aligned} \mathbf{C} &= \mathbf{I} - \frac{1}{n} \cdot \mathbf{J} = \mathbf{I} - \mathbf{j}(\mathbf{j}'\mathbf{j})^{-1}\mathbf{j}', \quad \text{noting } \mathbf{j}'\mathbf{j} = n \\ &= \mathbf{I} - \mathbf{P}_{\mathcal{W}} = \mathbf{P}_{\mathcal{W}^\perp}, \end{aligned}$$

where $\mathcal{W} = C(\mathbf{j})$ denotes the subspace spanned by $\mathbf{j}$; $\mathcal{W}^\perp$ is the subspace perpendicular to $\mathcal{W}$; and $\mathbf{P}_{\mathcal{W}}$ and $\mathbf{P}_{\mathcal{W}^\perp}$ are their respective projection matrices. Namely, matrix $\mathbf{C}$ is the project matrix on the subspace that is perpendicular to the subspace spanned by $\mathbf{j}$. It follows immediately that

$$(\mathbf{I} - \frac{1}{n} \cdot \mathbf{J})\mathbf{j} = \mathbf{0} \text{ and } \mathbf{j}'\mathbf{X}_c = \mathbf{0} \tag{3.45}$$

Using (3.45), the least squared estimators of $(\alpha, \boldsymbol{\beta}_1)$ are given by,

$$\begin{aligned} \begin{pmatrix} \hat{\alpha} \\ \widehat{\boldsymbol{\beta}}_1 \end{pmatrix} &= \{(\mathbf{j},\ \mathbf{X}_c)'(\mathbf{j},\ \mathbf{X}_c)\}^{-1} (\mathbf{j},\ \mathbf{X}_c)'\mathbf{y} = \begin{pmatrix} n & \mathbf{0}' \\ \mathbf{0} & \mathbf{X}_c'\mathbf{X}_c \end{pmatrix}^{-1} \begin{pmatrix} n\bar{y} \\ \mathbf{X}_c'\mathbf{y} \end{pmatrix} \\ &= \begin{pmatrix} 1/n & \mathbf{0}' \\ \mathbf{0} & (\mathbf{X}_c'\mathbf{X}_c)^{-1} \end{pmatrix} \begin{pmatrix} n\bar{y} \\ \mathbf{X}_c'\mathbf{y} \end{pmatrix} \\ &= \begin{pmatrix} \bar{y} \\ (\mathbf{X}_c'\mathbf{X}_c)^{-1}\mathbf{X}_c'\mathbf{y} \end{pmatrix}. \end{aligned}$$

Thus, $\widehat{\boldsymbol{\beta}}_1$ is the same as in the ordinary least squares estimator $\widehat{\boldsymbol{\beta}}_1$ and

$$\hat{\beta}_0 = \bar{y} - \bar{\mathbf{x}}'\widehat{\boldsymbol{\beta}}_1 \tag{3.46}$$

in view of (3.42) and uniqueness of LSE.

Using the centered form, many interesting properties of the least squares estimation can be easily obtained. First, the LS fitted regression plane satisfies

$$y - \hat{\alpha} = y - \bar{y} = (\mathbf{x} - \bar{\mathbf{x}})'\widehat{\boldsymbol{\beta}}_1$$

and hence must pass through the center of the data $(\bar{\mathbf{x}}, \bar{y})$.

Denote $\mathcal{V}_c = C(\mathbf{X}_c)$. Since $\mathcal{W} = C(\mathbf{j}) \perp \mathcal{V}_c$ using (3.45),

$$\mathcal{V} = C(\mathbf{X}) = \mathcal{W} \oplus \mathcal{V}_c.$$

The vector fitted values is

$$\widehat{\mathbf{y}} = \mathbf{P}_{\mathcal{V}}\mathbf{y} = \mathbf{P}_{\mathcal{W}}\mathbf{y} + \mathbf{P}_{\mathcal{V}_c}\mathbf{y} = \bar{y}\mathbf{j} + \mathbf{X}_c(\mathbf{X}_c'\mathbf{X}_c)^{-1}\mathbf{X}_c'\mathbf{y} = \bar{y}\mathbf{j} + \mathbf{X}_c\widehat{\beta}_1 \tag{3.47}$$

and the residual vector is

$$\boldsymbol{e} = (\mathbf{I} - \mathbf{P}_{\mathcal{W}} - \mathbf{P}_{\mathcal{V}_c})\mathbf{y} = (\mathbf{P}_{\mathcal{W}^\perp} - \mathbf{P}_{\mathcal{V}_c})\mathbf{y}. \tag{3.48}$$

Consider the sum of squared error (SSE), which becomes

$$\begin{aligned}
\text{SSE} &= \| \mathbf{y} - \widehat{\mathbf{y}} \|^2 = \boldsymbol{e}'\boldsymbol{e} \\
&= \mathbf{y}'(\mathbf{P}_{\mathcal{W}^\perp} - \mathbf{P}_{\mathcal{V}_c})\mathbf{y} = \mathbf{y}'\mathbf{P}_{\mathcal{W}^\perp}\mathbf{y} - \mathbf{y}'\mathbf{P}_{\mathcal{V}_c}\mathbf{y} \\
&= \sum_{i=1}^{n}(y_i - \bar{y})^2 - \widehat{\boldsymbol{\beta}}_1'\mathbf{X}_c'\mathbf{y} = \text{SST} - \widehat{\boldsymbol{\beta}}_1'\mathbf{X}_c'\mathbf{y}.
\end{aligned} \tag{3.49}$$

Namely, the sum of squares regression (SSR) is $\text{SSR} = \widehat{\boldsymbol{\beta}}_1'\mathbf{X}_c'\mathbf{y}$.

The leverage $h_i = \mathbf{x}_i'(\mathbf{X}'\mathbf{X})^{-1}\mathbf{x}_i$ can also be reexpressed for better interpretation using the centered form. Letting $\mathbf{x}_{1i} = (x_{i1}, x_{2i}, \ldots, x_{ik})'$,

$$\begin{aligned}
h_i &= (1,\ \mathbf{x}_{1i}' - \bar{\mathbf{x}}')\left\{(\mathbf{j},\ \mathbf{X}_c)'(\mathbf{j},\ \mathbf{X}_c)\right\}^{-1}\begin{pmatrix} 1 \\ \mathbf{x}_{1i} - \bar{\mathbf{x}} \end{pmatrix} \\
&= (1,\ \mathbf{x}_{1i}' - \bar{\mathbf{x}}')\begin{pmatrix} 1/n & \mathbf{0}' \\ \mathbf{0} & (\mathbf{X}_c'\mathbf{X}_c)^{-1} \end{pmatrix}\begin{pmatrix} 1 \\ \mathbf{x}_{1i} - \bar{\mathbf{x}} \end{pmatrix} \\
&= \frac{1}{n} + (\mathbf{x}_{1i} - \bar{\mathbf{x}})'(\mathbf{X}_c\mathbf{X}_c)^{-1}(\mathbf{x}_{1i} - \bar{\mathbf{x}}).
\end{aligned} \tag{3.50}$$

Note that

$$\mathbf{X}_c\mathbf{X}_c = (n-1)\mathbf{S}_{xx}, \tag{3.51}$$

where

$$\mathbf{S}_{xx} = \begin{pmatrix} s_1^2 & s_{12} & \cdots & s_{1k} \\ s_{21} & s_2^2 & \cdots & s_{2k} \\ \vdots & \vdots & & \vdots \\ s_{k1} & s_{k2} & \cdots & s_k^2 \end{pmatrix} \text{ with } \begin{cases} s_j^2 &= \sum_{i=1}^n (x_{ij} - \bar{x}_j)^2 \\ s_{jj'} &= \sum_{i=1}^n (x_{ij} - \bar{x}_j)(x_{ij'} - \bar{x}_{j'}) \end{cases}$$

is the sample variance-covariance matrix for $\mathbf{x}$ vectors. Therefore, h_i in (3.50) is

$$h_i = \frac{1}{n} + \frac{(\mathbf{x}_{1i} - \bar{\mathbf{x}})'\mathbf{S}_{xx}^{-1}(\mathbf{x}_{1i} - \bar{\mathbf{x}})}{n-1}. \tag{3.52}$$

Clearly, the term $(\mathbf{x}_{1i} - \bar{\mathbf{x}})'\mathbf{S}_{xx}^{-1}(\mathbf{x}_{1i} - \bar{\mathbf{x}})$ gives the Mahalanobis distance between $\mathbf{x}_{1i}$ and the center of the data $\bar{\mathbf{x}}$, which renders h_i an important diagnostic measure for assessing how outlying an observation is in terms of its predictor values.

Furthermore, both $\hat{\beta}_0$ and $\widehat{\boldsymbol{\beta}}_1$ can be expressed in terms of the sample variances and covariances. Let $\mathbf{s}_{yx}$ denote the covariance vector between Y and X_j's. Namely,

$$\mathbf{s}_{yx} = (s_{y1}, s_{y2}, \ldots, s_{yk})', \tag{3.53}$$

where

$$s_{yj} = \frac{\sum_{i=1}^n (x_{ij} - \bar{x}_j) \cdot (y_i - \bar{y})}{n-1} = \frac{\sum_{i=1}^n (x_{ij} - \bar{x}_j) \cdot y_i}{n-1}.$$

It can be easily seen that

$$(n-1) \cdot \mathbf{s}_{yx} = \mathbf{X}'\mathbf{y}. \tag{3.54}$$

Using equations (3.51) and (3.54), we have

$$\widehat{\boldsymbol{\beta}}_1 = \left(\frac{\mathbf{X}_c'\mathbf{X}_c}{n-1}\right)^{-1} \frac{\mathbf{X}_c'\mathbf{y}}{n-1} = \mathbf{S}_{xx}^{-1}\mathbf{s}_{yx} \tag{3.55}$$

and

$$\hat{\beta}_0 = \bar{y} - \widehat{\boldsymbol{\beta}}_1'\bar{\mathbf{x}} = \bar{y} - \mathbf{s}_{yx}'\mathbf{S}_{xx}^{-1}\bar{\mathbf{x}}. \tag{3.56}$$

The above forms are now analogous to those formulas for $\hat{\beta}_1$ and $\hat{\beta}_0$ in simple linear regression.

Besides, the coefficient of determination R^2 can also be expressed in terms of $\mathbf{S}_{xx}$ and $\mathbf{s}_{yx}$ as below

$$\begin{aligned} R^2 = \frac{\text{SSR}}{SST} &= \frac{\widehat{\boldsymbol{\beta}}_1'\mathbf{X}_c'\mathbf{X}_c\widehat{\boldsymbol{\beta}}_1}{\sum_{i=1}^n (y_i - \bar{y})^2} \\ &= \frac{\mathbf{s}_{yx}'\mathbf{S}_{xx}^{-1}(n-1)\mathbf{S}_{xx}\mathbf{S}_{xx}^{-1}\mathbf{s}_{yx}}{\sum_{i=1}^n (y_i - \bar{y})^2} \\ &= \frac{\mathbf{s}_{yx}'\,\mathbf{S}_{xx}^{-1}\,\mathbf{s}_{yx}}{s_y^2}. \end{aligned} \tag{3.57}$$

3.19 Numerical Computation of LSE via QR Decomposition

According to earlier derivation, the least squares estimator $\widehat{\boldsymbol{\beta}}$ is obtained by solving the normal equations

$$\mathbf{X}'\mathbf{X}\boldsymbol{\beta} = \mathbf{X}'\mathbf{y}. \tag{3.58}$$

Nevertheless, the approach is not very computationally attractive because it can be difficult to form matrices in (3.58) to a great numerical accuracy. Instead, computation of LSE, as well as many other related quantities, is carried out through QR decomposition of the design matrix $\mathbf{X}$. The basic idea of this approach utilizes a successive orthogonalization process on the predictors to form an orthogonal basis for the linear space $\mathcal{V} = C(\mathbf{X})$.

3.19.1 *Orthogonalization*

To motivate, we first consider the simple regression (Y versus X) with design matrix $\mathbf{X} = (\mathbf{j}, \mathbf{x})$. The LSE of β_1 is given by

$$\hat{\beta}_1 = \frac{\sum_{i=1}^n (x_i - \bar{x}) \cdot (y_i - \bar{y})}{\sum_{i=1}^n (x_i - \bar{x})^2} = \frac{\sum_{i=1}^n (x_i - \bar{x}) \cdot y_i}{\sum_{i=1}^n (x_i - \bar{x})^2} = \frac{\langle \mathbf{x} - \bar{x}\mathbf{j}, \mathbf{y}\rangle}{\langle \mathbf{x} - \bar{x}\mathbf{j}, \mathbf{x} - \bar{x}\mathbf{j}\rangle},$$

where $\langle \mathbf{x}, \mathbf{y}\rangle = \mathbf{x}^t\mathbf{y}$ denotes the inner product between $\mathbf{x}$ and $\mathbf{y}$. The above estimate $\hat{\beta}_1$ can be obtained in two steps, either applying a simple linear regression without intercept. In step 1, regress $\mathbf{x}$ on $\mathbf{j}$ without intercept and obtain the residual $\boldsymbol{e} = \mathbf{x} - \bar{x}\mathbf{j}$; and in step 2, regress $\mathbf{y}$ on the residual $\boldsymbol{e}$ without intercept to produce $\hat{\beta}_1$.

Note that regressing $\mathbf{u}$ on $\mathbf{v}$ without intercept by fitting model

$$u_i = \gamma v_i + \varepsilon_i$$

gives

$$\hat{\gamma} = \frac{\langle \mathbf{u}, \mathbf{v} \rangle}{\langle \mathbf{v}, \mathbf{v} \rangle} \quad \text{and residual vector} \quad \boldsymbol{e} = \mathbf{u} - \hat{\gamma}\mathbf{v}, \tag{3.59}$$

which is exactly the step in linear algebra taken to orthogonalize one vector $\mathbf{u}$ with respect to another vector $\mathbf{v}$. The key point is to ensure residual vector $\boldsymbol{e}$ to be orthogonal to $\mathbf{v}$, i.e., $\boldsymbol{e} \perp \mathbf{v}$ or $\langle \boldsymbol{e}, \mathbf{v} \rangle = 0$.

Orthogonality often provides great convenience and efficiency in designed experiments. It is easy to show, for example, that if the k predictors $\mathbf{x}_1, \mathbf{x}_2, \ldots, \mathbf{x}_k$ in a multiple linear regression model are orthogonal to each other, then the LSE of the j-th slope equals

$$\hat{\beta}_j = \frac{\langle \mathbf{x}_j, \mathbf{y} \rangle}{\langle \mathbf{x}_j, \mathbf{x}_j \rangle},$$

which is the same as the slope estimator obtained in a simple linear regression model that regresses $\mathbf{y}$ on $\mathbf{x}_j$. This implies that orthogonal predictors have no confounding effect on each other at all.

In the two-step approach, the subspace $C(\mathbf{j}, \mathbf{x})$ spanned by $(\mathbf{j}, \mathbf{x})$ is the same as the subspace spanned by the orthogonal basis $(\mathbf{j}, \boldsymbol{e})$. This idea can be generalized to multiple linear regression, which leads to the algorithm outlined below. This is the well-known Gram-Schmidt procedure for constructing an orthogonal basis from an arbitrary basis. Given a design matrix $\mathbf{X} = (\mathbf{x}_0 = \mathbf{j}, \mathbf{x}_1, \ldots, \mathbf{x}_k)$ with columns $\mathbf{x}_j$, the result of the algorithm is an orthogonal basis $(\boldsymbol{e}_0, \boldsymbol{e}_1, \ldots, \boldsymbol{e}_k)$ for the column subspace of $\mathbf{X}$, $\mathcal{V} = C(\mathbf{X})$.

Algorithm 9.1: Gram-Schmidt Algorithm for Successive Orthogonalization.

- Set $\boldsymbol{e}_0 = \mathbf{j}$;
- Compute $\gamma_{01} = \langle \mathbf{x}_1, \boldsymbol{e}_0 \rangle / \langle \boldsymbol{e}_0, \boldsymbol{e}_0 \rangle$ and $\boldsymbol{e}_1 = \mathbf{x}_1 - \gamma_{01}\, \boldsymbol{e}_0$;
- Compute $\gamma_{02} = \langle \mathbf{x}_2, \boldsymbol{e}_0 \rangle / \langle \boldsymbol{e}_0, \boldsymbol{e}_0 \rangle$ and $\gamma_{12} = \langle \mathbf{x}_2, \boldsymbol{e}_1 \rangle / \langle \boldsymbol{e}_1, \boldsymbol{e}_1 \rangle$ and obtain $\boldsymbol{e}_2 = \mathbf{x}_2 - (\gamma_{02}\, \boldsymbol{e}_0 + \gamma_{12}\, \boldsymbol{e}_1)$.

$\vdots$

- Continue the process up to $\mathbf{x}_k$, which involves computing $(\gamma_{0k}, \gamma_{1k}, \ldots, \gamma_{(k-1)k})$ with $\gamma_{jk} = \langle \mathbf{x}_k, \boldsymbol{e}_j \rangle / \langle \boldsymbol{e}_j, \boldsymbol{e}_j \rangle$ for $j = 0, 1, \ldots, (k-1)$ and then obtaining $\boldsymbol{e}_k = \mathbf{x}_k - (\gamma_{0k}\, \boldsymbol{e}_0 + \gamma_{1k}\, \boldsymbol{e}_1 + \cdots + \gamma_{(k-1)k}\, \boldsymbol{e}_{k-1})$.

Note that $\boldsymbol{e}_j \perp \boldsymbol{e}_{j'}$ for $j \neq j'$. It is interesting and insightful to take a few observations, as listed below. First, the slope estimate obtained by regressing $\mathbf{y}$ on e_k without intercept is the same as the LS slope estimate for $\mathbf{x}_k$ in multiple linear model of $\mathbf{y}$ versus $(\mathbf{x}_0, \mathbf{x}_1, \ldots, \mathbf{x}_k)$. That is,

$$\hat{\beta}_k = \frac{\langle \mathbf{y}, e_k \rangle}{\langle e_k, e_k \rangle} = \frac{\langle \mathbf{y}, e_k \rangle}{\| e_k \|^2}. \tag{3.60}$$

This can be verified by using the fact that e_j's form an orthogonal basis for the column space of $\mathbf{X}$ and $\mathbf{x}_k$ is only involved in $e_k = \mathbf{x}_k - \sum_{j=0}^{k-1} \gamma_{jk} e_j$, with coefficient 1.

Secondly, since $(e_0, e_1, \ldots e_{k-1})$ spans the same subspace as $(\mathbf{x}_0, \mathbf{x}_1, \ldots, \mathbf{x}_{k-1})$ does, the residual vector e_k is identical to the residual vector obtained by regressing $\mathbf{x}_k$ versus $(\mathbf{x}_0, \mathbf{x}_1, \ldots, \mathbf{x}_{k-1})$. This is the result that motivates the partial regression plots, in which the residuals obtained from regressing $\mathbf{y}$ on $(\mathbf{x}_0, \mathbf{x}_1, \ldots, \mathbf{x}_{k-1})$ are plotted versus e_k.

Thirdly, the same results clearly hold for any one of the predictors if one rearranges it to the last position. The general conclusion is that the j-th slope $\hat{\beta}_j$ can be obtained by fitting a simple linear regression of $\mathbf{y}$ on the residuals obtained from regressing $\mathbf{x}_j$ on other predictors $(\mathbf{x}_0, \mathbf{x}_1, \ldots, \mathbf{x}_{j-1}, \mathbf{x}_{j+1}, \ldots, \mathbf{x}_k)$. Thus, $\hat{\beta}_j$ can be interpreted as the additional contribution of $\mathbf{x}_j$ on $\mathbf{y}$ after $\mathbf{x}_j$ has been adjusted for other predictors. Furthermore, from (3.60) the variance of $\hat{\beta}_k$ is

$$\text{Var}(\hat{\beta}_k) = \frac{\sigma^2}{\| e_k \|^2}. \tag{3.61}$$

In other words, e_k, which represents how much of $\mathbf{x}_p$ is unexplained by other predictors, plays an important role in estimating β_k.

Fourthly, if $\mathbf{x}_p$ is highly correlation with some of the other predictors, a situation to which multicolinearity is referred, then the residual vector e_k will be close to zero in length $|e|$. From (3.60), the estimate $\hat{\beta}_k$ would be very unstable. From (3.61), the precision of the estimate would be poor as well. The effect due to multicolinearity is clearly true for all predictors in the correlated predictors.

3.19.2 *QR Decomposition and LSE*

The Gram-Schmidt algorithm can be represented in matrix form. It can be easily verified that

$$\gamma_{jl} = \frac{\langle \mathbf{x}_l, e_j \rangle}{\langle e_j, e_j \rangle} = \begin{cases} 1 & \text{if } j = l \\ 0 & \text{if } j < l \end{cases} \tag{3.62}$$

for $j, l = 0, 1, \ldots, k$. Denoting

$$\mathbf{\Gamma} = (\gamma_{jl}) = \begin{pmatrix} 1 & \gamma_{01} & \gamma_{02} & \cdots & \gamma_{0(k-1)} & \gamma_{0k} \\ 0 & 1 & \gamma_{12} & \cdots & \gamma_{1(k-1)} & \gamma_{1k} \\ \vdots & \vdots & \vdots & & \vdots & \vdots \\ 0 & 0 & 0 & \cdots & 1 & \gamma_{(k-1)k} \\ 0 & 0 & 0 & \cdots & 0 & 1 \end{pmatrix} \quad \text{and} \quad \mathbf{E} = (\boldsymbol{e}_0, \boldsymbol{e}_1, \ldots, \boldsymbol{e}_k),$$

we have

$$\mathbf{X} = \mathbf{E\Gamma} = (\mathbf{ED}^{-1})(\mathbf{D\Gamma}) = \mathbf{QR}, \tag{3.63}$$

where $\mathbf{D} = \text{diag}(d_{jj})$ with $d_{jj} = \| \boldsymbol{e}_{j-1} \|$ for $j = 1, \ldots, (k+1)$. The form given in (3.63) is the so-called QR decomposition of $\mathbf{X}$. The matrix

$$\mathbf{Q} = \mathbf{ED}^{-1} \tag{3.64}$$

is an orthogonal matrix of dimension $n \times (k+1)$ satisfying $\mathbf{Q}'\mathbf{Q} = \mathbf{I}$ and the matrix

$$\mathbf{R} = \mathbf{D\Gamma} \tag{3.65}$$

is a $(k+1) \times (k+1)$ upper triangular matrix.

Using the decomposition in (3.63), the normal equations becomes

$$\begin{aligned} \mathbf{X}^t\mathbf{X}\boldsymbol{\beta} &= \mathbf{X}^t\mathbf{y} \\ \Longrightarrow \mathbf{R}'\mathbf{Q}^t\mathbf{QR}\boldsymbol{\beta} &= \mathbf{R}'\mathbf{Q}'\mathbf{y} \\ \Longrightarrow \qquad \mathbf{R}\boldsymbol{\beta} &= \mathbf{Q}\mathbf{y}, \end{aligned} \tag{3.66}$$

which are easy to solve since $\mathbf{R}$ is upper triangular. This leads to the LSE

$$\widehat{\boldsymbol{\beta}} = \mathbf{R}^{-1}\mathbf{Q}'\mathbf{y}. \tag{3.67}$$

Its variance-covariance matrix can be expressed as

$$\text{Cov}(\widehat{\boldsymbol{\beta}}) = \sigma^2 \cdot (\mathbf{X}^t\mathbf{X})^{-1} = \sigma^2 \cdot (\mathbf{R}^t\mathbf{R})^{-1} = \sigma^2 \cdot \mathbf{R}^{-1}(\mathbf{R}^t)^{-1}. \tag{3.68}$$

To compute, $\mathbf{R}^{-1}$ is needed. Since $\mathbf{R}$ is upper triangular, $\mathbf{R}^{-1} = \mathbf{W}$ can be easily obtained with back-substitution in the system of linear equations

$$\mathbf{R}\,\mathbf{W} = \mathbf{I}. \tag{3.69}$$

Various other desired quantities in linear regression including the F test statistic for linear hypotheses can also be computed using the QR decomposition. For example, the predicted vector is

$$\widehat{\mathbf{y}} = \mathbf{X}\widehat{\boldsymbol{\beta}} = \mathbf{QQ}^t\mathbf{y},$$

the residual vector is

$$\boldsymbol{e} = \mathbf{y} - \widehat{\mathbf{y}} = (\mathbf{I} - \mathbf{QQ}^t)\mathbf{y}$$

and the sum of squared errors (SSE) is

$$\text{SSE} = \boldsymbol{e}^t\boldsymbol{e} = \mathbf{y}^t(\mathbf{I} - \mathbf{QQ}^t)\mathbf{y} = \| \mathbf{y} \|^2 - \| \mathbf{Q}^t\mathbf{y} \|^2 .$$

3.20 Analysis of Regression Residual

3.20.1 *Purpose of the Residual Analysis*

Definition 3.5. The residual of the linear regression model $\boldsymbol{y} = \boldsymbol{X}\beta + \varepsilon$ is defined as the difference between observed response variable $\boldsymbol{y}$ and the fitted value $\hat{\boldsymbol{y}}$, i.e., $\boldsymbol{e} = \boldsymbol{y} - \hat{\boldsymbol{y}}$.

The regression error term ε is unobservable and the residual is observable. Residual is an important measurement of how close the calculated response from the fitted regression model to the observed response. The purposes of the residual analysis are to detect model mis-specification and to verify model assumptions. Residuals can be used to estimate the error term in regression model, and the empirical distribution of residuals can be utilized to check the normality assumption of the error term (QQ plot), equal variance assumption, model over-fitting, model under-fitting, and outlier detection. Overall, residual analysis is useful for assessing a regression model.

Simple statistical properties of the regression residual can be discussed. The ith residual of the linear regression model can be written as

$$e_i = y_i - \hat{y}_i = y_i - \boldsymbol{x}_i b = y_i - \boldsymbol{x}_i(\boldsymbol{X}^{'}\boldsymbol{X})^{-1}\boldsymbol{X}^{'}\boldsymbol{y}.$$

Regression residual can be expressed in a vector form

$$\boldsymbol{e} = \boldsymbol{y} - \boldsymbol{X}(\boldsymbol{X}^{'}\boldsymbol{X})^{-1}\boldsymbol{X}^{'}\boldsymbol{y} = (I - \boldsymbol{X}(\boldsymbol{X}^{'}\boldsymbol{X})^{-1}\boldsymbol{X}^{'})\boldsymbol{y} = (I - H)\boldsymbol{y}, \quad (3.70)$$

where $H = \boldsymbol{X}(\boldsymbol{X}^{'}\boldsymbol{X})^{-1}\boldsymbol{X}^{'}$ is called the HAT matrix. Note that $I - H$ is symmetric and idempotent, i.e., $I - H = (I - H)^2$. The covariance matrix of the residual $\boldsymbol{e}$ is given by:

$$\text{Cov}(\boldsymbol{e}) = (I - H)\text{Var}(\boldsymbol{y})(I - H)^{'} = (I - H)\sigma^2.$$

Denote the hat matrix $H = (h_{ij})$ we have

$$\text{Var}(e_i) = (1 - h_{ii})\sigma^2$$

and

$$\text{Cov}(e_i, e_j) = -h_{ij}\sigma^2.$$

The HAT matrix contains useful information for detecting outliers and identifying influential observations.

Table 3.7 United States Population Data (in Millions)

Year	Population	Year	Population	Year	Population	Year	Population
1790	3.929	1800	5.308	1810	7.239	1820	9.638
1830	12.866	1840	17.069	1850	23.191	1860	31.443
1870	39.818	1880	50.155	1890	62.947	1900	75.994
1910	91.972	1920	105.710	1930	122.775	1940	131.669
1950	151.325	1960	179.323	1970	203.211		

3.20.2 *Residual Plot*

A plot of residuals e_i's against the fitted values $\hat{y}_i$'s is residual plot, which is a simple and convenient tool for regression model diagnosis. The residuals evenly distributed on both sides of $y = 0$ imply that the assumptions $E(\varepsilon) = 0$ and constant variance $\text{Var}(\varepsilon_i) = \sigma^2$ are appropriate. A curvature appearance in residual plot implies that some higher order terms in regression model may be missing. A funnel shape of residual plot indicates heterogeneous variance and violation of model assumption $\text{Var}(\varepsilon) = \sigma^2$. In addition, periodical and curvature residuals may indicate that the possible regression model may be piecewise and some higher order terms in the model may be missing. The following Fig. 3.3 illustrate different situations of the residuals in regression model. Figure (a) displays residuals evenly distributed about 0, (b) shows residuals with uneven variances, (c) displays residuals with curvature pattern, and (d) displays periodic and curvature residuals. The classical regression model assumptions $E(\boldsymbol{\varepsilon}) = 0$ and $\text{Var}(\varepsilon_i) = \sigma^2$ are satisfied only when residuals are evenly distributed about 0. The other residual plots imply some deviations from the classical regression model assumptions. When model assumptions are violated the model is no longer valid and statistical inference based on model is not reliable anymore.

We now discuss how to use residual plot to improve regression model. The illustrative example is the regression model of the populations of the United States from Year 1790 to Year 1970. We will show how to improve the regression model based on residual diagnosis. The population data (in millions) are presented in Table 3.7.

Example 3.2. First, we fit the data to the simple linear regression model

$$\text{Population} = \beta_0 + \beta_1 \text{Year} + \varepsilon.$$

The estimates of the model parameters are presented in Table 3.8. We then compute the residuals and plot the residuals against the fitted values

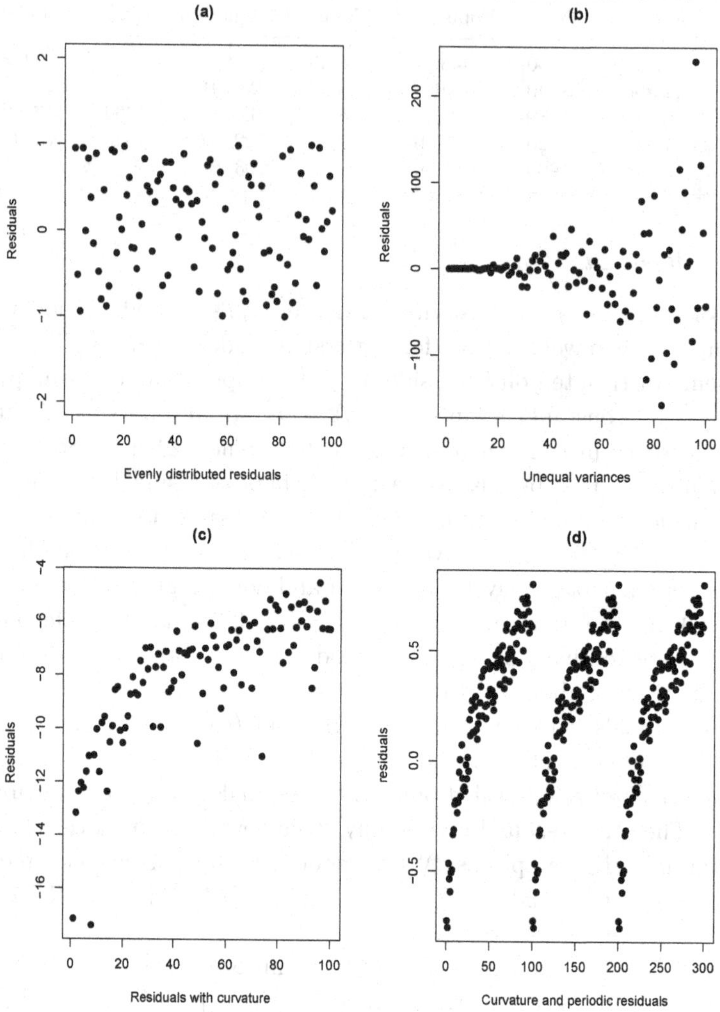

Fig. 3.3 Various Shapes of Residual Plots

$\hat{y}$ for regression model Population $= \beta_0 + \beta_1 \text{Year} + \varepsilon$. The residual plot is presented in Fig. 3.4 (b). The curvature appearance of the residual plot implies that the proposed regression model may be under-fitted. i.e., some necessary higher order terms in the regression model may be missing.

Table 3.8 Parameter Estimates for Model Population=Year

MODEL	TYPE	DEPVAR	RMSE	Intercept	year
MODEL1	PARMS	population	18.1275	-1958.37	1.0788
MODEL1	STDERR	population	18.1275	142.80	0.0759
MODEL1	T	population	18.1275	-13.71	14.2082
MODEL1	PVALUE	population	18.1275	0.00	0.0000
MODEL1	L95B	population	18.1275	-2259.66	0.9186
MODEL1	U95B	population	18.1275	-1657.08	1.2390

Table 3.9 Parameter Estimates for Model Population=Year+Year^2

TYPE	DEPVAR	RMSE	Intercept	Year	Year^2
PARMS	population	2.78102	20450.43	-22.7806	0.0063
STDERR	population	2.78102	843.48	0.8978	0.0002
T	population	2.78102	24.25	-25.3724	26.5762
PVALUE	population	2.78102	0.00	0.0000	0.0000
L95B	population	2.78102	18662.35	-24.684	0.0058
U95B	population	2.78102	22238.52	-20.8773	0.0069

The curvature of a quadratic appearance in the residual plot suggests that a quadratic term in the model may be missing. We then add a term $Year^2$ into the model and fit the data to the following regression model:

$$\text{Population} = \beta_0 + \beta_1 \text{Year} + \beta_2 \text{Year}^2 + \varepsilon.$$

The estimates of the model parameters are presented in Table 3.9. The residual plot of above regression model is presented in Fig. 3.4 (c) and the shape of the residual plot is clearly better than the residual plot Fig. 3.4 (b), since residuals become more evenly distributed on both sides of $y = 0$. If we take a closer look at the residual plot, we still observe that two residuals, which are at years 1940 and 1950, are far from the line $y = 0$. We know that in the history these are the years during the World War II. We think there might be a shift of populations due to the war. So we try to add a dummy variable z into the model. The dummy variable z takes value 1 at years 1940 and 1950, and 0 elsewhere. The regression model with mean shift term z can be written as

$$\text{Population} = \beta_0 + \beta_1 \text{Year} + \beta_2 \text{Year}^2 + \beta_3 z + \varepsilon.$$

We then fit the US population data to the above model. The estimates of the model parameters are presented in Table 3.10. The residual plot for

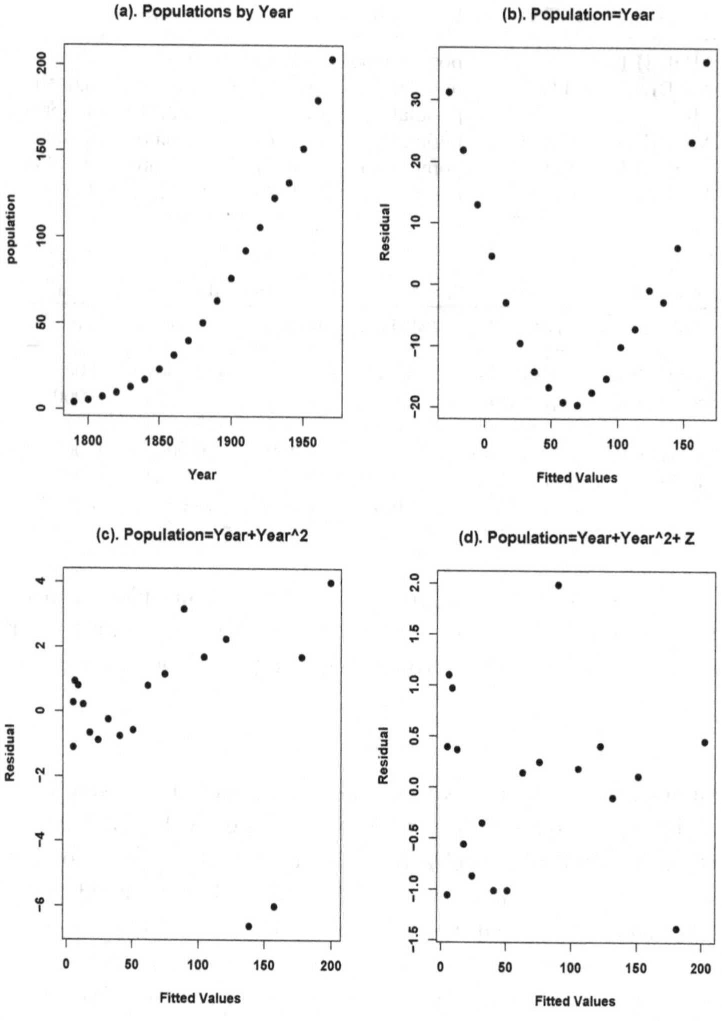

Fig. 3.4 Residual Plots of the Regression Model of the US Population

this model is presented in Fig 3.4 (d) and it is clearly improved, since the residuals are much more evenly distributed on both sides of $y = 0$, including the residuals at Years 1940 and 1950.

The SAS program for generating analysis results above is provided below for illustrative purpose.

Table 3.10 Parameter Estimates for Regression Model Population=β_0 +β_1 Year+ β_2 Year^2+z

TYPE	DEPVAR	RMSE	Intercept	Year	Year^2	z
PARMS	population	0.93741	20982.75	-23.3664	0.0065	-8.7415
STDERR	population	0.93741	288.25	0.3071	0.0001	0.7793
T	population	0.93741	72.79	-76.0838	79.5883	-11.2170
PVALUE	population	0.93741	0.00	0.0000	0.0000	0.0000
L95B	population	0.93741	20368.37	-24.0210	0.0063	-10.4026
U95B	population	0.93741	21597.14	-22.7118	0.0067	-7.0805

```
data pop; set pop;
yyear=year*year;
if year in (1940, 1950) then z=1;
else z=0;
run;

proc reg data=pop outest=out1 tableout;
     model population=year;
     output out=out2
     p=yhat r=yresid student=sresid;
run;

proc reg data=pop outest=out3 tableout;
     model population=year yyear;
     output out=out4
     p=yhat r=yresid student=sresid;
run;

proc reg data=pop outest=out5 tableout;
     model population=year yyear z;
     output out=out6
     p=yhat r=yresid student=sresid;
run;

proc gplot data=out2; symbol v=dot h=1;
     plot yresid*yhat/caxis=red ctext=blue vref=0;
title "population=year";

proc gplot data=out4; symbol v=dot h=1;
```

```
      plot yresid*yhat/caxis=red ctext=blue vref=0;
title "population=year+year*year";

proc gplot data=out6; symbol v=dot h=1;
      plot yresid*yhat/caxis=red ctext=blue vref=0;
title "population=year+year*year +Z";
run;
```

The above regression analysis of the US population over years can be performed using the free software R. One advantage of software R over SAS is that R generates regression diagnosis graphs relatively easily. We present the following R code that perform the regression analysis of the US population over years and generate all regression diagnosis plots in postscript format for different regression models.

```
year<-c(1790,1800,1810,1820,1830,1840,1850,1860,1870,1880,
        1890,1900,1910,1920,1930,1940,1950,1960,1970)
pop<-c(3.929,5.308,7.239,9.638,12.866,17.069,23.191,31.443,
       39.818,50.155, 62.947,75.994,91.972,105.710,122.775,
       131.669,151.325,179.323,203.211)
postscript("C:\\uspop.eps",horizontal=FALSE, onefile= FALSE,
           print.it=FALSE)

par(mfrow=c(2, 2))
plot(pop~year, pch=20, font=2, font.lab=2,
     ylab="population",xlab="Year",
     main="populations by Year")

fit<-lm(pop~year)
fitted<-fit$fitted
resid<-fit$residual
plot(fitted, resid, pch=20, cex=1.5, font=2, font.lab=2,
     ylab="Residual", xlab="Fitted Values",
     main="Population=Year")

yyear<-year*year
fit1<-lm(pop ~ year + yyear)
fitted1<-fit1$fitted
resid1<-fit1$residual
```

```
plot(fitted1, resid1, pch=20, cex=1.5, font=2, font.lab=2,
     ylab="Residual", xlab="Fitted Values",
     main="population=Year+Year^2")

z<-ifelse((year==1940)|(year==1950), 1, 0)
fit2<-lm(pop ~ year + yyear +z)
fitted2<-fit2$fitted
resid2<-fit2$residual
plot(fitted2, resid2, pch=20, cex=1.5, font=2, font.lab=2,
     ylab="Residual", xlab="Fitted Values",
     main="population=Year+Year^2+ Z")
dev.off()
```

3.20.3 *Studentized Residuals*

Without normalization the usual residual $e_i = y_i - \hat{y}_i$ is subject to the scale of the response y_i. It is inconvenient when several regression models are discussed together. We then consider the normalized residual. Since $\text{Var}(e_i) = (1 - h_{ii})\sigma^2$ the normalized regression residual can be defined as

$$r_i = \frac{e_i}{s\sqrt{1 - h_{ii}}}. \tag{3.71}$$

This normalized residual is called the studentized residual. Note that σ is unknown and it can be estimated by s. The studentized residual is scale-free and can be used for checking model assumption. Also it can be used for model diagnosis. If several regression models need to be compared the scale-free studentized residuals is a better measurement for model comparison.

3.20.4 *PRESS Residual*

The PRESS residual is the leave-one-out residual. To obtain the PRESS residual we fit the regression model without using the ith observation and calculate the fitted value from that model

$$\hat{y}_{i,-i} = \boldsymbol{x}_i b_{-i},$$

where b_{-i} is the least squares estimate of regression parameters without using the ith observation. $\hat{y}_{i,-i}$ is the fitted value calculated from the regression model without using the ith observation. The ith PRESS residual is defined as

$$e_{i,-i} = y_i - \hat{y}_{i,-i} = y_i - x_i b_{-i}. \tag{3.72}$$

The PRESS residual is the measurement of influential effect of the ith observation on the regression model. If the ith observation has a small influence on the regression model then $\hat{y}_i$ should be fairly close to $\hat{y}_{i,-i}$, therefore, the PRESS residual $e_{i,-i}$ should be close to the usual residual e_i. In order to discuss the PRESS residual and establish the relationship between usual the residual e_i and the PRESS residual $e_{i,-i}$ we first introduce the following the theorem (see Rao, 1973).

Theorem 3.16. *Let A be a nonsingular square $p \times p$ matrix and z be a p-dimensional column vector. The matrix $(A - zz')^{-1}$ is given by*

$$(A - zz')^{-1} = A^{-1} + \frac{A^{-1}zz'A^{-1}}{1 - z'A^{-1}z}. \tag{3.73}$$

The proof of the theorem is to directly show that $A - zz'$ multiply the matrix on the right side of the above formula yields an identity matrix. This theorem will be used later to establish the relationship between the PRESS residual and the ordinary residual. For regression model $y = \boldsymbol{X}\beta + \varepsilon$, write $\boldsymbol{X}$ as $(1, \boldsymbol{x}_2, \boldsymbol{x}_2, \cdots, \boldsymbol{x}_p)$, where $\boldsymbol{x}_i$ is an n-dimensional vector. It is easy to verify that

$$\boldsymbol{X}'\boldsymbol{X} = \begin{pmatrix} n & 1'\boldsymbol{x}_1 & 1'\boldsymbol{x}_2 & \cdots & 1'\boldsymbol{x}_p \\ 1'\boldsymbol{x}_1 & \boldsymbol{x}_1'\boldsymbol{x}_1 & \boldsymbol{x}_1'\boldsymbol{x}_2 & \cdots & \boldsymbol{x}_1'\boldsymbol{x}_p \\ 1'\boldsymbol{x}_2 & \boldsymbol{x}_1'\boldsymbol{x}_1 & \boldsymbol{x}_2'\boldsymbol{x}_2 & \cdots & \boldsymbol{x}_2'\boldsymbol{x}_p \\ \cdots \\ 1'\boldsymbol{x}_p & \boldsymbol{x}_2'\boldsymbol{x}_p & \boldsymbol{x}_3'\boldsymbol{x}_2 & \cdots & \boldsymbol{x}_p'\boldsymbol{x}_p \end{pmatrix}$$

$$= \begin{pmatrix} n & \sum_j x_{1j} & \sum_j x_{2i} & \cdots & \sum_j x_{pj} \\ \sum_j x_{1j} & \sum_j x_{1j}^2 & \sum_j x_{1j}x_{2j} & \cdots & \sum_j x_{1j}x_{pj} \\ \sum_j x_{2j} & \sum_j x_{2j}x_{2j} & \sum_j x_{2j}^2 & \cdots & \sum_j x_{2j}x_{pj} \\ \cdots \\ \sum_j x_{pj} & \sum_j x_{pj}x_{pj} & \sum_j x_{pj}x_{2j} & \cdots & \sum_j x_{pj}^2 \end{pmatrix}.$$

Remove the ith observation from $\boldsymbol{X}$ and perform the matrix multiplication of $\boldsymbol{X}'_{-i}\boldsymbol{X}_{-i}$ we have

$$\boldsymbol{X}'_{-i}\boldsymbol{X}_{-i} = \begin{pmatrix} n-1 & \sum_{j\neq i} x_{1j} & \sum_{j\neq i} x_{2j} & \cdots & \sum_{j\neq i} x_{pj} \\ \sum_{j\neq i} x_{1j} & \sum_{j\neq i} x_{1j}^2 & \sum_{j\neq i} x_{1j}x_{2j} & \cdots & \sum_{j\neq i} x_{1j}x_{pj} \\ \sum_{j\neq i} x_{2j} & \sum_{j\neq i} x_{2j}x_{2j} & \sum_{j\neq i} x_{2j}^2 & \cdots & \sum_{j\neq i} x_{2j}x_{pj} \\ \cdots \\ \sum_{j\neq i} x_{pj} & \sum_{j\neq i} x_{pj}x_{pj} & \sum_{j\neq i} x_{pj}x_{2j} & \cdots & \sum_{j\neq i} x_{pj}^2 \end{pmatrix}$$

$$= \boldsymbol{X}'\boldsymbol{X} - \boldsymbol{x}_i\boldsymbol{x}_i'.$$

Thus, we establish that

$$\boldsymbol{X}'_{-i}\boldsymbol{X}_{-i} = \boldsymbol{X}'\boldsymbol{X} - \boldsymbol{x}_i\boldsymbol{x}_i'.$$

Using the formula above and set $A = \boldsymbol{X}'\boldsymbol{X}$ we find

$$(\boldsymbol{X}'_{-i}\boldsymbol{X}_{-i})^{-1} = (\boldsymbol{X}'\boldsymbol{X} - \boldsymbol{x}_j\boldsymbol{x}_i')^{-1}$$

$$= (\boldsymbol{X}'\boldsymbol{X})^{-1} + \frac{(\boldsymbol{X}'\boldsymbol{X})^{-1}\boldsymbol{x}_i\boldsymbol{x}_i'(\boldsymbol{X}'\boldsymbol{X})^{-1}}{1 - \boldsymbol{x}_i'(\boldsymbol{X}'\boldsymbol{X})^{-1}\boldsymbol{x}_i}$$

$$= (\boldsymbol{X}'\boldsymbol{X})^{-1} + \frac{(\boldsymbol{X}'\boldsymbol{X})^{-1}\boldsymbol{x}_i\boldsymbol{x}_i'(\boldsymbol{X}'\boldsymbol{X})^{-1}}{1 - h_{ii}}$$

The following theorem gives the relationship between the PRESS residual and the usual residual.

Theorem 3.17. *Let regression model be* $\boldsymbol{y} = \boldsymbol{X\beta} + \boldsymbol{\varepsilon}$. *The relationship between the* i*th PRESS residual* $e_{i,-i}$ *and the ordinary* i*th residual* e_i *is given by*

$$e_{i,-i} = \frac{e_i}{1 - h_{ii}} \tag{3.74}$$

$$Var(e_{i,-i}) = \frac{\sigma^2}{1 - h_{ii}}. \tag{3.75}$$

Proof. For the regression model without using the ith observation the residual is

$$
\begin{aligned}
e_{i,-i} &= \boldsymbol{y}_i - \boldsymbol{x}_i^{'} b_{-i} = \boldsymbol{y}_i - \boldsymbol{x}_i^{'}(\boldsymbol{X}_{-i}^{'}\boldsymbol{X}_{-i})^{-1}\boldsymbol{X}_{-i}^{'}\boldsymbol{y}_{-i} \\
&= \boldsymbol{y}_i - \boldsymbol{x}_i^{'}\Big[(\boldsymbol{X}^{'}\boldsymbol{X})^{-1} + \frac{(\boldsymbol{X}^{'}\boldsymbol{X})^{-1}\boldsymbol{x}_i\boldsymbol{x}_i^{'}(\boldsymbol{X}^{'}\boldsymbol{X})^{-1}}{1-h_{ii}}\Big]\boldsymbol{X}_{-i}^{'}\boldsymbol{y}_{-i} \\
&= \frac{(1-h_{ii})\boldsymbol{y}_i - (1-h_{ii})\boldsymbol{x}_i^{'}(\boldsymbol{X}^{'}\boldsymbol{X})^{-1}\boldsymbol{X}_{-i}^{'}\boldsymbol{y}_{-i} - h_{ii}\boldsymbol{x}_i^{'}(\boldsymbol{X}^{'}\boldsymbol{X})^{-1}\boldsymbol{X}_{-i}^{'}\boldsymbol{y}_{-i}}{1-h_{ii}} \\
&= \frac{(1-h_{ii})\boldsymbol{y}_i - \boldsymbol{x}_i^{'}(\boldsymbol{X}^{'}\boldsymbol{X})^{-1}\boldsymbol{X}_{-i}^{'}\boldsymbol{y}_{-i}}{1-h_{ii}}
\end{aligned}
$$

Note that $\boldsymbol{X}_{-i}^{'}\boldsymbol{y}_{-i} + \boldsymbol{x}_i\boldsymbol{y}_i = \boldsymbol{X}^{'}\boldsymbol{y}$ we have

$$
\begin{aligned}
e_{i,-i} &= \frac{(1-h_{ii})\boldsymbol{y}_i - \boldsymbol{x}_i^{'}(\boldsymbol{X}^{'}\boldsymbol{X})^{-1}(\boldsymbol{X}^{'}\boldsymbol{y} - \boldsymbol{x}_i\boldsymbol{y}_i)}{1-h_{ii}} \\
&= \frac{(1-h_{ii})\boldsymbol{y}_i - \boldsymbol{x}_i^{'}(\boldsymbol{X}^{'}\boldsymbol{X})^{-1}\boldsymbol{X}^{'}\boldsymbol{y} + \boldsymbol{x}_i^{'}(\boldsymbol{X}^{'}\boldsymbol{X})^{-1}\boldsymbol{x}_i\boldsymbol{y}_i}{1-h_{ii}} \\
&= \frac{(1-h_{ii})\boldsymbol{y}_i - \hat{\boldsymbol{y}}_i + h_{ii}\boldsymbol{y}_i}{1-h_{ii}} = \frac{\boldsymbol{y}_i - \hat{\boldsymbol{y}}_i}{1-h_{ii}} = \frac{e_i}{1-h_{ii}}
\end{aligned}
$$

For variance of PRESS residual $\text{Var}(e_{i,-i})$ we have

$$
\text{Var}(e_{i,-i}) = \text{Var}(e_i)\frac{1}{(1-h_{ii})^2} = [\sigma^2(1-h_{ii})]\frac{1}{(1-h_{ii})^2} = \frac{\sigma^2}{1-h_{ii}} \qquad \square
$$

The ith standardized PRESS residual is

$$
\frac{e_{i,-i}}{\sigma_{i,-i}} = \frac{e_i}{\sigma\sqrt{1-h_{ii}}}. \tag{3.76}
$$

3.20.5 *Identify Outlier Using PRESS Residual*

The standardized PRESS residual can be used to detect outliers since it is related to the ith observation and is scale free. If the ith PRESS residual is large enough then the ith observation may be considered as a potential outlier. In addition to looking at the magnitude of the ith PRESS residual, according to the relationship between the PRESS residual $e_{i,-i}$ and the regular residual e_i, the ith observation may be a potential outlier if the leverage h_{ii} is close to 1.

We now discuss how to deal with outlier in regression model. First, what is an outlier? An outlier is an observation at which the fitted value is not close enough to the observed response. i.e., there is breakdown in the model at the ith observation such that the location of the response is shifted. In this situation, the ith data point could be a potential outlier. To mathematically formulate this mean shift or model breakdown, we can write $E(\varepsilon_i) = \Delta \neq 0$. i.e., there is a non-zero mean shift in error term at the ith observation. If we believe that the choice and model assumptions are appropriate, it is suspectable that the ith observation might be an outlier in terms of the shift of the response from the model at that observation.

Another aspect of an outlier is that at the ith data point the $\text{Var}(\varepsilon)$ exceeds the error variance at other data points. i.e., there might be an inflation in variance at the ith observation. If the equal variance assumption is appropriate we may consider the ith observation as an outlier if the variance is inflated at the ith observation. So, outlier could be examined by checking both the mean response shift and the variance inflation at the ith data point. If equal variance assumption is no longer appropriate in the regression model we can use the generalized least squares estimate where the equal variance assumption is not required. The generalized least squares estimate will be discussed later.

A convenient test statistic used to detect outlier in regression model is the ith PRESS residual

$$e_{i,-i} = y_i - \hat{y}_{i,-i}.$$

If there is a mean shift at the ith data point, then we have

$$E(y_i - \hat{y}_{i,-i}) = E(e)i, -i = \Delta_i > 0.$$

Similarly, if there is a variance inflation at the ith data point we would like to use the standardized PRESS residual

$$\frac{e_{i,-i}}{\sigma_{i,-i}} = \frac{e_i}{\sigma\sqrt{1-h_{ii}}}$$

to detect a possible outlier. Since σ is unknown, we can replace σ by its estimate s to calculate the standardized PRESS residual. Note that in the presence of a mean shift outlier s is not an ideal estimate of true standard deviation of σ. If we consider the situation where there is a mean shift outlier, the sample standard deviation s is biased upward, and is not an ideal estimate of standard error σ. One way to cope with it is to leave the

ith observation out and calculate the leave-one-out sum of squared residuals s_{-i}. It can be shown that the relationship between s_{-i} and regular s is

$$s_{-i} = \sqrt{\frac{(n-p)s^2 - e_i^2/(1-h_{ii})}{n-p-1}}. \tag{3.77}$$

Replacing σ with s_{-i} we can construct a test statistic

$$t_i = \frac{y_i - \hat{y}_i}{s_{-i}\sqrt{1-h_{ii}}} \sim t_{n-p-1}. \tag{3.78}$$

Under the null hypothesis $H_0 : \Delta_i = 0$, the above test statistic has the centralized t distribution with degrees of freedom $n-p-1$, where n is the sample size and $p+1$ is the total number of parameters in the regression model. This test statistic can be used to test the hypothesis $H_0 : \Delta_i = 0$ versus the alternative $H_1 : \Delta_i \neq 0$. The above statistic is often called the R-student statistic. It tends larger if the ith data point is a mean shift outlier. Note that the two-tailed t-test should be used to test a mean shift outlier using the R-student statistic.

The R-student statistic can also be used to test variance inflation at the ith observation. If there is inflation in variance at the ith observation we should have $\text{Var}(\varepsilon_i) = \sigma^2 + \sigma_i^2$. Here σ_i^2 represents the increase in variance at the ith data point. The hypothesis may be defined as $H_0 : \sigma_i^2 = 0$ versus $H_1 : \sigma_i^2 \neq 0$. Note that the two-tailed t-test should be used as well.

3.20.6 *Test for Mean Shift Outlier*

Example 3.3. The coal-cleansing data will be used to illustrate the mean shift outlier detection in multiple regression. The data set has three independent variables. Variable x_1 is the percent solids in the input solution; x_2 is the pH value of the tank that holds the solution; and x_3 is the flow rate of the cleansing polymer in ml/minute. The response variable y is the measurement of experiment efficiency. The data set is presented in Table 3.11.

We first fit the coal-cleansing data to the multiple regression model:

$$y = \beta_0 + \beta_1 x_1 + \beta_2 x_2 + \beta_3 x_3 + \varepsilon.$$

The SAS procedure REG is used to calculate the parameter estimates, HAT matrix, ordinary residuals, and R-student residuals for the above regression model. The program is presented as follows:

Table 3.11 Coal-cleansing Data

Experiment	x_1	x_2	x_3	y
1	1.5	6.0	1315	243
2	1.5	6.0	1315	261
3	1.5	9.0	1890	244
4	1.5	9.0	1890	285
5	2.0	7.5	1575	202
6	2.0	7.5	1575	180
7	2.0	7.5	1575	183
8	2.0	7.5	1575	207
9	2.5	9.0	1315	216
10	2.5	9.0	1315	160
11	2.5	6.0	1890	104
12	2.5	6.0	1890	110

```
proc reg data=coal outest=out1 tableout;
    model y=x1 x2 x3;
    output out=out2
    p=yhat r=resid h=hat rstudent=Rresid;
    run;
```

The fitted regression model is found to be

$$\hat{y} = 397.087 - 110.750x_1 + 15.5833x_2 - 0.058x_3.$$

The estimates of the regression parameters and the corresponding P-values are presented in Table 3.12.

Table 3.12 Parameter Estimates for Regression Model for Coal–Cleansing Data

Type	RMSE	Intercept	x1	x2	x3
PARMS	20.8773	397.087	-110.750	15.5833	-0.05829
STDERR	20.8773	62.757	14.762	4.9208	0.02563
T	20.8773	6.327	-7.502	3.1668	-2.27395
PVALUE	20.8773	0.000	0.000	0.0133	0.05257
L95B	20.8773	252.370	-144.792	4.2359	-0.11741
U95B	20.8773	541.805	-76.708	26.9308	0.00082

Before the analysis there was suspicion by the experimental engineer that the 9th data point was keyed in erroneously. We first fit the model without deleting the 9th data point. The fitted responses, residuals, values of diagonal elements in the HAT matrix, and values of the R-student statistic associated with each observation are calculated and listed in Table 3.13. The largest residual is the 9th residual ($e_9 = 32.192$) and the corresponding R-student statistic value is 2.86951, which implies that the 9th residual is greater than zero statistically. This finding would support the suspicion that the 9th data point was originally keyed in incorrectly.

Table 3.13 Residuals

Experiment	y_i	$\hat{y}_i$	e_i	h_{ii}	t_i
1	243	247.808	-4.8080	0.45013	-0.29228
2	261	247.808	13.1920	0.45013	0.83594
3	244	261.040	-17.0400	0.46603	-1.13724
4	285	261.040	23.9600	0.46603	1.76648
5	202	200.652	1.3480	0.08384	0.06312
6	180	200.652	-20.6520	0.08384	-1.03854
7	183	200.652	-17.6520	0.08384	-0.86981
8	207	200.652	6.3480	0.08384	0.29904
9*	216	183.808	32.1920*	0.45013	2.86951*
10	160	183.808	-23.8080	0.45013	-1.71405
11	104	103.540	0.4600	0.46603	0.02821
12	110	103.540	6.4600	0.46603	0.40062

Note that the statistical analysis only confirms that the 9th data point does not fit the proposed regression model well. Therefore, it may be a potential mean shift outlier. The decision on whether or not keeping this data point in the model has to be made jointly by regression model diagnosis, rechecking the experimental data, and consulting with the engineer who collected the data.

In the example above the mean shift outlier is tested individually. If there are multiple mean shift outliers, we can test these mean shift outliers simultaneously. To do so the threshold is calculated by the t distribution with degrees of freedom $n-p-1$ and test level α is chosen to be $0.025/m$, where n=total number of observations, $p+1$=number of regression parameters in the model, and m is the number of potential outliers that need to be tested. For small data set one may choose $m = n$. The Manpower data will be used to illustrate the simultaneous test for multiple mean shift outliers. The data were collected from 25 office sites by U.S. Navy. The purpose of the regression analysis is to determine the needs for the manpower in Bachelor Officers Quarters. The 7 independent variables and the response variable y in the data set are

x_1: Average daily occupancy
x_2: Monthly average numbers of check-ins
x_3: Weekly hours of service desk operation
x_4: Square feet of common use area
x_5: Number of building wings
x_6: Operational berthing capacity
x_7: Number of rooms
y : Monthly man-hours

The data set is presented in Table 3.14:

Table 3.14 Manpower Data

Site	x1	x2	x3	x4	x5	x6	x7	y
1	2.00	4.00	4	1.26	1	6	6	180.23
2	3.00	1.58	40	1.25	1	5	5	182.61
3	16.60	23.78	40	1.00	1	13	13	164.38
4	7.00	2.37	168	1.00	1	7	8	284.55
5	5.30	1.67	42.5	7.79	3	25	25	199.92
6	16.50	8.25	168	1.12	2	19	19	267.38
7	25.89	3.00	40	0	3	36	36	999.09
8	44.42	159.75	168	0.60	18	48	48	1103.24
9	39.63	50.86	40	27.37	10	77	77	944.21
10	31.92	40.08	168	5.52	6	47	47	931.84
11	97.33	255.08	168	19.00	6	165	130	2268.06
12	56.63	373.42	168	6.03	4	36	37	1489.50
13	96.67	206.67	168	17.86	14	120	120	1891.70
14	54.58	207.08	168	7.77	6	66	66	1387.82
15	113.88	981.00	168	24.48	6	166	179	3559.92
16	149.58	233.83	168	31.07	14	185	202	3115.29
17	134.32	145.82	168	25.99	12	192	192	2227.76
18	188.74	937.00	168	45.44	26	237	237	4804.24
19	110.24	410.00	168	20.05	12	115	115	2628.32
20	96.83	677.33	168	20.31	10	302	210	1880.84
21	102.33	288.83	168	21.01	14	131	131	3036.63
22	274.92	695.25	168	46.63	58	363	363	5539.98
23	811.08	714.33	168	22.76	17	242	242	3534.49
24	384.50	1473.66	168	7.36	24	540	453	8266.77
25	95.00	368.00	168	30.26	9	292	196	1845.89

Data Source: Procedure and Analyses for Staffing Standards: Data Regression Analysis handbook (San Diego, California: Navy Manpower and Material Analysis Center, 1979).

The SAS program for the simultaneous outlier detection is provided as follows. In this example we choose $m = n = 25$ since data set is not too large and we can test all observations simultaneously in the data set.

```
proc reg data=manpow outest=out1 tableout;
     model y=x1 x2 x3 x4 x5 x6 x7;
     output out=out2
     p=yhat r=e h=h RSTUDENT=t ;
run;

data out2; set out2;
cutoff=-quantile('T', 0.025/50, 25-8-1);
if abs(t)> cutoff then outlier="Yes";
else outlier="No";
run;
```

The output with information on multiple mean shift outliers is presented in Table 3.15. Note that in this example we tested all data points ($n = 25$) simultaneously. The cutoff for identifying multiple mean shift outlier is $\alpha/2n = 0.025/25$ quantile from the t distribution with degrees of freedom $n - 1 -$ number of parameters $= 25 - 1 - 8 = 16$. In the output, "No" indicates that the corresponding observation is not a mean shift outlier and "Yes" means a mean shift outlier.

In Table 3.15, we detect outlier using all data points as a whole. This approach is based on rather conservative Bonferroni inequality, i.e., set the critical value to be $t_{\alpha/2n,n-p-1}$, where n is the total number of observations to be tested and p is the total number of parameters in the regression model. We use this approach in situation where individual outlier detection and residual plot do not provide us enough information on model fitting. Detection of outlier as a whole may tell us that even individually there is no evidence to identify an outlier, but as compare to other residuals in the overall data set, one residual may be more extreme than other. The idea behind this approach is that when we fit data to a model we would expect the model can provide satisfactory fitted values for all data points as a whole.

In this example we set $\alpha = 0.05$, $n = 25$, and $p = 8$. The cutoff for the test statistic is

$$-t_{\alpha/2n,n-p-1} = -t_{0.05/50,16} = 3.686155.$$

Table 3.15 Simultaneous Outlier Detection

Obs	y_i	$\hat{y}_i$	e_i	h_{ii}	t_i	Outlier
1	180.23	209.98	-29.755	0.25729	-0.07360	No
2	182.61	213.80	-31.186	0.16088	-0.07257	No
3	164.38	360.49	-196.106	0.16141	-0.45944	No
4	284.55	360.11	-75.556	0.16311	-0.17621	No
5	199.92	380.70	-180.783	0.14748	-0.41961	No
6	267.38	510.37	-242.993	0.15890	-0.57043	No
7	999.09	685.17	313.923	0.18288	0.75320	No
8	1103.24	1279.30	-176.059	0.35909	-0.47199	No
9	944.21	815.47	128.744	0.28081	0.32464	No
10	931.84	891.85	39.994	0.12954	0.09139	No
11	2268.06	1632.14	635.923	0.12414	1.55370	No
12	1489.50	1305.18	184.323	0.20241	0.44258	No
13	1891.70	1973.42	-81.716	0.08020	-0.18179	No
14	1387.82	1397.79	-9.966	0.09691	-0.02235	No
15	3559.92	4225.13	-665.211	0.55760	-2.51918	No
16	3115.29	3134.90	-19.605	0.40235	-0.05406	No
17	2227.76	2698.74	-470.978	0.36824	-1.33105	No
18	4804.24	4385.78	418.462	0.44649	1.25660	No
19	2628.32	2190.33	437.994	0.08681	1.00741	No
20	1880.84	2750.91	-870.070	0.36629	-2.86571	No
21	3036.63	2210.13	826.496	0.07039	2.05385	No
22	5539.98	5863.87	-323.894	0.78537	-1.60568	No
23	3534.49	3694.77	-160.276	0.98846	-5.24234	Yes
24	8266.77	7853.50	413.265	0.87618	3.20934	No
25	1845.89	1710.86	135.029	0.54674	0.42994	No

We then compare the absolute value of each t_i with this cutoff to determine whether the corresponding observation is a possible outlier. Assuming that the regression model is correctly specified, the comparison between this cutoff and each observation in the data set alerts that the 23th observation might be an outlier.

The following example demonstrates the detection of multiple mean shift outliers. The magnitude of mean shift at different data point may be different. The technique for multiple outlier detection is to create variables which take value 1 at these suspicious data point and 0 elsewhere. We need to create as many such columns as the number of suspicious outliers if we believe there are different mean shifts at those data points. This way, we can take care of different magnitudes of mean shift for all possible outliers. If we think some outliers are of the same mean shift then for these outlier we should create a dummy variable that takes value 1 at these outliers and

0 elsewhere.

In the following example, the data points with a larger value of variable x_2 are suspicious and we would like to consider multiple data points 15, 18, 22, 23, and 24 as possible multiple outliers. We first create 5 dummy variables D_1, D_2, D_3, D_4, and D_5 that takes value 1 at observations 15, 18, 22, 23, and 24, and value 0 for all other observations in the data set. We then include these dummy variables in the regression model. The SAS program for detecting the mean shift outliers is provided below.

```
data manpow; set manpow;
if _n_=15 then D15=1; else D15=0;
if _n_=18 then D18=1; else D18=0;
if _n_=22 then D22=1; else D22=0;
if _n_=23 then D23=1; else D23=0;
if _n_=24 then D24=1; else D24=0;
run;

Proc reg data=manpow;
     model y=x1 x2 x3 x4 x5 x6 x7 D15 D18 D22 D23 D24;
run;
```

The output is presented in Table 3.16. The identified outlier is the 23th observation since the corresponding P-value is $0.0160 < 0.05$. Note that this time we identified the same outlier via multiple outlier detection approach.

Table 3.16 Detection of Multiple Mean Shift Outliers

Variable	df	b_i	std	t_i	$P > \|t\|$
Intercept	1	142.1511	176.0598	0.81	0.4351
x1	1	23.7437	8.8284	2.69	0.0197
x2	1	0.8531	0.8608	0.99	0.3412
x3	1	-0.2071	1.7234	-0.12	0.9063
x4	1	9.5700	16.0368	0.60	0.5618
x5	1	12.7627	21.7424	0.59	0.5681
x6	1	-0.2106	6.9024	-0.03	0.9762
x7	1	-6.0764	12.5728	-0.48	0.6376
Data15	1	723.5779	914.0654	0.79	0.4440
Data18	1	139.5209	611.3666	0.23	0.8233
Data22	1	-592.3845	900.1980	-0.66	0.5229
Data23*	1	-15354	5477.3308	-2.80	0.0160*
Data24	1	262.4439	1386.053	0.19	0.8530

3.21 Check for Normality of the Error Term in Multiple Regression

We now discuss how to check normality assumption on error term of a multiple regression model. It is known that the sum of squared residuals, divided by $n-p$, is a good estimate of the error variance, where n is the total number of observations and p is the number of parameters in the regression model, The residual vector in a multiple linear regression is given by

$$\boldsymbol{e} = (I - H)y = (I - H)(\boldsymbol{X\beta} + \boldsymbol{\varepsilon}) = (I - H)\boldsymbol{\varepsilon},$$

where H is the HAT matrix for this regression model. Each component $e_i = \varepsilon_i - \sum_{j=1}^{n} h_{ij}\varepsilon_j$. Therefore, the normality of residual is not simply the normality of the error term in the multiple regression model. Note that

$$\text{Cov}(\boldsymbol{e}) = (I - H)\sigma^2(I - H)^{'} = (I - H)\sigma^2.$$

Hence we can write $\text{Var}(e_i) = (1 - h_{ii})\sigma^2$. If sample size is much larger than the number of the model parameters, i.e., $n >> p$, or sample size n is large enough, h_{ii} will be small as compared to 1, then $\text{Var}(e_i) \approx \sigma^2$. Thus, a residual in multiple regression model behaves like error if sample size is large. However, it is not true for small sample size. We point out that it is unreliable to check normality assumption using the residuals from a multiple regression model when sample size is small.

3.22 Example

In this section we provide some illustrative examples of multiple regression using SAS. The following SAS program is for calculating confidence intervals on regression mean and regression prediction. The Pine Tree data set used in this example is presented in Table 3.17.

The corresponding estimated of regression parameters for the model including all independent variables x_1, x_2, x_3 and the model including x_3, x_2 are presented in Tables 3.18 and 3.19. The confidence intervals on regression mean and regression prediction for the model including all variables and the model including x_1, x_2 are presented in Tables 3.20 and 3.21.

Table 3.17 Stand Characteristics of Pine Tree Data

Age	HD	N	MDBH
19	51.5	500	7.0
14	41.3	900	5.0
11	36.7	650	6.2
13	32.2	480	5.2
13	39.0	520	6.2
12	29.8	610	5.2
18	51.2	700	6.2
14	46.8	760	6.4
20	61.8	930	6.4
17	55.8	690	6.4
13	37.3	800	5.4
21	54.2	650	6.4
11	32.5	530	5.4
19	56.3	680	6.7
17	52.8	620	6.7
15	47.0	900	5.9
16	53.0	620	6.9
16	50.3	730	6.9
14	50.5	680	6.9
22	57.7	480	7.9

Data Source: Harold E, et al. "Yield of Old-field Loblolly Pine Plantations", Division of Forestry and Wildlife Resources Pub. FWS-3-72, Virginia Polytechnic Institute and State University, Blacksburg, Virginia, 1972.

```
data pinetree; set pinetree;
x1= HD;
x2=age*N;
x3=HD/N;
run;

proc reg data=pinetree outest=out tableout;
model MDBH=x1 x2 x3/all;
run;

*Calculation of partial sum;
proc reg data=pinetree;
model MDBH=x1 x2 x3;
run;
```

Table 3.18 Parameter Estimates and Confidence Intervals Using x_1, x_2 and x_3

MODEL	TYPE	DEPVAR	RMSE	Intercept	x1	x2	x3
MODEL1	PARMS	MDBH	0.29359	3.23573	0.09741	-0.00017	3.4668
MODEL1	STDERR	MDBH	0.29359	0.34666	0.02540	0.00006	8.3738
MODEL1	T	MDBH	0.29359	9.33413	3.83521	-2.79003	0.4140
MODEL1	PVALUE	MDBH	0.29359	0.00000	0.00146	0.01311	0.6844
MODEL1	L95B	MDBH	0.29359	2.50085	0.04356	-0.00030	-14.2848
MODEL1	U95B	MDBH	0.29359	3.97061	0.15125	-0.00004	21.2185

Table 3.19 Parameter Estimates and Confidence Intervals after Deleting x_3

MODEL	TYPE	DEPVAR	RMSE	Intercept	x_1	x_2
MODEL1	PARMS	MDBH	0.28635	3.26051	0.1069	-0.00019
MODEL1	STDERR	MDBH	0.28635	0.33302	0.0106	0.00003
MODEL1	T	MDBH	0.28635	9.79063	10.1069	-5.82758
MODEL1	PVALUE	MDBH	0.28635	0.00000	0.0000	0.00002
MODEL1	L95B	MDBH	0.28635	2.55789	0.0846	-0.00026
MODEL1	U95B	MDBH	0.28635	3.96313	0.1292	-0.00012

```
proc reg data=pinetree outest=out tableout;
model MDBH=x1 x2;
run;

*Calculate fitted values and residuals;
proc reg data=pinetree;
model MDBH=x1 x2 x3;
output out=out  p=yhat  r=yresid  student=sresid
       LCLM=L_mean  UCLM=U_mean
       LCL=L_pred   UCL=U_pred;
run;

*Calculate fitted values and residuals after deleting X3;
proc reg data=pinetree;
model MDBH=x1 x2;
output out=out p=yhat r=yresid  student=sresid
       LCLM=L_mean  UCLM=U_mean
       LCL=L_pred   UCL=U_pred;
run;
```

If collinearity exists the regression analysis become unreliable. Although

Table 3.20 Confidence Intervals on Regression Mean and Prediction Without Deletion

Obs	MDBH	yhat	*Lmean*	*Umean*	*Lpred*	*Upred*	yresid	sresid
1	7.0	7.00509	6.69973	7.31046	6.31183	7.69835	-0.00509	-0.01991
2	5.0	5.29011	4.99935	5.58088	4.60316	5.97707	-0.29011	-1.11762
3	6.2	5.79896	5.53676	6.06115	5.12360	6.47431	0.40104	1.50618
4	5.2	5.55111	5.23783	5.86439	4.85433	6.24789	-0.35111	-1.38404
5	6.2	6.15311	5.92449	6.38173	5.49007	6.81615	0.04689	0.17171
6	5.2	5.07177	4.76288	5.38066	4.37695	5.76659	0.12823	0.50309
7	6.2	6.34891	6.17875	6.51908	5.70369	6.99414	-0.14891	-0.52731
8	6.4	6.21119	5.98863	6.43376	5.55021	6.87217	0.18881	0.68863

Table 3.21 Confidence Intervals on Regression Mean and Prediction After Deleting x_3

Obs	MDBH	yhat	*Lmean*	*Umean*	*Lpred*	*Upred*	yresid	sresid
1	7.0	6.96391	6.74953	7.17829	6.32286	7.60495	0.03609	0.13482
2	5.0	5.28516	5.00400	5.56632	4.61880	5.95151	-0.28516	-1.12512
3	6.2	5.82751	5.61623	6.03878	5.18749	6.46752	0.37249	1.38853
4	5.2	5.51907	5.26001	5.77813	4.86173	6.17641	-0.31907	-1.23345
5	6.2	6.14741	5.92731	6.36751	5.50443	6.79039	0.05259	0.19721
6	5.2	5.05755	4.76617	5.34892	4.38681	5.72828	0.14245	0.56791
7	6.2	6.34360	6.18055	6.50665	5.71785	6.96935	-0.14360	-0.52082
8	6.4	6.24510	6.10990	6.38030	5.62602	6.86418	0.15490	0.55504

we can identify highly dependent regressors and include one of them in the regression model to eliminate collinearity. In many applications, often it is rather difficulty to determine variable deletion. A simple way to combat collinearity is to fit the regression model using centralized data. The following example illustrates how to perform regression analysis on the centralized data using SAS. The regression model for centralized data is given by

$$y_i = \beta_0 + \beta_1(x_{1i} - \bar{x}_1) + \beta_2(x_{2i} - \bar{x}_2) + \varepsilon_i.$$

We then create the centralize variables, $(x_{1i} - \bar{x}_1)$ and $(x_{2i} - \bar{x}_2)$, before performing the regression analysis. The following SAS code is for regression analysis using centralized data. The regression parameter estimators using the centralized data are presented in Table 3.22.

```
data example;
input yield temp time @@;
datalines;
77  180  1 79  160  2 82  165  1 83  165  2
```

```
85  170  1 88  170  2 90  175  1 93  175  2;
run;
*Centralize data;
proc means data=example noprint;
var temp time;
output out=aa mean=meantemp meantime;
run;

data aa; set aa;
call symput('mtemp', meantemp);
call symput('mtime', meantime);
run;

*Created centralized data ctime and ctemp;
data example; set example;
ctemp=temp-&mtemp;
ctime=time-&mtime;
run;

proc reg data=example outest=out1 tableout;
    model yield=ctemp ctime/noprint;
run;
```

Table 3.22 Regression Model for Centralized Data

Obs	MODEL	TYPE	DEPVAR	RMSE	Intercept	ctemp	ctime
1	MODEL1	PARMS	yield	5.75369	84.6250	0.37000	4.1000
2	MODEL1	STDERR	yield	5.75369	2.0342	0.36390	4.4568
3	MODEL1	T	yield	5.75369	41.6003	1.01678	0.9199
4	MODEL1	PVALUE	yield	5.75369	0.0000	0.35591	0.3998
5	MODEL1	L95B	yield	5.75369	79.3958	-0.56542	-7.3566
6	MODEL1	U95B	yield	5.75369	89.8542	1.30542	15.5566

The final multiple regression model is

$$\text{yield} = 84.625 + 0.37(\text{temperature} - 170) + 4.10(\text{time} - 1.5)$$

The test for linear hypothesis is useful in many applications. For linear regression models, SAS procedures GLM and MIXED are often used. The following SAS program uses the procedure GLM for testing linear hypothesis. Note that SAS procedure REG can also be used for testing linear hypothesis. The variable GROUP in the following example is a class variable. The results of the linear hypothesis tests are presented in Tables 3.23 and 3.24.

```
data example; input group weight HDL;
datalines;
 1 163.5 75.0
 ...
 1 144.0 63.5
 2 141.0 49.5
 ...
 2 216.5 74.0
 3 136.5 54.5
 ...
 3 139.0 68.0
 ;
 run;

*Regression analysis by group;
proc sort data=example;
by group;
run;

proc reg data=example outest=out1 tableout;
     model HDL=weight/noprint;
     by group;
run;

*Test for linear hypothesis of equal slopes;
proc glm data=example outstat=out1;
     class group;
     model HDL=group weight group*weight/ss3;
run;

proc print data=out1;
var _SOURCE_  _TYPE_  DF  SS  F PROB;
run;
```

We use the Pine Trees Data in Table 3.17 to illustrate how to test for

Table 3.23 Test for Equal Slope Among 3 Groups

SOURCE	TYPE	DF	SS	F	PROB
error	error	20	1712.36		
group	SS3	2	697.20	4.07157	0.03285
weight	SS3	1	244.12	2.85124	0.10684
weight*group	SS3	2	505.05	2.94946	0.07542

Table 3.24 Regression by Group

Group	Model	Type	Depvar	Rmse	Intercept	Weight
1	MODEL1	PARMS	HDL	7.2570	23.054	0.24956
1	MODEL1	STDERR	HDL	7.2570	25.312	0.15733
1	MODEL1	T	HDL	7.2570	0.911	1.58629
1	MODEL1	PVALUE	HDL	7.2570	0.398	0.16377
1	MODEL1	L95B	HDL	7.2570	-38.883	-0.13540
1	MODEL1	U95B	HDL	7.2570	84.991	0.63452
2	MODEL1	PARMS	HDL	10.3881	14.255	0.25094
2	MODEL1	STDERR	HDL	10.3881	17.486	0.11795
2	MODEL1	T	HDL	10.3881	0.815	2.12741
2	MODEL1	PVALUE	HDL	10.3881	0.446	0.07749
2	MODEL1	L95B	HDL	10.3881	-28.532	-0.03769
2	MODEL1	U95B	HDL	10.3881	57.042	0.53956
3	MODEL1	PARMS	HDL	9.6754	76.880	-0.08213
3	MODEL1	STDERR	HDL	9.6754	16.959	0.10514
3	MODEL1	T	HDL	9.6754	4.533	-0.78116
3	MODEL1	PVALUE	HDL	9.6754	0.002	0.45720
3	MODEL1	L95B	HDL	9.6754	37.773	-0.32458
3	MODEL1	U95B	HDL	9.6754	115.987	0.16032

linear hypothesis. The following SAS code test the linear hypothesis (a) H_0: $\beta_0 = \beta_1 = \beta_2 = \beta_3 = 0$ versus H_1: at least one $\beta_i \neq \beta_j$. (b) H_0: $\beta_1 = \beta_2$ versus H_1: $\beta_1 \neq \beta_2$, both at a level 0.05 (default).

```
proc reg data=pinetree alpha=0.05;
model MDBH=x1 x2 x3;
test  intercept=0,
      x1=0,
      x2=0,
      x3=0;
test  x1=x2;
run;
```

The first hypothesis test (a) is the multiple test for checking if all parameters are zero and the observed value of the corresponding F test statistic is 2302.95 with the p-value $< .0001$. Thus, we cannot confirm H_0. For the second hypothesis test (b) the observed value of the corresponding F test statistic is 14.69 with the p-value 0.0015. Since the p-value is less than the significance level we cannot confirm H_0 either.
Again, we use the Pine Trees Data to illustrate how to find the least squares estimation under linear restrictions. The following SAS code compute the least squares estimates under the linear restrictions $\beta_0 = 3.23$ and $\beta_1 = \beta_2$.

```
proc reg data=pinetree;
model MDBH=x1 x2 x3;
restrict intercept=3.23, x1=x2/print;
run;
```

It is noted that the least squares estimates from the regression model without any linear restrictions are $b_0 = 3.2357$, $b_1 = 0.09741$, $b_2 = -0.000169$ and $b_3 = 3.4668$. The least squares estimates with the linear restrictions $\beta_0 = 3.23$ and $\beta_1 = \beta_2$ are $b_0 = 3.23$, $b_1 = b_2 = 0.00005527$ and $b_3 = 33.92506$.

Problems

1. Using the matrix form of the simple linear regression to show the unbiasness of the $\boldsymbol{b}$. Also, calculate the covariance of $\boldsymbol{b}$ using the matrix format of the simple linear regression.

2. Let $\boldsymbol{X}$ be a matrix of $n \times m$ and $\boldsymbol{X} = (\boldsymbol{X}_1, \boldsymbol{X}_2)$, where $\boldsymbol{X}_1$ is $n \times k$ matrix and $\boldsymbol{X}_2$ is $n \times (m-k)$ matrix. Show that
   (a). The matrices $\boldsymbol{X}(\boldsymbol{X}'\boldsymbol{X})^{-1}\boldsymbol{X}'$ and $\boldsymbol{X}_1(\boldsymbol{X}_1'\boldsymbol{X}_1)^{-1}\boldsymbol{X}_1'$ are idempotent.
   (b). The matrix $\boldsymbol{X}(\boldsymbol{X}'\boldsymbol{X})^{-1}\boldsymbol{X}' - \boldsymbol{X}_2(\boldsymbol{X}_2'\boldsymbol{X}_2)^{-1}\boldsymbol{X}_2'$ is idempotent.
   (c). Find the rank of the matrix $\boldsymbol{X}(\boldsymbol{X}'\boldsymbol{X})^{-1}\boldsymbol{X}' - \boldsymbol{X}_2(\boldsymbol{X}_2'\boldsymbol{X}_2)^{-1}\boldsymbol{X}_2'$.

3. The least squares estimators of the regression model $Y = \boldsymbol{X\beta} + \varepsilon$ are linear function of the y-observations. When $(\boldsymbol{X}'\boldsymbol{X})^{-1}$ exists the least squares estimators of $\boldsymbol{\beta}$ is $\boldsymbol{b} = (\boldsymbol{X}'\boldsymbol{X})^{-1}\boldsymbol{X}\boldsymbol{y}$. Let A be a constant matrix. Using $\text{Var}(Ay) = A\text{Var}(\boldsymbol{y})A'$ and $\text{Var}(\boldsymbol{y}) = \sigma^2 I$ to show that $\text{Var}(\boldsymbol{b}) = \sigma^2(\boldsymbol{X}'\boldsymbol{X})^{-1}$.

4. Show that the HAT matrix in linear regression model has the property $tr(H) = p$ where p is the total numbers of the model parameters.
5. Let h_{ii} be the ith diagonal elements of the HAT matrix. Prove that

   (a). For a multiple regression model with a constant term $h_{ii} \geq 1/n$.

   (b). Show that $h_{ii} \leq 1$. (Hint: Use the fact that the HAT matrix is idempotent.)

6. Assume that the data given in Table 3.25 satisfy the model

$$y_i = \beta_0 + \beta_1 x_{1i} + \beta_2 x_{2i} + \varepsilon_i,$$

where ε_i's are iid $N(0, \sigma^2)$.

Table 3.25 Data Set for Calculation of Confidence Interval on Regression Prediction

y	12.0	11.7	9.3	11.9	11.8	9.5	9.3	7.2	8.1	8.3	7.0	6.5	5.9
x_1	3	4	5	6	7	8	9	10	11	12	13	14	15
x_2	6	4	2	1	0	1	2	1	-1	0	-2	-1	-3

Data Source: Franklin A. Grabill, (1976), Theory and Application of the linear model. p. 326.

   (a). Find 80 percent, 90 percent, 95 percent, and 99 percent confidence interval for y_0, the mean of one future observation at $x_1 = 9.5$ and $x_2 = 2.5$.

   (b). Find a 90 percent confidence interval for $\bar{y}_0$, the mean of six observations at $x_1 = 9.5$ and $x_2 = 2.5$.

7. Consider the general linear regression model $\boldsymbol{y} = \boldsymbol{X\beta} + \varepsilon$ and the least squares estimate $\boldsymbol{b} = (\boldsymbol{X}'\boldsymbol{X})^{-1}\boldsymbol{X}'\boldsymbol{y}$. Show that

$$\boldsymbol{b} = \boldsymbol{\beta} + R\varepsilon,$$

where $R = (\boldsymbol{X}'\boldsymbol{X})^{-1}\boldsymbol{X}'$.

8. A scientist collects experimental data on the radius of a propellant grain (y) as a function of powder temperature, x_1, extrusion rate, x_2, and die temperature, x_3. The data is presented in Table 3.26.

   (a). Consider the linear regression model

$$y_i = \beta_0^\star + \beta_1(x_{1i} - \bar{x}_1) + \beta_2(x_{2i} - \bar{x}_2) + \beta_3(x_{3i} - \bar{x}_3) + \epsilon_i.$$

   Write the vector $\boldsymbol{y}$, the matrix $\boldsymbol{X}$, and vector $\boldsymbol{\beta}$ in the model $\boldsymbol{y} = \boldsymbol{X\beta} + \varepsilon$.

Table 3.26 Propellant Grain Data

Grain Radius	Powder Temp (x_1)	Extrusion Rate (x_2)	Die Temp (x_3)
82	150	12	220
92	190	12	220
114	150	24	220
124	150	12	250
111	190	24	220
129	190	12	250
157	150	24	250
164	190	24	250

(b). Write out the normal equation $(\boldsymbol{X}'\boldsymbol{X})\boldsymbol{b} = \boldsymbol{X}'\boldsymbol{y}$. Comment on what is special about the $\boldsymbol{X}'\boldsymbol{X}$ matrix. What characteristic in this experiment do you suppose to produce this special form of $\boldsymbol{X}'\boldsymbol{X}$.

(c). Estimate the coefficients in the multiple linear regression model.

(d). Test the hypothesis $H_0 : L\beta_1 = 0$, $H_0 : \beta_2 = 0$ and make conclusion.

(e). Compute $100(1-\alpha)\%$ confidence interval on $E(y|x)$ at each of the locations of x_1, x_2, and x_3 described by the data points.

(f). Compute the HAT diagonals at eight data points and comment.

(g). Compute the variance inflation factors of the coefficients b_1, b_2, and b_3. Do you have any explanations as to why these measures of damage due to collinearity give the results that they do?

9. For the data set given in Table 3.27

Table 3.27 Data Set for Testing Linear Hypothesis

y	x_1	x_2
3.9	1.5	2.2
7.5	2.7	4.5
4.4	1.8	2.8
8.7	3.9	4.4
9.6	5.5	4.3
19.5	10.7	8.4
29.3	14.6	14.6
12.2	4.9	8.5

(a). Find the linear regression model.

(b). Use the general linear hypothesis test to test

$$H_0 : \beta_1 = \beta_2 = 0$$

and make your conclusion. Use full and restricted model residual sums of squares.

10. Consider the general linear regression model $\boldsymbol{y} = \boldsymbol{X}\boldsymbol{\beta} + \varepsilon$ and the least squares estimate $\boldsymbol{b} = (\boldsymbol{X}^{'}\boldsymbol{X})^{-1}\boldsymbol{X}^{'}\boldsymbol{y}$. Show that

$$\boldsymbol{b} = \boldsymbol{\beta} + R\varepsilon,$$

where $R = (\boldsymbol{X}^{'}\boldsymbol{X})^{-1}\boldsymbol{X}^{'}$.

11. In an experiment in the civil engineering department of Virginia Polytechnic Institute and State University in 1988, a growth of certain type of algae in water was observed as a function of time and dosage of copper added into the water. The collected data are shown in Table 3.28.

(a). Consider the following regression model

$$y_i = \beta_0 + \beta_1 x_{1i} + \beta_2 x_{2i} + \beta_{12} x_{1i} x_{2i} + \varepsilon_i$$

Estimate the coefficients of the model, using multiple linear regression.

(b). Test $H_0 : \beta_{12} = 0$ versus $H_1 : \beta_{12} \neq 0$. Do you have any reason to change the model given in part (a).

(c). Show a partitioning of total degrees of freedom into those attributed to regression, pure error, and lack of fit.

(d). Using the model you adopted in part (b), make a test for lack of fit and draw conclusion.

(e). Plot residuals of your fitted model against x_1 and x_2 separately, and comment.

Table 3.28 Algae Data

y(unit of algae)	x_1(copper, mg)	x_2(days)
.3	1	5
.34	1	5
.2	2	5
.24	2	5
.24	2	5
.28	3	5
.2	3	5
.24	3	5
.02	4	5
.02	4	5
.06	4	5
0	5	5
0	5	5
0	5	5
.37	1	12
.36	1	12
.30	2	12
.31	2	12
.30	2	12
.30	3	12
.30	3	12
.30	3	12
.14	4	12
.14	4	12
.14	4	12
.14	5	12
.15	5	12
.15	5	12
.23	1	18
.23	1	18
.28	2	18
.27	2	18
.25	2	18

Table 3.28 Cont'd

y(unit of algae)	x_1(mg copper)	x_2(days)
.27	3	18
.25	3	18
.25	3	18
.06	4	18
.10	4	18
.10	4	18
.02	5	18
.02	5	18
.02	5	18
.36	1	25
.36	1	25
.24	2	25
.27	2	25
.31	2	25
.26	3	25
.26	3	25
.28	3	25
.14	4	25
.11	4	25
.11	4	25
.04	5	25
.07	5	25
.05	5	25

Chapter 4

Detection of Outliers and Influential Observations in Multiple Linear Regression

After data is collected, and cleaned the next step is to fit the data to a selected statistical model so that the relationship between response variable and independent variables may be established. Choosing a statistical model largely depends on the nature of data from the experiment and the scientific questions that need to be answered. If linear regression model is selected, determination of the model is to solve for the regression parameters. After that it is needed to perform hypothesis tests for those parameters and it is necessary to perform model diagnosis.

We have already stated that for classical regression model the method for solving regression parameters is the least squares method. If the error term in the regression model is normally distributed the least squares estimates of the regression parameters are the same as the maximum likelihood estimates. After the estimates of the linear regression model are obtained, the next question is to answer whether or not this linear regression model is reasonably and adequately reflect the true relationship between response variable and independent variables. This falls into the area of regression diagnosis. There are two aspects of the regression diagnosis. One is to check if a chosen model is reasonable enough to reflect the true relationship between response variable and independent variables. Another is to check if there are any data points that deviate significantly from the assumed model. The first question belongs to model diagnosis and the second question is to check outliers and influential observations. We focus on the detection of outliers and influential observations in this section.

Identifying outliers and influential observations for a regression model is based on the assumption that the regression model is correctly specified. That is, the selected regression model adequately specifies the relationship between response variable and independent variables. Any data points that fit well the assumed model are the right data points for the assumed model. Sometimes, however, not all data points fit the model equally well. There may be some data points that may deviate significantly from the assumed model. These data points may be considered as outliers if we believe that the selected model is correct. Geometrically, linear regression is a line (for simple linear regression) or a hyperplane (for multiple regression). If a regression model is appropriately selected, most data points should be fairly close to regression line or hyperplane. The data points which are far away from regression line or hyperplane may not be "ideal" data points for the selected model and could potentially be identified as the outliers for the model. An outlier is the data point that is statistically far away from the chosen model if we believe that the selected regression model is correct. An influential observation is one that has a relatively larger impact on the estimates of one or more regression parameters. i.e., including or excluding it in the regression model fitting will result in unusual change in one or more estimates of regression parameters. We will discuss the statistical procedures for identifying outliers and influential observations.

4.1 Model Diagnosis for Multiple Linear Regression

4.1.1 *Simple Criteria for Model Comparison*

Before checking a regression model, one has to understand the following:

- Scientific questions to be answered
- Experiment from which the data is collected
- Statistical model and model assumptions

It should be understood that the model selection may be rather complicated and sometimes it is difficult to find a "perfect model" for your data. David R. Cox made famous comment on model selection, "all models are wrong, but some are useful". This implies that no "optimal scheme" is ever in practice optimal. The goal of statistical modeling is to find a reasonable and useful model upon which we can answer desired scientific questions based on data obtained from well-designed scientific experiment. This can be done through model comparison. The following are some basic criteria that are commonly used for regression model diagnosis:

1. Coefficients of determination $R^2 = 1 - \frac{SS_{Res}}{SS_{Total}}$. The preferred model

would be the model with R^2 value close to 1. If the data fit well the regression model then it should be expected that y_i is close enough to $\hat{y}_i$. Hence, SS_{Res} should be fairly close to zero. Therefore, R^2 should be close to 1.

2. Estimate of error variance s^2. Among a set of possible regression models a preferred regression model should be one that has a smaller value of s^2 since this corresponds to the situation where the fitted values are closer to the response observations as a whole.
3. Adjusted $\bar{R}^2$. Replace SS_{Res} and SS_{Total} by their means:

$$\bar{R}^2 = 1 - \frac{SS_{Res}/n-p}{SS_{Total}/n-1} = 1 - \frac{s^2(n-1)}{SS_{Total}}$$

Note that SS_{Total} is the same for all models and the ranks of R^2 for all models are the same as the ranks of s^2 for all models. We would like to choose a regression model with adjusted $\bar{R}^2$ close to 1.

4.1.2 *Bias in Error Estimate from Under-specified Model*

We now discuss the situation where the selected model is under-specified. i.e.,

$$\boldsymbol{y} = \boldsymbol{X}_1\boldsymbol{\beta}_1 + \boldsymbol{\varepsilon}. \tag{4.1}$$

Here, the under-specified model is a model with inadequate regressors. In other words, if more regressors are added into the model, the linear combination of the regressors can better predict the response variable. Sometimes, the under-specified model is also called as reduced model. Assuming that the full model is

$$\boldsymbol{y} = \boldsymbol{X}_1\boldsymbol{\beta}_1 + \boldsymbol{X}_2\boldsymbol{\beta}_2 + \boldsymbol{\varepsilon}, \tag{4.2}$$

and the number of parameters in the reduced model is p, let s_p^2 be the error estimate based on the under-specified regression model. The fitted value from the under-specified regression model is

$$\hat{\boldsymbol{y}} = \boldsymbol{X}_1(\boldsymbol{X}_1^{'}\boldsymbol{X}_1)^{-1}\boldsymbol{X}_1^{'}\boldsymbol{y}$$

and error estimate from the under-specified regression model is

$$s_p^2 = \boldsymbol{y}^{'}(I - \boldsymbol{X}_1(\boldsymbol{X}_1^{'}\boldsymbol{X}_1)^{-1}\boldsymbol{X}_1^{'})\boldsymbol{y}.$$

We then compute the expectation of the error estimate s_p^2.

$$E(s_p^2) = \sigma^2 + \frac{\boldsymbol{\beta}_2^{'}[\boldsymbol{X}_2^{'}\boldsymbol{X}_2 - \boldsymbol{X}_2^{'}\boldsymbol{X}_1(\boldsymbol{X}_1^{'}\boldsymbol{X}_1)^{-1}\boldsymbol{X}_2^{'}\boldsymbol{X}_1]\boldsymbol{\beta}_2}{n-p}. \tag{4.3}$$

The above derivations need to use the formula of inverse matrix of partitioned matrix and the fact that $E(\boldsymbol{y}) = \boldsymbol{X}_1\boldsymbol{\beta}_1 + \boldsymbol{X}_2\boldsymbol{\beta}_2$. Note that s_p^2 based on the under-specified regression model is a biased estimate of the error variance. The bias due to under-specification of a regression model is a function of deleted parameters $\boldsymbol{\beta}_2$ and its covariance matrix.

4.1.3 *Cross Validation*

Suppose that y_i is the ith response and $(x_{i1}, x_{i2}, \cdots, x_{ip})$ is the ith regressor observation vector. The data can be written in the following matrix format:

$$(\boldsymbol{y}, \boldsymbol{X}) = \begin{pmatrix} y_1 & x_{1,1} & \cdots & x_{1,p} \\ y_2 & x_{2,1} & \cdots & x_{2,p} \\ & \vdots & & \\ y_{n_1} & x_{n_11} & \cdots & x_{n_1,p} \\ y_{n_1+1} & x_{n_1+1,1} & \cdots & x_{n_1+1,p} \\ & \vdots & & \\ y_n & x_{n,1} & \cdots & x_{n,p} \end{pmatrix} = \begin{pmatrix} \boldsymbol{y}_1 & \boldsymbol{X}_1 \\ \boldsymbol{y}_2 & \boldsymbol{X}_2 \end{pmatrix}, \tag{4.4}$$

where $1 < n_1 < n$. The cross validation approach is to select n_1 data points out of n data points to fit regression model and use the rest $n - n_1$ data points to assess the model. This way, the first n_1 data points serve as the learning sample and the remaining $n - n_1$ data points are the testing sample. To assess the model using the rest $n - n_1$ data points we use sum of prediction errors $\sum_{i=n_1+1}^{n}(y_i - \hat{y}_i)^2$ as assessment criterion. Note that the prediction $\hat{y}_i$ is computed using the regression model from the first n_1 data points and the observations in the remaining $n - n_1$ data points. The idea behind this model checking approach is that the model parameters are obtained using the first n_1 data points. If this is an appropriate model for the whole data set then it should also yield a smaller value of the sum of the squared prediction errors from the remaining $n - n_1$ observations. Note that this criterion is harder than the criterion we have previously discussed, where the mean square error criterion is used rather than the criterion of prediction error. Furthermore, it should be noted that the prediction has a larger variance than the fitted value.

The above cross validation approach can be modified by randomly choosing a subset of size ν to fit a regression model, then the remaining $n - \nu$ data points are used to calculate the prediction error. This can be done k times, where k is a pre-defined integer. Suppose that $\boldsymbol{\beta}_1, \boldsymbol{\beta}_2, \cdots, \boldsymbol{\beta}_k$ are the least squares estimates each uses the randomly selected ν data points. We can use the average $\boldsymbol{\beta} = \frac{1}{k}\sum_{i=1}^{k}\boldsymbol{\beta}_i$ as the estimate of the regression parameters and check the regression model using the average prediction errors. This average of prediction errors can be used as the criterion of regression model diagnosis. This technique is called *ν-fold cross validation* method and is often better than the cross validation method of choosing only one subset as a learning sample.

When the cross validation approach is used, the entire data set is split into learning sample and testing sample to fit the regression model and to test the regression model. If selected regression model is appropriate through model checking using the testing samples mentioned above, the final regression model parameters should be estimated by the entire data set. Finally, studied by Snee (1977), as a rule of thumb, the total number of observations required for applying the cross validation approach is $n \geq 2p + 20$, where p is the number of parameters in the regression model.

4.2 Detection of Outliers in Multiple Linear Regression

If the difference between fitted value $\hat{y}_i$ and response y_i is large, we may suspect that the ith observation is a potential outlier. The purpose of outlier detection in regression analysis is tempted to eliminate some observations that have relatively larger residuals so that the model fitting may be improved. However, eliminating observations cannot be made solely based on statistical procedure. Determination of outlier should be made jointly by statisticians and subject scientists. In many situations, eliminating observations is never made too easy. Some observations that do not fit the model well may imply a flaw in the selected model and outlier detection might actually result in altering the model.

Regression residuals carry the most useful information for model fitting. There are two types of residuals that are commonly used in the outlier detection: the standardized residual and the studentized residual.

Definition 4.1. Let s be the mean square error of a regression model. The standardized residual is defined as

$$z_i = \frac{y_i - \hat{y}_i}{s} \tag{4.5}$$

The standardized residual is simply the normalized residual or the z-score of the residual.

Definition 4.2. Let s be the mean square error of a regression model. The studentized residual is defined as

$$z_i = \frac{y_i - \hat{y}_i}{s\sqrt{1 - h_{ii}}} \tag{4.6}$$

where h_{ii} is the ith diagonal element of the HAT matrix (also called leverage). The studentized residuals are so named because they follow approximately the t-distribution.

According to Chebyshev's inequality, for any random variable X, if $E(X)$ exists, $P(|X - E(X)| \leq k\sigma) \geq 1 - 1/k^2$. If $k = 3$, the probability of a random variable within 3 times of standard deviation from its mean is not less than $1 - 1/3^2 = 8/9 \approx 88.9\%$. One would like to define an observation as a potential outlier if it is 3 times of the standard deviation away from its mean value.

Definition 4.3. If the residual of an observation is larger than 3 times of the standard deviation (or standardized residual is larger than 3) then the observation may be considered as an *outlier*.

Note that this outlier definition considers outlier individually. If we consider multiple outliers simultaneously then the simplest way is to use t-distribution and the Bonferroni multiplicity adjustment.

4.3 Detection of Influential Observations in Multiple Linear Regression

4.3.1 *Influential Observation*

It is known that data points may have different influences on the regression model. Different influences mean that some data points may have larger impact on the estimates of regression coefficients than that of other data points.

An influential observation is the data point that causes a significant change in the regression parameter estimates if it is deleted from the whole data set. Based on this idea we remove one observation at a time to fit the same regression model then calculate the fitted value $\hat{y}_{i,-i}$ from the regression model without using the ith data point. The fitted value of y without using the ith data point is defined as

$$\hat{y}_{i,-i} = \boldsymbol{x}_i \boldsymbol{b}_{-i},$$

where $\boldsymbol{x}_i$ is the ith regressor vector, and b_{-i} is the least squares estimate based on the same regression model using all data points except for the response y_i and the regressor vector $\boldsymbol{x}_i$. To measure the difference between response y_i and $\hat{y}_{i,-i}$ we introduce the following PRESS residual.

Definition 4.4. The PRESS residual is defined as

$$e_{i,-i} = y_i - \hat{y}_{i,-i} \tag{4.7}$$

The PRESS residual measures the impact of the ith observation on the regression model. An observation with a larger PRESS residual may be a possible influential observation. i.e., the fitted regression model based on the data excluding this observation is quite different from the fitted regression model based on the data including this observation. The sum of the squared PRESS residuals is defined as the PRESS statistic.

Definition 4.5. The PRESS statistic is defined as

$$PRESS = \sum_{i=1}^{n} (y_i - \hat{y}_{i,-i})^2 = \sum_{i=1}^{n} (e_{i,-i})^2 \tag{4.8}$$

The PRESS residual can be expressed by the usual residual. It can be shown that

$$\begin{aligned} PRESS &= \sum_{i=1}^{n} (e_{i,-i})^2 \\ &= \sum_{i=1}^{n} \Big(\frac{y_i - \hat{y}_i}{1 - x_i'(\boldsymbol{X}'\boldsymbol{X})^{-1}x_i} \Big)^2 = \sum_{i=1}^{n} \Big(\frac{e_i}{1 - h_{ii}} \Big)^2, \end{aligned} \tag{4.9}$$

where h_{ii} is the ith diagonal element of the HAT matrix. This relationship makes calculation of the PRESS residual easier. Note that h_{ii} is always between 0 and 1. If h_{ii} is small (close to 0), even a larger value of the ordinary residual e_i may result in a smaller value of the PRESS residual. If h_{ii} is larger (close to 1), even a small value of residual e_i could result in a larger value of the PRESS residual. Thus, an influential observation is determined not only by the magnitude of residual but also by the corresponding value of leverage h_{ii}.

The PRESS statistic can be used in the regression model selection. A better regression model should be less sensitive to each individual observation. In other words, a better regression model should be less impacted by excluding one observation. Therefore, a regression model with a smaller value of the PRESS statistic should be a preferred model.

4.3.2 Notes on Outlier and Influential Observation

What is outlier? If there is breakdown in the model at the ith data point such that the mean is shifted. Typically, it is the phenomenon that $E(\epsilon_i) = \Delta \neq 0$. i.e., there is a mean shift for the ith observation. If we believe that the regression model is correctly specified, the ith observation may be a possible outlier when there is a mean shift at the ith observation.

Another aspect of an outlier is that the variance of ϵ is larger at the ith observation. i.e., there might be an inflation in variance at ith observation. If we believe that equal variance assumption is appropriate, the ith observation may be a potential outlier when its variance is inflated. So, an outlier could be examined by looking at both mean shift and variance inflation at a particular data point. In addition, the ith PRESS residual $e_{i,-i} = y_i - \hat{y}_{i,-i}$ can be used to determine outlier in regression model. If there is a mean shift at the ith data point, then we have

$$E(y_i - \hat{y}_{i,-i}) = \Delta_i.$$

Similarly, if there is a variance inflation at the ith data point we would like to use the standardized PRESS residual

$$\frac{e_{i,-i}}{\sigma_{i,-i}} = \frac{e_i}{\sigma\sqrt{1 - h_{ii}}} \tag{4.10}$$

to detect a possible outlier. Since σ is unknown, we can replace σ by its estimate s in the calculation of the standardized PRESS residual. However, in the presence of a mean shift outlier s is not an ideal estimate of true standard deviation of σ.

If there is a mean shift outlier at the ith observation, then s is biased upward. Hence, s is not an ideal estimate of σ. One way to deal with this situation is to leave the ith observation out and calculate the standard deviation with using the ith observation, denoted by s_{-i}. s_{-i} can be easily calculated based on the relationship between s_{-i} and s:

$$s_{-i} = \sqrt{\frac{(n-p)s^2 - e_i^2/(1 - h_{ii})}{n - p - 1}}. \tag{4.11}$$

If equal variance assumption is suspicious for a regression model, we can use the generalized least squares estimate where the equal variance assumption is not required. To conclude the discussion on outlier and influential observation in regression model, we point out that each observation can be examined under the following four groups.

(1). Usual observation: this is the observation which has equal effects on the fitted values. Regression model will not be affected with or without this observation. This type of observations should be used to estimate the model parameters.
(2). Outlier: an outlier is an observation for which the studentized residual is large in magnitude as compared to the studentized residuals of other observations in the data set. Outlier may indicate the violation of model assumptions or may imply the need for an alternative model.
(3). High leverage point: it is an observation which is far from the center of X space compared to other observations. Note that the concept of leverage is only related to the regressors, not the response variable.
(4). Influential observation: influential observation is the observation that has more impact on the regression coefficients or the estimate of variance as compared to other observations in the data set.

The relationship among outlier, influential observation, and high-leverage observation can be summarized as follows:

(1). An outlier needs not be an influential observation.
(2). An influential observation need not be an outlier.
(3). There is a tendency for a high-leverage observation to have small residuals and to influence the regression fit disproportionately.
(4). A high-leverage observation may not necessarily be an influential observation and an influential observation may not necessarily be a high-leverage observation.
(5). An observation might be an outlier, a high-leverage point, or an influential observation, simultaneously.

4.3.3 *Residual Mean Square Error for Over-fitted Regression Model*

In this section we discuss the consequence of under-fitting and over-fitting in regression analysis. Let's assume that the correct model should include all regressors $\boldsymbol{X}_1$ and $\boldsymbol{X}_2$, where $\boldsymbol{X}_1$ and $\boldsymbol{X}_2$ have p_1 and p_2 independent variables, respectively. We thus write the following full regression model:

$$\boldsymbol{y} = \boldsymbol{X}_1\boldsymbol{\beta}_1 + \boldsymbol{X}_2\boldsymbol{\beta}_2 + \boldsymbol{\varepsilon}.$$

Assume that the under-fitted model only includes $\boldsymbol{X}_1$ and is given by

$$\boldsymbol{y} = \boldsymbol{X}_1\boldsymbol{\beta}_1 + \boldsymbol{\varepsilon}^*.$$

In other words, an under-fitted regression model is the model from which some necessary regressors are omitted. Based on the under-fitted model we can solve for the least squares estimate of $\boldsymbol{\beta}_1$:

$$b_1 = (\boldsymbol{X}_1'\boldsymbol{X}_1)^{-1}\boldsymbol{X}_1'\boldsymbol{y}.$$

It can be verified that b_1 is a biased estimate of $\boldsymbol{\beta}_1$:

$$\begin{aligned} E(b_1) &= (\boldsymbol{X}_1'\boldsymbol{X}_1)^{-1}\boldsymbol{X}_1'E(\boldsymbol{y}) \\ &= (\boldsymbol{X}_1'\boldsymbol{X}_1)^{-1}\boldsymbol{X}_1'(\boldsymbol{X}_1\boldsymbol{\beta}_1 + \boldsymbol{X}_2\boldsymbol{\beta}_2) \\ &= \boldsymbol{\beta}_1 + (\boldsymbol{X}_1'\boldsymbol{X}_1)^{-1}\boldsymbol{X}_1'\boldsymbol{X}_2\boldsymbol{\beta}_2 \\ &= \boldsymbol{\beta}_1 + A\boldsymbol{\beta}_2. \end{aligned}$$

Thus, an under-fitted model has a bias $(\boldsymbol{X}_1'\boldsymbol{X}_1)^{-1}\boldsymbol{X}_1'\boldsymbol{X}_2\boldsymbol{\beta}_2$. Suppose that we over-fit a regression model with m parameters

$$\boldsymbol{y} = \boldsymbol{X}_1\boldsymbol{\beta}_1 + \boldsymbol{X}_2\boldsymbol{\beta}_2 + \boldsymbol{\varepsilon}$$

when the right model would be the one with $\boldsymbol{\beta}_2 = 0$. The residual mean square of the over-fitted model is

$$s_m^2 = \frac{\boldsymbol{y}'(I - \boldsymbol{X}(\boldsymbol{X}'\boldsymbol{X})^{-1}\boldsymbol{X})\boldsymbol{y}}{n-m}. \tag{4.12}$$

We have

$$E(s_m^2) = \sigma^2 + \frac{1}{n-m}\boldsymbol{\beta}_1'\boldsymbol{X}_1'(I - \boldsymbol{X}(\boldsymbol{X}'\boldsymbol{X})^{-1}\boldsymbol{X})\boldsymbol{X}_1\boldsymbol{\beta}_1. \tag{4.13}$$

It is known that the linear space spanned by the columns of $(I - \boldsymbol{X}(\boldsymbol{X}'\boldsymbol{X})^{-1}\boldsymbol{X})$ is perpendicular to the linear space spanned by the column $\boldsymbol{X}$, therefore, to the linear space spanned by the columns of $\boldsymbol{X}_1$. Thus, we have

$$\boldsymbol{X}_1'(I - \boldsymbol{X}(\boldsymbol{X}'\boldsymbol{X})^{-1}\boldsymbol{X})\boldsymbol{X}_1 = 0.$$

This yields $E(s_m^2) = \sigma^2$ which implies that s_m^2 is an unbiased estimate of the true variance σ^2 regardless of over-fitting. Furthermore, under the normality assumption of the model error term we have

$$\frac{(n-m)s_m^2}{\sigma^2} \sim \chi_{n-m}^2. \tag{4.14}$$

According to the χ^2 distribution we have

$$\text{Var}(s_m^2) = \frac{\sigma^4}{(n-m)^2}\text{Var}(\chi_{n-m}^2) = \frac{2\sigma^4}{n-m}. \tag{4.15}$$

Note that if the model is over-fitted the degrees of freedom is $n-m$, which is smaller than that from an ideal regression model. Therefore, for an over-fitted regression model, even s_m^2 is an unbiased estimate of the true variance σ^2, the variance of s_m^2 tends to be larger. We will further discuss the effect of under-fitting and over-fitting for multiple regression model in the next chapter.

4.4 Test for Mean-shift Outliers

When the ith observation is suspected to be an outlier, we can consider the so-called mean-shift outlier model

$$\boldsymbol{y} = \boldsymbol{X\beta} + e_i\gamma + \varepsilon, \tag{4.16}$$

where e_i denotes the ith $n \times 1$ unit vector. The problem to test whether or not the ith observation is a possible outlier can be formulated into the following hypothesis testing problem in the mean-shift outlier model:

$$H_0: \ \gamma = 0 \ \text{ versus } \ H_1: \gamma \neq 0.$$

We now discuss how to test mean shift outlier individually using the t-test statistic. When the ith observation is a potential mean shift outlier s^2 is a biased estimate of the error variance σ^2; therefore, we would prefer to use the leave-one-out s_{-i} and replacing σ by s_{-i} to construct the following test statistic

$$t_i = \frac{y_i - \hat{y}_i}{s_{-i}\sqrt{1-h_{ii}}} \sim t_{n-p-1}. \tag{4.17}$$

Under the null hypothesis $H_0: \Delta_i = 0$, the above test statistic has the centralized t distribution with degrees of freedom of $n-p-1$. It can be used to test the hypothesis $H_0 : \Delta_i = 0$ versus the alternative $H_1 : \Delta_i \neq 0$. The above statistic is often called the R-student statistic. It tends to be larger if the ith observation is a mean shift outlier. Note that the two-tailed t-test should be used. The rejection of H_0 implies that the ith observation may be a possible mean shift outlier. Note also that this test requires prior knowledge about a chosen observation being a potential outlier. If such knowledge is not available, then we can weaken the alternative hypothesis to "there is at least one outlier in the model". One may use the test statistic

$$t_{\max} = \max_{1 \le i \le n} t_i$$

and reject the null hypothesis H_0 at level α if

$$|t_{\max}| = |\max_{1 \le i \le n} t_i| > t_{n-p-1,\, \alpha/2n}. \tag{4.18}$$

This is equivalent to testing each individual observation and rejecting the null hypothesis if at least one test is significant. Note that we have to choose the test level to be α/n, which is the Bonferroni multiplicity adjustment, in order to ensure that the probability of the Type I error for the overall test is bounded by α.

The R-student statistic can also be used to test variance inflation at the ith observation. If there is a variance inflation at the ith observation then we should have $\mathrm{Var}(\epsilon_i) = \sigma^2 + \sigma_i^2$. Here σ_i^2 represents the increase in variance at ith observation. The hypothesis is $H_0 : \sigma_i^2 = 0$ versus $H_1 : \sigma_i^2 \neq 0$. Note that the two-tailed t-test should be used.

We use the following example to illustrate how to compute R-student statistic to test mean shift outlier. The Coal-cleansing Data is used in the example. There are 3 variables in the data set. Variable x_1 is the percent solids in the input solution; x_2 is the pH of the tank that holds the solution; and x_3 is the flow rate of the cleansing polymer in ml/minute. The response variable y is the polymer used to clean coal and suspended solids (mg/l). There are a total of 15 experiments performed. The data set is presented in Table 4.1.

```
proc reg data=coal outest=out1 tableout;
     model y=x1 x2 x3;
     output out=out2
     p=yhat r=yresid h=h RSTUDENT=rstudent;
     run;
```

Table 4.1 Coal-cleansing Data

Experiment	x1 Percent Solids	x2 pH Value	x3 Flow Rate	y Cleansing Polymer
1	1.5	6.0	1315	243
2	1.5	6.0	1315	261
3	1.5	9.0	1890	244
4	1.5	9.0	1890	285
5	2.0	7.5	1575	202
6	2.0	7.5	1575	180
7	2.0	7.5	1575	183
8	2.0	7.5	1575	207
9	2.5	9.0	1315	216
10	2.5	9.0	1315	160
11	2.5	6.0	1890	104
12	2.5	6.0	1890	110

Data Source: Data were generated by the Mining Engineering Department and analyzed by the Statistical Consulting Center, Virginia Polytechnic Institute and State University, Blacksburg, Virginia, 1979.

Table 4.2 Parameter Estimates for $y = b_0 + b_1x_1 + b_2x_2 + b_3x_3$

TYPE	RMSE	Intercept	x1	x2	x3
PARMS	20.8773	397.087	-110.750	15.5833	-0.05829
STDERR	20.8773	62.757	14.762	4.9208	0.02563
T	20.8773	6.327	-7.502	3.1668	-2.27395
PVALUE	20.8773	0.000	0.000	0.0133	0.05257
L95B	20.8773	252.370	-144.792	4.2359	-0.11741
U95B	20.8773	541.805	-76.708	26.9308	0.00082

The fitted regression model is found to be

$$\hat{y} = 397.087 - 110.750x_1 + 15.5833x_2 - 0.058x_3.$$

Before the analysis there was suspicion by experimental engineer that the 9th data point was keyed in erroneously. We then fit the model without deleting the 9th data point. The fitted responses, residuals, values of diagonal elements in the HAT matrix, and values of the R-student statistic associated with each observation are calculated and listed in Table 4.3. The largest residual is the 9th residual, $e_9 = 32.192$, and the corresponding value of the R-student statistic is 2.86951 which is statistically significant when test level α is 0.05. This adds suspicion that the 9th observation was originally keyed in erroneously.

Note that the statistical analysis only confirms that the 9th observation dose not fit the model well and may be a possible outlier. The decision on

Table 4.3 Residuals, Leverage, and t_i

Experiment	y_i	$\hat{y}_i$	e_i	h_{ii}	t_i
1	243	247.808	-4.8080	0.45013	-0.29228
2	261	247.808	13.1920	0.45013	0.83594
3	244	261.040	-17.0400	0.46603	-1.13724
4	285	261.040	23.9600	0.46603	1.76648
5	202	200.652	1.3480	0.08384	0.06312
6	180	200.652	-20.6520	0.08384	-1.03854
7	183	200.652	-17.6520	0.08384	-0.86981
8	207	200.652	6.3480	0.08384	0.29904
9*	216	183.808	32.1920*	0.45013	2.86951*
10	160	183.808	-23.8080	0.45013	-1.71405
11	104	103.540	0.4600	0.46603	0.02821
12	110	103.540	6.4600	0.46603	0.40062

whether or not this data point should be deleted must be made jointly by statistical model diagnosis and scientific judgement.

4.5 Graphical Display of Regression Diagnosis

4.5.1 *Partial Residual Plot*

In simple regression model, the visualization of the outlier is the residual plot which is the plot of regression residuals e_i against fitted values $\hat{y}_i$. However, in the multiple regression setting, there are more than one regressors and the plot of residuals against fitted values may not be able to tell which regressor is the cause of outlier. We may consider to plot y against individual regressor, but these plots may not highlight the effect of each regressor on the predictor since regressors may be dependent. The problem of outlier detection using residual plot in the multiple regression can be resolved by introducing the partial regression plot. To this end, we rearrange the $\boldsymbol{X}$ as $(\boldsymbol{x}_j, \boldsymbol{X}_{-j})$, where $\boldsymbol{x}_j$ is the jth column in the $\boldsymbol{X}$ and $\boldsymbol{X}_{-j}$ is the remaining $\boldsymbol{X}$ after deleting the jth column. If we regress $\boldsymbol{y}$ against $\boldsymbol{X}_{-j}$ the vector of residuals is

$$\boldsymbol{y} - \boldsymbol{X}_{-j}(\boldsymbol{X}'_{-j}\boldsymbol{X}_{-j})^{-1}\boldsymbol{X}'_{-j}\boldsymbol{y} = e_{y|X_{-j}}.$$

These residuals are not impacted by variable $\boldsymbol{x}_j$. We further consider the regression of $\boldsymbol{x}_j$ against $\boldsymbol{X}_{-j}$. The corresponding vector of residuals is

$$\boldsymbol{x}_j - \boldsymbol{X}_{-j}(\boldsymbol{X}'_{-j}\boldsymbol{X}_{-j})^{-1}\boldsymbol{X}'_{-j}\boldsymbol{x}_j = \boldsymbol{e}_{x_j|\boldsymbol{X}_{-j}}$$

This regression is to assess how much of $\boldsymbol{x}_j$ can be explained by the other independent variables $\boldsymbol{X}_{-j}$. The *partial regression plot* is the plot of

$$\boxed{\boldsymbol{e}_{\boldsymbol{y}|\boldsymbol{X}_{-j}} \text{ against } \boldsymbol{e}_{\boldsymbol{x}_j|\boldsymbol{X}_{-j}}}$$

The partial regression plot was proposed by Mosteller and Turkey (1977). To understand what is expected in the partial regression plot we proceed with the following derivations:

$$\boldsymbol{y} = \boldsymbol{X}\boldsymbol{\beta} + \boldsymbol{\varepsilon} = \boldsymbol{X}_{-j}\boldsymbol{\beta}_{-j} + \boldsymbol{x}_j\beta_j + \boldsymbol{\varepsilon}$$

Multiply both sides by $I - H_{-j}$, where $H_{-j} = \boldsymbol{X}_{-j}(\boldsymbol{X}'_{-j}\boldsymbol{X}_{-j})^{-1}\boldsymbol{X}'_{-j}$, the HAT matrix without using the ith column in $\boldsymbol{X}$,

$$(I - H_{-j})\boldsymbol{y} = (I - H_{-j})(\boldsymbol{X}_{-j}\boldsymbol{\beta}_{-j} + \boldsymbol{x}_j\beta_j + \boldsymbol{\varepsilon})$$

Since $(I - H_{-j})\boldsymbol{X}_{-j} = 0$ we have

$$\boldsymbol{e}_{y|\boldsymbol{X}_{-j}} = \beta_j \boldsymbol{e}_{\boldsymbol{x}_j|\boldsymbol{X}_{-j}} + \boldsymbol{\varepsilon}^*,$$

where $\boldsymbol{\varepsilon}^* = (I - H_{-j})\boldsymbol{\varepsilon}$. Thus, we expect that the partial regression plot is almost a line passing through the origin with the slope β_j. For any multiple regression model with p regressors there are a total of p partial residual plots to be produced.

The following example illustrates how to create a partial regression plot using SAS. Note that the ID statement is introduced to indicate the data points in the partial regression plot so that it helps identifying the observations that may be misrepresented in the model. We leave discussions on variable transformation to later chapter. For partial residual plot, the navy functional activity data is used in the example. There are 4 variables in the data set: man-hours is the response variable; the independent variables are site and two workload factors x_1 and x_2 which have substantial influence on man-hours.

The starting model in the example is

$$y = \beta_0 + \beta_1 x_1 + \beta_2 x_2 + \epsilon.$$

We generate two partial plots $\boldsymbol{e}_{\boldsymbol{y}|\boldsymbol{X}_{-1}}$ versus $\boldsymbol{e}_{\boldsymbol{x}_1|\boldsymbol{X}_{-1}}$ and $\boldsymbol{e}_{\boldsymbol{y}|\boldsymbol{X}_{-2}}$ versus $\boldsymbol{e}_{\boldsymbol{x}_1|\boldsymbol{X}_{-2}}$. The partial regression plot of $\boldsymbol{e}_{\boldsymbol{y}|\boldsymbol{X}_{-1}}$ versus $\boldsymbol{e}_{\boldsymbol{x}_1|\boldsymbol{X}_{-1}}$ shows a

Table 4.4 Navy Manpower Data

site	y(man-hours)	x1	x2
A	1015.09	1160.33	9390.83
B	1105.18	1047.67	14942.33
C	1598.09	4435.67	14189.08
D	1204.65	5797.72	4998.58
E	2037.35	15409.60	12134.67
F	2060.42	7419.25	20267.75
G	2400.30	38561.67	11715.00
H	3183.13	36047.17	18358.29
I	3217.26	40000.00	20000.00
J	2776.20	35000.00	15000.00

Data Source: SAS Institute Inc. SAS User's Guide: Statistics, Version 5 (Carry, North Carolina, SAS Institute Inc. 1985.

flat pattern in the last four sites G, H, I and J. This suggests that adding a cubic term of x_1 into the model may improve model fitting. We then fit the regression model with a term $x_1^{\frac{1}{3}}$ added:

$$y = \beta_0 + \beta_1 x_1^{\frac{1}{3}} + \beta_2 x_2 + \epsilon.$$

The partial regression plot for the model with the transformed term $x_1^{\frac{1}{3}}$ is clearly improved. The SAS program is provided as follows.

```
data navy; set navy;
label y ="Manpower hours"
      x1="Manpower factor 1"
      x2="Manpower factor 2"
      X3="Cubic Root of x1";
x3=x1**(1/3);
run;

proc reg data=navy;
model y=x1 x2/partial;
id site;
output out=out1 p=pred r=resid;
run;

proc reg data=navy;
model y=x3 x2/partial;
id site;
```

```
output out=out2 p=pred r=resid;
run;
```

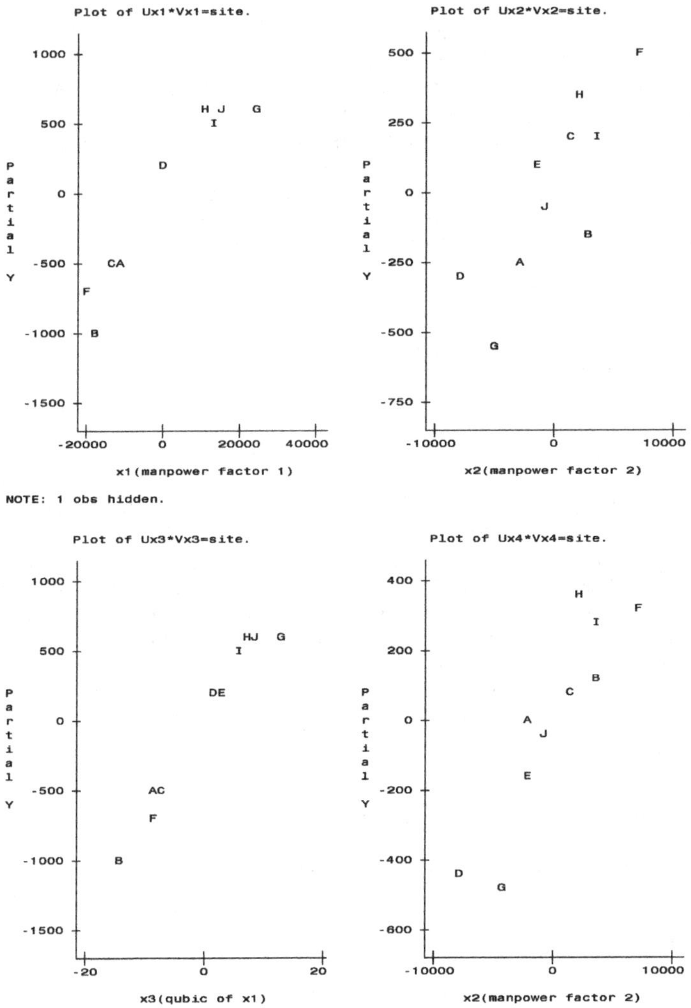

Fig. 4.1 Partial Residual Plots for Regression Models

Table 4.5 Residuals for Model With Regressors x_1 and x_2

site	y	x_1	x_2	$\hat{y}_i$	e_i
A	1015.09	1160.33	9390.83	1110.76	-95.672
B	1105.18	1047.67	14942.33	1433.61	-328.431
C	1598.09	4435.67	14189.08	1520.50	77.589
D	1204.65	5797.72	4998.58	1031.58	173.072
E	2037.35	15409.60	12134.67	1824.66	212.690
F	2060.42	7419.25	20267.75	1994.40	66.015
G	2400.30	38561.67	11715.00	2697.09	-296.787
H	3183.13	36047.17	18358.29	2991.22	191.915
I	3217.26	40000.00	20000.00	3241.16	-23.896
J	2776.20	35000.00	15000.00	2752.69	23.507

$R^2 = 0.9468$ $s = 216.672$

Table 4.6 Residuals for Model With Regressors $x_1^{\frac{1}{3}}$ and x_2

site	y	x_1	x_2	$x_1^{\frac{1}{3}}$	$\hat{y}_i$	e_i
A	1015.09	1160.33	9390.83	10.5082	896.57	118.525
B	1105.18	1047.67	14942.33	10.1564	1203.15	-97.966
C	1598.09	4435.67	14189.08	16.4306	1590.56	7.529
D	1204.65	5797.72	4998.58	17.9647	1148.60	56.052
E	2037.35	15409.60	12134.67	24.8846	2050.64	-13.287
F	2060.42	7419.25	20267.75	19.5038	2164.54	-104.124
G	2400.30	38561.67	11715.00	33.7846	2638.86	-238.557
H	3183.13	36047.17	18358.29	33.0337	2982.99	200.136
I	3217.26	40000.00	20000.00	34.1995	3161.15	56.108
J	2776.20	35000.00	15000.00	32.7107	2760.62	15.584

$R^2 = 0.9776$ $s = 140.518$

4.5.2 *Component-plus-residual Plot*

The component-plus-residual (CPR) plot is one of regression diagnosis plots. It is useful for assessing nonlinearity in independent variables in the model. The CPR plot is the scatter plot of

$$\boxed{e_{y|X} + x_j b_j \text{ against } x_j}$$

This plot was introduced by Ezekiel (1924), and was recommended later by Larsen and McCleary (1972). This scatter plot has the least squares slope of b_j and the abscissa x_j and is effective when one wants to find nonlinearity of x_j in regression model.

4.5.3 *Augmented Partial Residual Plot*

The augmented partial residual (APR) plot is another graphical display of regression diagnosis. The APR plot is the plot of

$$\boxed{e_{y|Xx_j^2} + x_j b_j + x_j^2 b_{jj} \text{ against } x_j}$$

where b_j and b_{jj} are the least squares estimates from model

$$\boldsymbol{y} = \boldsymbol{X}\boldsymbol{\beta} + \beta_j \boldsymbol{x}_j + \beta_{jj}\boldsymbol{x}_j^2 + \boldsymbol{\varepsilon}.$$

This plot was suggested by Mallows (1986) to explore whether or not a transformation of x_j is needed in the linear multiple regression model. Note that the APR plot does not just intend to detect the need for a quadratic term in the regression model. The introduced quadratic term is really a truncated version of the potential nonlinear form of x_j.

The CPR plot is not very sensitive in finding the transformation for an entering variable or in detecting nonlinearity. The APR plot appears to be more sensitive to nonlinearity. In the absence of nonlinearity the CPR plot is similar to the APR plot. However, if the nonlinearity is present in a variable for a regression model (it may not be necessarily quadratic), in general, the APR plot gives a better picture to show the nonlinearity of the variable than the CPR plot.

4.6 Test for Inferential Observations

Outliers and Influential Observations are easy to understand in simple regression model $y = a + bx + \varepsilon$ since there is only one independent variable in the model. We can simply plot the fitted regression line, a data point which lies far away from the regression line (and thus has a large residual value) is known as a potential outlier. Such point may represent erroneous data, or may indicate a poorly fitting of the regression model. If a data point lies far away from the other data points in the horizontal direction, it is known as potential influential observation since this data point may have a significant impact on the slope of the regression line. If the data point is far away from other data points but the residual at this data point is small then this data point may not be an influential observation. However, if the residual is large at this observation then including or excluding this data point may have great impact on the regression slope. Hence, this data point may be an influential observation for the regression model.

The outlier detection for linear regression analysis is to find observations with larger residuals which are larger than it would actually be produced by chance. i.e., the possible error in y-direction. A data analyst may also want to identify the extent whether or not these observations influence predicted values, estimators of regression coefficients, and performance criteria of a regression model. This is the detection of inferential observations.

Source of influential observation to the regression model depends on the location of the observation in either y-direction or x-direction. If a data point is far away from the rest of the data points in x-direction this data point may be an influential observation since it may affect the slope of the simple regression line. If a data point is far away from the rest of the data points in y-direction then it may not be an influential observation to the regression slope. We now briefly discuss the quantities that could affect the regression model in various ways.

(1) **HAT diagonal elements**: A data point with a large HAT diagonal element may or may not be an influential observation to the regression coefficients, but it could be an outlier to affect the variance of the estimate of regression coefficients. The HAT diagonal element h_{ii} can be used to detect a possible outlier if

$$h_{ii} = \boldsymbol{x}_i^{'}(\boldsymbol{X}^{'}\boldsymbol{X})^{-1}\boldsymbol{x}_i > 2p/n \qquad (4.19)$$

where n is the total number of observations and p is the number of independent variables in the regression model.

(2) **R-student residual**: Any R-student residual deviates from zero by 2 times of the estimated standard error could be a potential outlier.

(3) **DFITTS** : The DFITTS is defined as the difference between the standardized fitted values with and without the ith observation $\boldsymbol{x}_i$:

$$\text{DFITTS}_i = \frac{\hat{y}_i - \hat{y}_{i,-i}}{s_{-i}\sqrt{h_{ii}}} \qquad (4.20)$$

It can be shown that

$$\text{DFITTS}_i = \text{R-student}_i \Big[\frac{h_{ii}}{1-h_{ii}}\Big]^{1/2} \qquad (4.21)$$

It can be seen from (4.21) that a larger value of R-student$_i$ or a near-one value of h_{ii} may cause the ith observation to be a potential outlier.

DFITTS$_i$ is useful in application and it tells how much changes you can expect in the response variable with deletion of the ith observation.

To take into account for sample size, as suggested by some authors, the "yardstick cutoff values" are $2\sqrt{p/n}$ for DFFITS$_i$ to detect possible outliers. In situations of moderate and small sample size, a yardstick cutoff of ± 2 can be used to detect possible outliers.

(4) **DEBETAS**: The DEBETAS is the difference of regression estimate b_j between with the ith observation included and with the ith observation excluded. Namely it is given by

$$\text{DFBETAS}_{j,-i} = \frac{b_j - b_{j,-i}}{s_{-i}\sqrt{c_{jj}}}, \tag{4.22}$$

where c_{jj} is the diagonal element of $(\boldsymbol{X}'\boldsymbol{X})^{-1}$. A larger value of DFBETAS$_{j,-i}$ implies that the ith observation may be a potential outlier. DEBETAS$_{j,-i}$ is useful in application and it tells the change of regression parameter estimate b_j due to the deletion of the ith observation.

To take into account for sample size, as suggested by some authors, the "yardstick cutoff values" are $2/\sqrt{n}$ for DEBETAS$_{j,-i}$ to detect possible outliers. In situations of moderate and small sample size, a yardstick cutoff of ± 2 can be used to detect possible outliers.

(5) **Cook's D**: The Cook's D is the distance between the least squares estimates of regression coefficients with $\boldsymbol{x}_i$ included and excluded. The composite measurement Cook's D is defined as

$$D_i = \frac{(\boldsymbol{b} - \boldsymbol{b}_{-i})'(\boldsymbol{X}'\boldsymbol{X})(\boldsymbol{b} - \boldsymbol{b}_{-i})}{ps^2}, \tag{4.23}$$

where $\boldsymbol{b}$ is the least squares estimate of the parameter β_j using all observations and $\boldsymbol{b}_{-i}$ is the least squares estimate of the parameter β_j with the ith observation deleted.

(6) **Generalized Variance (GV)**: The GV provides a scalar measure of the variance-covariance properties of the regression coefficients. It is defined as

$$GV = |(\boldsymbol{X}'\boldsymbol{X})^{-1}\sigma^2| \tag{4.24}$$

(7) **Covariance Ratio (COVRATIO)**: The covariance ratio is the ratio of GVs with $\boldsymbol{x}_i$ included and excluded. It is defined as

$$(\text{COVARATIO})_i = \frac{|(\boldsymbol{X}'_{-i}\boldsymbol{X}_{-i})^{-1}s^2_{-i}|}{|(\boldsymbol{X}'\boldsymbol{X})^{-1}s^2|} \tag{4.25}$$

It can be shown that

$$(\text{COVARATIO})_i = \frac{(s^2_{-i})^{2p}}{s^{2p}}\Big(\frac{1}{1-h_{ii}}\Big) \tag{4.26}$$

A rough "yardstick cutoff value" for the COVARATIO is suggested by Belsley, Kuh and Welsch (1980). The ith observation may be identified as a possible outlier if

$$(\text{COVARATIO})_i > 1 + (3p/n) \tag{4.27}$$

or

$$(\text{COVARATIO})_i < 1 - (3p/n) \tag{4.28}$$

where p is the number of parameters in regression model and n is the total number of observations.

There are much overlap in the information provided by the regression diagnosis. For example, if the Cook's D produces an unusually high result, then at least we can expect that one of $\text{DEBETAS}_{j,i}$, or one of DIFFITS_i will provide strong result as well. So, information provided by regression diagnosis may overlap. Usually, an experienced analyst would briefly look at the HAT diagnosis and R-student residuals to explore possible outliers. As always, it should be very cautious to remove outliers. Sometimes adding one or more mean shift variables may be a good approach to take.

4.7 Example

In this section we give an illustrative example to calculate residuals, PRESS residuals, and PRESS statistic for several selected regression models using the SAS. The values of the PRESS statistic for different regression models are computed and compared in order to select a better regression model.

The following example is based on the Sales of Asphalt Roofing Shingle Data (see Table 4.7). The variables in the data set are district code, number

Table 4.7 Sales Data for Asphalt Shingles

District	x_1 Promotional Accounts	x_2 Active Accounts	x_3 Competing Brands	x_4 Sale Potential	y Sales in Thousands of Dollars
1	5.5	31	10	8	79.3
2	2.5	55	8	6	200.1
3	8.67	12	9	1	63.2
4	3.0	50	7	16	200.1
5	3.0	38	8	15	146.0
6	2.9	71	12	17	177.7
7	8.0	30	12	8	30.9
8	9.0	56	5	10	291.9
9	4.0	42	8	4	160.0
10	6.5	73	5	16	339.4
11	5.5	60	11	7	159.6
12	5.0	44	12	12	86.3
13	6.0	50	6	6	237.5
14	5.0	39	10	4	107.2
15	3.5	55	10	4	155.0

Data Source: Raymond H. Myers, *Classical and Modern Regression with Applications*. Duxbury, p. 174.

of promotional accounts (x_1), number of active accounts (x_2), number of competing brands (x_3), and sale potential for district (x_4). The response variable is the sales in thousands of dollars (y).

The SAS program for computing values of the PRESS statistic for various regression models is presented as follows:

```
proc reg data=sale;
model y=x2 x3;
output out=out1
r=r12 press=press12 H=h12;

proc reg data=sale;
model y=x1 x2 x3;
output out=out2
r=r123 press=press123 H=h123;

proc reg data=sale;
model y=x1 x2 x3 x4;
output out=out3
r=r1234 press=presss1234 H=h1234;
```

```
run;

proc sort data=out1; by district;
proc sort data=out2; by district;
proc sort data=out3; by district;
run;

data all; merge out1 out2 out3;
by district;
p12=press12*press12;
p123=press123*press123;
p1234=press1234*press1234;
run;

proc means data=all;
var p12 p123 p1234;
output out=summary sum=p12 p123 p1234;
run;
```

Table 4.8 Values of PRESS Statistic for Various Regression Models

	Model includes x_2, x_3	Model includes x_1, x_2, x_3	Model includes x_1, x_2, x_3, x_4
PRESS	782.19	643.36	741.76

Data Source: Asphalt Roofing Shingle Data

After calculating the values of the PRESS statistic for different regression models, we select the model with the regressors x_1, x_2, x_3 because it has the smallest value of the PRESS statistic.

Problems

1. Let the ith residual of the regression model be $e_i = y_i - \hat{y}_i$. Prove that $\text{Var}(e_i) = s^2(1 - h_{ii})$, where s^2 is the mean square error of the regression model and h_{ii} is the ith diagonal element of the HAT matrix.

Table 4.9 PRESS Residuals and Leverages for Models Including x_1, x_2 (press12, h12) and Including x_1, x_2, x_3 (press123, h123)

district	yresid12	press12	h12	yresid123	press123	h123
1	1.5744	2.0192	0.22030	0.64432	0.8299	0.22360
2	-7.8345	-8.5519	0.08389	-2.11269	-2.6705	0.20889
3	1.5332	2.1855	0.29847	-4.59006	-8.2204	0.44163
4	-12.2631	-13.8695	0.11582	-7.54144	-9.4379	0.20094
5	-1.2586	-1.4852	0.15253	3.29538	4.2893	0.23171
6	1.7565	2.7618	0.36402	6.39862	11.5561	0.44630
7	1.2924	1.8385	0.29705	-5.08116	-9.2747	0.45215
8	13.5731	18.1046	0.25030	5.87488	11.2237	0.47657
9	-1.5353	-1.7361	0.11564	0.94022	1.0921	0.13904
10	0.3972	0.6555	0.39405	-1.92210	-3.2833	0.41459
11	0.6429	0.7600	0.15403	-0.17736	-0.2103	0.15660
12	6.7240	8.0934	0.16919	6.76184	8.1389	0.16920
13	2.8625	3.4654	0.17397	1.35393	1.6565	0.18266
14	0.9210	1.0521	0.12456	1.09787	1.2542	0.12468
15	-8.3857	-9.1765	0.08619	-4.94225	-5.6903	0.13146

2. Show the following relationship between the leave-one-out residual and the ordinary residual:

$$PRESS = \sum_{i=1}^{n}(e_{i,-i})^2 = \sum_{i=1}^{n}\Big(\frac{y_i - \hat{y}_i}{1 - x_i'(\boldsymbol{X}'\boldsymbol{X})^{-1}x_i}\Big)^2 = \sum_{i=1}^{n}\Big(\frac{e_i}{1-h_{ii}}\Big)^2,$$

where h_{ii} is the ith diagonal elements of the HAT matrix.

3. Consider the data set in Table 4.10. It is known a prior that in observations 10, 16, and 17 there were some difficulties in measuring the response y. Apply the mean shift model and test simultaneously

$$H_0: \ \Delta_{10} = 0, \ \Delta_{16} = 0, \ \Delta_{17} = 0.$$

Make your conclusions (hint: Use indicator variables).

4. In the partial residual plot discussed in this chapter, show that the least squares slope of the elements $e_{y|\boldsymbol{X}_j}$ regressed against $e_{x_j|\boldsymbol{X}_{-j}}$ is b_j, the slope of x_j in the multiple regression of y on $\boldsymbol{X}$.

5. The variance inflation model can be written

$$y_i = \boldsymbol{x}_i'\boldsymbol{\beta} + \epsilon_i \quad \text{for} i = 1, 2, \cdots, n$$

with model error being normally distributed and

$$E(\epsilon_j) = \sigma^2 + \sigma_\Delta^2 \ \ \text{and} E(\epsilon_i) = \sigma^2 \ \text{ for all } \ i \neq j$$

and $E(\epsilon_i) = 0$.

Table 4.10 Data Set for Multiple Mean Shift Outliers

Observation	x_1	x_2	x_3	y
1	12.980	0.317	9.998	57.702
2	14.295	2.028	6.776	59.296
3	15.531	5.305	2.947	56.166
4	15.133	4.738	4.201	55.767
5	15.342	7.038	2.053	51.722
6	17.149	5.982	-0.055	60.466
7	15.462	2.737	4.657	50.715
8	12.801	10.663	3.408	37.441
9	13.172	2.039	8.738	55.270
10	16.125	2.271	2.101	59.289
11	14.340	4.077	5.545	54.027
12	12.923	2.643	9.331	53.199
13	14.231	10.401	1.041	41.896
14	15.222	1.22	6.149	63.264
15	15.74	10.612	-1.691	45.798
16	14.958	4.815	4.111	58.699
17	14.125	3.153	8.453	50.086
18	16.391	9.698	-1.714	48.890
19	16.452	3.912	2.145	62.213
20	13.535	7.625	3.851	45.625
21	14.199	4.474	5.112	53.923
22	16.565	8.546	8.974	56.741
23	13.322	8.598	4.011	43.145
24	15.945	8.290	-0.248	50.706
25	14.123	0.578	-0.543	56.817

To identify variance inflation outlier it is needed to test the hypothesis $H_0 : \sigma^2_\Delta = 0$ and the corresponding test statistic is the R-student statistic. Thus, the R-student statistic is appropriate for testing regression outlier no matter which type of outlier exists.

(a). Show that the $(R\!-\!student)_j$ has the t-distribution under $H_0 : \sigma^2_\Delta = 0$.

(b). Find the distribution of $(R - student)^2_j$ under $H_0 : \sigma^2_\Delta \neq 0$.

6. The following data in Table 4.11 was collected from a study of the effect of stream characteristics on fish biomass. The regressor variables are

   x_1: Average depth of 50 cells
   x_2: Area of stream cover (i.e., undercut banks, logs, boulders, etc.)
   x_3: Percent of canopy cover (average of 12)
   x_4: Area $\geq$ 25cm in depth

   The response variable is y, the fish biomass. The data set is as follows:

Table 4.11 Fish Biomass Data

Observation	y	x_1	x_2	x_3	x_4
1	100	14.3	15.0	12.2	48.0
2	388	19.1	29.4	26.0	152.2
3	755	54.6	58.0	24.2	469.7
4	1288	28.8	42.6	26.1	485.9
5	230	16.1	15.9	31.6	87.6
6	0.0	10.0	56.4	23.3	6.9
7	551	28.5	95.1	13.0	192.9
8	345	13.8	60.6	7.5	105.8
9	0.0	10.7	35.2	40.3	0.0
10	348	25.9	52.0	40.3	116.6

Data Source: Raymond H. Myers: Classical and Modern Regression With Application, p. 201.

(a). Compute s^2, C_p, PRESS residuals, and PRESS statistic for the model $y = \beta_0 + \beta_1 x_1 + \beta_2 x_2 + \beta_3 x_3 + \beta_4 x_4 + \epsilon$.

(b). Compute s^2, C_p, PRESS residuals, and PRESS statistic for the model $y = \beta_0 + \beta_1 x_1 + \beta_2 x_3 + \beta_3 x_4 + \epsilon$.

(c). Compare the regression models from parts (a) and (b) and comments on the prediction of fish biomass.

Chapter 5

Model Selection

In previous chapters, we have proceeded as if predictors included in the model, as well as their functional forms (i.e., linear), are known. This is certainly not the case in reality. Practically, model misspecification occurs in various ways. In particular, the proposed model may underfit or overfit the data, which respectively correspond to situations where we mistakenly exclude important predictors or include unnecessary predictors. Thus, there involves an issue of variable selection. Furthermore, the simple yet naive linear functions might be neither adequate nor proper for some predictors. One needs to evaluate the adequacy of linearity for predictors, and if inadequacy identified, seek appropriate functional forms that fit better.

In this chapter, we mainly deal with variable selection, while deferring the latter issues, linearity assessment and functional form detection, to model diagnostics. To gain motivation of and insight into variable selection, we first take a look at the adverse effects on model inference that overfitting or underfitting may cause. Then we discuss two commonly used methods for variable selection. The first is the method of all possible regressions, which is most suitable for situations where total number of predictors, p, is small or moderate. The second type is stepwise algorithmic procedures. There are many other concerns that warrant caution and care and new advances newly developed in the domain of model selection. We will discuss some important ones at the end of the chapter.

5.1 Effect of Underfitting and Overfitting

Consider the linear model $\mathbf{y} = \mathbf{X}\boldsymbol{\beta} + \boldsymbol{\varepsilon}$, where $\mathbf{X}$ is $n \times (k+1)$ of full column rank $(k+1)$ and $\boldsymbol{\varepsilon} \sim \text{MVN}\{\mathbf{0},\ \sigma^2 \cdot I\}$. To facilitate a setting for underfitting and overfitting, we rewrite it in a partitioned form

$$\begin{aligned} \mathbf{y} &= (\mathbf{X_1}\ \mathbf{X_2}) \begin{pmatrix} \boldsymbol{\beta_1} \\ \boldsymbol{\beta_2} \end{pmatrix} + \boldsymbol{\varepsilon} \\ &= \mathbf{X_1}\boldsymbol{\beta_1} + \mathbf{X_2}\boldsymbol{\beta_2} + \boldsymbol{\varepsilon}, \end{aligned} \tag{5.1}$$

where $\mathbf{X}_1$ is $n \times (p+1)$ of full rank $(p+1)$ and $\mathbf{X}_2$ is $n \times (k-p)$. Then the following two situations may be encountered: *underfitting* if $\mathbf{X_2}\boldsymbol{\beta_2}$ is left out when it should be included and *overfitting* if $\mathbf{X_2}\boldsymbol{\beta_2}$ is included when it should be left out, both related to another model of reduced form

$$\mathbf{y} = \mathbf{X_1}\boldsymbol{\beta}_1^\star + \boldsymbol{\varepsilon}^\star, \tag{5.2}$$

with $\boldsymbol{\varepsilon}^\star \sim \text{MVN}\{\mathbf{0},\ \sigma_\star^2 \cdot I\}$. Let $\hat{\boldsymbol{\beta}} = (\mathbf{X}'\mathbf{X})^{-1}\mathbf{X}'\mathbf{y}$ from model (5.1) partitioned as

$$\hat{\boldsymbol{\beta}} = \begin{pmatrix} \hat{\boldsymbol{\beta}}_1 \\ \widehat{\boldsymbol{\beta}}_2 \end{pmatrix}.$$

Let $\widehat{\boldsymbol{\beta}}_1^\star = (\mathbf{X}_1'\mathbf{X}_1)^{-1}\mathbf{X}_1'\mathbf{y}$ denote the LSE of $\boldsymbol{\beta}_1^\star$ in model (5.2).

We shall next present two theorems that characterize the effects of underfitting and overfitting on least squares estimation and model prediction, respectively. In their proofs that follow, some matrix properties listed in the lemma below are useful.

Lemma 5.1.

(1) If $\mathbf{A}$ is a positive definite (p.d.) matrix, then $\mathbf{A}^{-1}$ is also positive definite. This can be proved by using the fact that a symmetric matrix $\mathbf{A}$ is p.d. if and only if there exists a nonsingular matrix $\mathbf{P}$ such that $\mathbf{A} = \mathbf{P}'\mathbf{P}$.

(2) If $\mathbf{A}$ is $p \times p$ p.d. and B is $k \times p$ with $k \leq p$, then $\mathbf{BAB}'$ is positive semidefinite (p.s.d.). Furthermore, if $\mathbf{B}$ is of rank k, then $\mathbf{BAB}'$ is p.d.

(3) If a symmetric p.d. matrix $\mathbf{A}$ is partitioned in the form

$$\mathbf{A} = \begin{pmatrix} \mathbf{A}_{11} & \mathbf{A}_{12} \\ \mathbf{A}_{21} & \mathbf{A}_{22} \end{pmatrix},$$

where $\mathbf{A}_{11}$ and $\mathbf{A}_{22}$ are square matrices, then both $\mathbf{A}_{11}$ and $\mathbf{A}_{22}$ are p.d. and the inverse of $\mathbf{A}$ is given by

$$\mathbf{A}^{-1} = \begin{pmatrix} \mathbf{A}_{11} + \mathbf{A}_{11}^{-1}\mathbf{A}_{12}\mathbf{B}^{-1}\mathbf{A}_{21}\mathbf{A}_{11}^{-1} & -\mathbf{A}_{11}^{-1}\mathbf{A}_{12}\mathbf{B}^{-1} \\ -\mathbf{B}\mathbf{A}_{21}\mathbf{A}_{11}^{-1} & \mathbf{B}^{-1} \end{pmatrix},$$

with $\mathbf{B} = \mathbf{A}_{22} - \mathbf{A}_{21}\mathbf{A}_{11}^{-1}\mathbf{A}_{12}$. *Besides, one can verify by straightforward algebra that*

$$(\,\mathbf{x}_1'\ \mathbf{x}_2'\,)\ \mathbf{A}^{-1}\begin{pmatrix}\mathbf{x}_1\\ \mathbf{x}_2\end{pmatrix} = \mathbf{x}_1'\mathbf{A}_{11}^{-1}\mathbf{x}_1 + \mathbf{b}'\mathbf{B}^{-1}\mathbf{b}, \tag{5.3}$$

where $\mathbf{b} = \mathbf{x}_2 - \mathbf{A}_{21}\mathbf{A}_{11}^{-1}\mathbf{x}_1$.

Theorem 5.1. *Suppose that Model* $\mathbf{y} = \boldsymbol{X}_1\boldsymbol{\beta}_1 + \boldsymbol{X}_2\boldsymbol{\beta}_2 + \varepsilon$ *is the true underlying model, but we actually fit model* $\mathbf{y} = \boldsymbol{X}_1\boldsymbol{\beta}_1^\star + \varepsilon^\star$, *which is thus an underfitted model. Then:*

(i) $E(\widehat{\boldsymbol{\beta}}_1^\star) = \boldsymbol{\beta}_1 + \mathbf{A}\boldsymbol{\beta}_2$, *where* $\mathbf{A} = (\mathbf{X}_1'\mathbf{X}_1)^{-1}\mathbf{X}_1'\mathbf{X}_2$.

(ii) $Cov(\widehat{\boldsymbol{\beta}}_1^\star) = \sigma^2(\mathbf{X}_1'\mathbf{X}_1)^{-1}$ *and*

$$Cov(\hat{\boldsymbol{\beta}}_1) = \sigma^2 \cdot \left\{(\mathbf{X}_1'\mathbf{X}_1)^{-1} + \mathbf{A}\mathbf{B}^{-1}\mathbf{A}'\right\},$$

where $\mathbf{B} = \mathbf{X}_2'\mathbf{X}_2 - \mathbf{X}_2'\mathbf{X}_1(\mathbf{X}_1'\mathbf{X}_1)^{-1}\mathbf{X}_1'\mathbf{X}_2$. *We have*

$$Cov(\hat{\boldsymbol{\beta}}_1) - Cov(\hat{\boldsymbol{\beta}}_1^\star) \geq \mathbf{0} \quad \textit{i.e. positive semi-definite.}$$

(iii) Let $\mathcal{V}_1$ *denote the column space of* $\mathbf{X}_1$ *and*

$$\mathbf{P}_{\mathcal{V}_1^\perp} = \mathbf{I} - \mathbf{P}_{\mathcal{V}_1} = \mathbf{I} - \mathbf{X}_1(\mathbf{X}_1'\mathbf{X}_1)^{-1}\mathbf{X}_1'$$

be the projection matric of subspace $\mathcal{V}_1^\perp$. *Then an estimator of* σ^2

$$\hat{\sigma}^2 = \frac{\|\, \mathbf{P}_{\mathcal{V}_1^\perp}\mathbf{y}\,\|^2}{n-p-1}$$

has the expected value of

$$E(\hat{\sigma}^2) = \sigma^2 + \frac{\|\, \mathbf{P}_{\mathcal{V}_1^\perp}(\mathbf{X}_2'\boldsymbol{\beta}_2)\,\|^2}{n-p-1}.$$

(iv) Given $\mathbf{x}_0 = \begin{pmatrix}\mathbf{x}_{01}\\ \mathbf{x}_{02}\end{pmatrix}$, *let* $\hat{y}^\star = \mathbf{x}_{01}'\widehat{\boldsymbol{\beta}}_1^\star$ *denote the estimate based on the working model and* $\hat{y}_0 = \mathbf{x}_0'\widehat{\boldsymbol{\beta}} = \begin{pmatrix}\hat{y}_{01}\\ \hat{y}_{02}\end{pmatrix}$ *denote the estimate based on the true model. Then it can be found that*

$$E(\hat{y}^\star) = \mathbf{x}_{01}'\,(\boldsymbol{\beta}_1 + \boldsymbol{A}\boldsymbol{\beta}_2)$$

and

$$Var(\hat{y}^\star) = \sigma^2 \cdot \mathbf{x}_{01}'(\mathbf{X}_1'\mathbf{X}_1)^{-1}\mathbf{x}_{01} \leq \min\{\,Var(\hat{y}_{01}),\, Var(\hat{y}_0)\},$$

where $Var(\hat{y}_{01}) = Var(\mathbf{x}_{01}'\widehat{\boldsymbol{\beta}}_1)$.

Proof.

(i) Consider

$$E(\boldsymbol{\beta}_1^\star) = E\{(\mathbf{X}_1'\mathbf{X}_1)^{-1}\mathbf{X}_1'\mathbf{y}\} = (\mathbf{X}_1'\mathbf{X}_1)^{-1}\mathbf{X}_1'E(\mathbf{y})$$
$$= (\mathbf{X}_1'\mathbf{X}_1)^{-1}\mathbf{X}_1'(\mathbf{X}_1\boldsymbol{\beta}_1 + \mathbf{X}_2\boldsymbol{\beta}_2)$$
$$= \boldsymbol{\beta}_1 + (\mathbf{X}_1'\mathbf{X}_1)^{-1}\mathbf{X}_1'\mathbf{X}_2\boldsymbol{\beta}_2$$

(ii) First

$$\mathrm{Cov}(\hat{\boldsymbol{\beta}}^\star) = \mathrm{cov}\{(\mathbf{X}_1'\mathbf{X}_1)^{-1}\mathbf{X}_1'\mathbf{y}\}$$
$$= (\mathbf{X}_1'\mathbf{X}_1)^{-1}\mathbf{X}_1'(\sigma^2\mathbf{I})\mathbf{X}_1(\mathbf{X}_1'\mathbf{X}_1)^{-1}$$
$$= \sigma^2(\mathbf{X}_1'\mathbf{X}_1)^{-1}$$

Next, to find $\mathrm{Cov}(\hat{\boldsymbol{\beta}}_1)$, consider

$$\mathrm{Cov}(\hat{\boldsymbol{\beta}}) = \mathrm{Cov}\begin{pmatrix}\hat{\boldsymbol{\beta}}_1\\ \hat{\boldsymbol{\beta}}_2\end{pmatrix} = \sigma^2\cdot(\mathbf{X}'\mathbf{X})^{-1} = \sigma^2\begin{pmatrix}\mathbf{X}_1'\mathbf{X}_1 & \mathbf{X}_1'\mathbf{X}_2\\ \mathbf{X}_2'\mathbf{X}_1 & \mathbf{X}_2'\mathbf{X}_2\end{pmatrix}^{-1}$$
$$= \sigma^2\begin{pmatrix}\mathbf{G}^{11} & \mathbf{G}^{12}\\ \mathbf{G}^{21} & \mathbf{G}^{22}\end{pmatrix},$$

where $\mathbf{G}^{ij}$ is the corresponding block of the partitioned inverse matrix $(\mathbf{X}'\mathbf{X})^{-1}$. It can be found that

$$\mathbf{G}^{11} = (\mathbf{X}_1'\mathbf{X}_1)^{-1} + \mathbf{A}\mathbf{B}^{-1}\mathbf{A}' \quad \text{and} \quad \mathbf{G}^{22} = \mathbf{B}^{-1}$$

Hence, $\mathrm{Cov}(\hat{\boldsymbol{\beta}}_1) - \mathrm{Cov}(\hat{\boldsymbol{\beta}}^\star{}_1) = \sigma^2\mathbf{A}\mathbf{B}^{-1}\mathbf{A}'$. Now $(\mathbf{X}'\mathbf{X})^{-1}$ is p.d., so is $\mathbf{G}^{22} = \mathbf{B}^{-1}$. Hence, matrix $\mathbf{A}\mathbf{B}^{-1}\mathbf{A}'$ is p.s.d. This implies that $\mathrm{Var}(\hat{\beta}_j) \geq \mathrm{Var}(\hat{\beta}_j^\star)$.

(iii) The sum of squared error, SSE_1, associated with model $\mathbf{y} = \boldsymbol{X}_1\boldsymbol{\beta}_1^\star + \varepsilon^\star$ is

$$SSE_1 = \|\, \mathbf{y} - \mathbf{X}_1\hat{\boldsymbol{\beta}}^\star \,\|^2 = \|\, \mathbf{y} - P_{\mathcal{V}_1}\mathbf{y} \,\|^2$$
$$= \|\, P_{\mathcal{V}_1^\perp}\mathbf{y} \,\|^2 = \mathbf{y}'P_{\mathcal{V}_1^\perp}\mathbf{y}.$$

We have

$$E(SSE_1) = E(\mathbf{y}'\mathbf{P}_{\mathcal{V}_1^\perp}\mathbf{y}) = \mathrm{tr}\{\mathbf{P}_{\mathcal{V}_1^\perp}\sigma^2\mathbf{I}\} + \boldsymbol{\beta}'\mathbf{X}'\mathbf{P}_{\mathcal{V}_1^\perp}\mathbf{X}\boldsymbol{\beta}$$
$$= (n-p-1)\sigma^2 + \|\, \mathbf{P}_{\mathcal{V}_1^\perp}(\mathbf{X}_1\boldsymbol{\beta}_1 + \mathbf{X}_2\boldsymbol{\beta}_2) \,\|^2$$
$$= (n-p-1)\sigma^2 + \|\, \mathbf{P}_{\mathcal{V}_1^\perp}(\mathbf{X}_2\boldsymbol{\beta}_2) \,\|^2, \tag{5.4}$$

since $\mathbf{P}_{\mathcal{V}_1^\perp}(\mathbf{X}_1\boldsymbol{\beta}_1) = \mathbf{0}$. Therefore,

$$\mathrm{E}(\hat{\sigma}^2) = \frac{\mathrm{E}(SSE_1)}{n-p-1} = \sigma^2 + \frac{\|\, \mathbf{P}_{\mathcal{V}_1^\perp}(\mathbf{X}_2'\boldsymbol{\beta}_2) \,\|^2}{n-p-1},$$

which is larger than or equal to σ^2.

(iv) First,

$$\begin{aligned} E(\hat{y}_0^\star) &= E(\mathbf{x}_{01}'\hat{\boldsymbol{\beta}}_1^\star = \mathbf{x}_{01}'(\boldsymbol{\beta}_1 + \mathbf{A}\boldsymbol{\beta}_2) \neq \mathbf{x}_{01}'\boldsymbol{\beta}_1 \\ &= \mathbf{x}_0'\boldsymbol{\beta} - (\mathbf{x}_{02} - \mathbf{A}'\mathbf{x}_{01})'\boldsymbol{\beta}_2 \neq \mathbf{x}_0'\boldsymbol{\beta}. \end{aligned}$$

Next, it can be found that

$$\mathrm{Var}(\hat{y}_0^\star) = \mathbf{x}_{01}'\mathrm{Cov}(\boldsymbol{\beta}_1^\star\mathbf{x}_{01}) = \sigma^2\mathbf{x}_{01}'(\mathbf{X}'\mathbf{X})^{-1}\mathbf{x}_{01}.$$

On the other hand,

$$\begin{aligned} \mathrm{Var}(\hat{y}_{01}) &= \mathrm{Var}(\mathbf{x}_{01}'\hat{\boldsymbol{\beta}}_1) = \mathbf{x}_{01}'\mathrm{Cov}(\hat{\boldsymbol{\beta}}_1)\mathbf{x}_{01} \\ &= \sigma^2 \cdot \mathbf{x}_{01}' \left\{(\mathbf{X}_1'\mathbf{X}_1)^{-1} + \mathbf{A}\mathbf{B}^{-1}\mathbf{A}'\right\}\mathbf{x}_{01} \\ &\geq \mathrm{Var}(\hat{y}_0^\star). \end{aligned}$$

Also,

$$\begin{aligned} \mathrm{Var}(\mathbf{x}_0'\hat{\boldsymbol{\beta}}) &= \sigma^2 \cdot \mathbf{x}_0'(\mathbf{X}'\mathbf{X})^{-1}\mathbf{x}_0 \\ &= \sigma^2 \cdot \mathbf{x}_{01}'(\mathbf{X}_1'\mathbf{X}_1)^{-1}\mathbf{x}_{01} + \sigma^2 \cdot \mathbf{b}'\mathbf{B}^{-1}\mathbf{b} \quad \text{by equation (5.3)}, \\ &\geq \sigma^2 \cdot \mathbf{x}_{01}'(\mathbf{X}_1'\mathbf{X}_1)^{-1}\mathbf{x}_{01} = \mathrm{Var}(\hat{y}_0^\star), \quad \text{as } \mathbf{B}^{-1} \text{ is p.d.,} \end{aligned}$$

where $\mathbf{b} = \mathbf{x}_{02} - \mathbf{A}'\mathbf{x}_{01}$. □

Theorem 5.1 indicates that underfitting would result in biased estimation and prediction, while leading to deflated variance. Subsequently, we consider the overfitting scenario.

Theorem 5.2. *Suppose that model* $\mathbf{y} = \boldsymbol{X}_1\boldsymbol{\beta}_1^\star + \varepsilon^\star$ *is the true underlying model, but we actually fit model* $\mathbf{y} = \boldsymbol{X}_1\boldsymbol{\beta}_1 + \boldsymbol{X}_2\boldsymbol{\beta}_2 + \varepsilon$, *an overfitted one. Then*

(i) $E(\widehat{\boldsymbol{\beta}}) = \begin{pmatrix} \boldsymbol{\beta}_1^\star \\ \mathbf{0} \end{pmatrix}$.

(ii) $Cov(\widehat{\boldsymbol{\beta}}) = \sigma_\star^2 \cdot (\mathbf{X}'\mathbf{X})^{-1}$ *and* $Cov(\widehat{\boldsymbol{\beta}}_1) = \sigma_\star^2 \cdot \left\{(\mathbf{X}_1'\mathbf{X}_1)^{-1} + \mathbf{A}\mathbf{B}^{-1}\mathbf{A}'\right\}$. *Again, compared to* $Cov(\hat{\boldsymbol{\beta}}^\star) = \sigma_\star^2 \cdot (\mathbf{X}_1'\mathbf{X}_1)^{-1}$, *we have* $Cov(\widehat{\boldsymbol{\beta}}_1) - Cov(\widehat{\boldsymbol{\beta}}_1^\star) \geq \mathbf{0}$, *i.e. positive semi-definite.*

(iii) An estimator of $\sigma_\star^2$,

$$\hat{\sigma}_\star^2 = \frac{\| \mathbf{P}_{\mathcal{V}^\perp}\mathbf{y} \|^2}{n - k - 1}$$

has the expected value of $\sigma_\star^2$.

(iv) Given $\mathbf{x}_0 = \begin{pmatrix} \mathbf{x}_{01} \\ \mathbf{x}_{02} \end{pmatrix}$, *let* $\hat{y}_0 = \mathbf{x}_0'\widehat{\boldsymbol{\beta}} = \begin{pmatrix} \hat{y}_{01} \\ \hat{y}_{02} \end{pmatrix}$ *denote the estimate based on the working model and* $\hat{y}^\star = \mathbf{x}_{01}'\widehat{\boldsymbol{\beta}}_1^\star$ *denote the estimate based on the true model. Then it can be found that* $E(\hat{y}) = \mathbf{x}_{01}'\boldsymbol{\beta}_1^\star$ *and* $Var(\hat{y}) = \sigma_\star^2 \cdot \mathbf{x}_0'(\mathbf{X}'\mathbf{X})^{-1}\mathbf{x}_0 \geq Var(\hat{y}^\star)$.

Proof. We will prove (i) and (iii) only, leaving the rest for exercise.

(i) Consider

$$\begin{aligned} E(\hat{\boldsymbol{\beta}}) &= E\{(\mathbf{X}'\mathbf{X})^{-1}\mathbf{X}'\mathbf{y}\} = (\mathbf{X}'\mathbf{X})^{-1}\mathbf{X}'\mathbf{X}_1\boldsymbol{\beta}_1^\star \\ &= (\mathbf{X}'\mathbf{X})^{-1}\mathbf{X}'\left(\mathbf{X}_1\ \mathbf{X}_1\right)\begin{pmatrix} \boldsymbol{\beta}_1^\star \\ \mathbf{0} \end{pmatrix} \\ &= (\mathbf{X}'\mathbf{X})^{-1}\mathbf{X}'\mathbf{X}\begin{pmatrix} \boldsymbol{\beta}_1^\star \\ \mathbf{0} \end{pmatrix} = \begin{pmatrix} \boldsymbol{\beta}_1^\star \\ \mathbf{0} \end{pmatrix} \end{aligned}$$

(iii) Consider the SSE associated with Model (5.1),

$$\begin{aligned} E(SSE) &= E(\mathbf{y}'\mathbf{P}_{\mathcal{V}^\perp}\mathbf{y}) = \text{tr}\{\mathbf{P}_{\mathcal{V}^\perp}\sigma_\star^2\mathbf{I}\} + \boldsymbol{\beta}_1^{\star\prime}\mathbf{X}_1'\mathbf{P}_{\mathcal{V}^\perp}\mathbf{X}_1\boldsymbol{\beta}_1^\star \\ &= (n-k-1)\sigma_\star^2, \end{aligned}$$

since $\mathbf{P}_{\mathcal{V}^\perp}(\mathbf{X}_1\boldsymbol{\beta}_1^\star) = \mathbf{0}$. Thus, $E(\hat{\sigma}_\star^2) = E\left\{SSE/(n-k-1)\right\} = \sigma_\star^2$. □

Theorem 5.2 implies that overfitting would yield unbiased estimation and prediction, however result in inflated variances.

Example (*A Simulation Study*) In some situations such as predictive data mining, prediction accuracy is the ultimate yardstick to judge the model performance. In order to see how overfitting or underfitting affects the generalization ability of a model to new observations, one may study the sum of squared prediction error (SSPE) within the similar theoretical framework as in Theorems 5.1 and 5.2. Alternatively, empirical evaluation can be made via simulation studies. Here, we illustrate the latter approach using a simple example. The data are generated from the following model

$$y = \beta_0 + \beta_1 x_1 + \cdots + \beta_{10}x_{10} + \varepsilon,$$

with $x_j \sim \text{uniform}(0,1)$ and $\varepsilon \sim N(0,1)$ independently. Each data set also includes 20 additional predictors, also from uniform(0,1), which are totally unrelated to the response. In each simulation run, we generated two data sets: a set containing 1000 observations and another independent

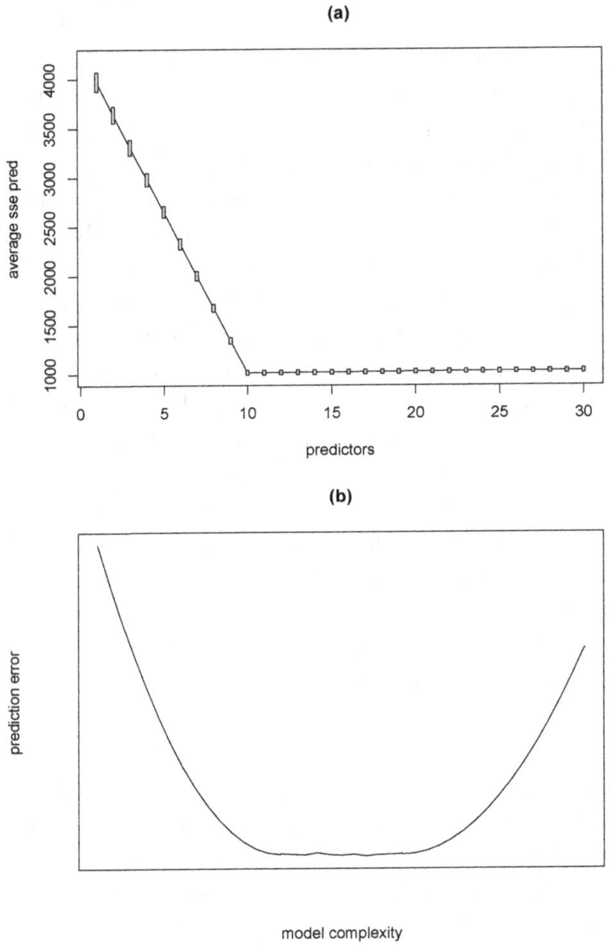

Fig. 5.1 (a) Plot of the Average and Standard Deviation of SSPE Prediction from 100 Simulation Runs; (b) A Hypothetical Plot of Prediction Error Versus Model Complexity in General Regression Problems.

set containing 1000 observations. For the training sample, we fit 30 nested models

$$y = \beta_0 + \beta_1 x_1 + \cdots + \beta_k x_k + \varepsilon,$$

where $k = 1, \ldots, 30$. Thus the first 9 models are underfitted ones while the last 20 models are overfitted ones. We then send down the test sample set to each fitted model and calculate its SSPE. The experiment is repeated for

100 simulation runs and we integrated the results by computing the average and standard deviation (sd) of SSPE values for each of the 30 models.

Figure 5.1 (a) gives a plot of the results. The line shows the average SSPE and the hight of the vertical bar at each point corresponds to the sd of the SSPE values for each model. It can be seen that both the average of SSPE reaches the minimum when the true model is fit. However, the averaged SSPE is inflated much more in the underfitting cases than in the overfitting ones. A similar observation can be drawn for the sd of SSPE. The pattern shown here about the prediction error is generally true in many regression problems. When the model complexity increases starting from the null model, the prediction error drops gradually to its bottom and then slowly increases after a flat valley, as shown in a hypothetical plot in Fig. 5.1 (b). The specific shape of the curve and length of the flat valley would vary depending on types of the regression methods being used, the signal-to-noise ratio in the data, and other factors.

In summary, both underfitting and overfitting cause concerns. Although a simplified scenario for model mis-specification has been employed to make inference, we, again, see that underfitting leads to biased estimation and overfitting leads to increased variances, which is the well-known *bias-variance tradeoff* principle. The task of variable selection is to seek an appropriate balance between this bias-variance tradeoff. □

The bias-variance tradeoff can also be viewed via the concept of mean square error (MSE). In statistics, MSE is a widely used criterion for measuring the quality of estimation. Consider a typical problem of estimating parameter θ, which may correspond to the true effect or sloe of a predictor, or prediction for a new observation. Let $\hat{\theta}$ be an estimator of θ. The MSE of $\hat{\theta}$ is defined as

$$MSE(\hat{\theta}) = E(\hat{\theta} - \theta)^2. \tag{5.5}$$

MSE provides a way, by applying the squared error loss, to quantify the amount by which an estimator differs from the true value of the parameter being estimated. Note that MSE is an expectation, a real number, not a random variable. It may be a function of the unknown parameter θ, but it does not involve any random quantity. It can be easily verified that

$$\begin{aligned} MSE(\hat{\theta}) &= E\left[\left\{\hat{\theta} - E(\hat{\theta})\right\} + \left\{E(\hat{\theta}) - \theta\right\}\right]^2 \\ &= E\left\{\hat{\theta} - E(\hat{\theta})\right\}^2 + \left\{E(\hat{\theta}) - \theta\right\}^2 \\ &= \text{Var}(\hat{\theta}) + \text{bias}^2(\hat{\theta}). \end{aligned} \tag{5.6}$$

Namely, MSE is the the sum of the variance and the squared bias of the estimator $\hat{\theta}$. An under-fitted model suffers from bias in both estimation of the regression coefficients and the regression prediction while an over-fitted model suffers from inflated variance. It is reasonable to expect that a good balance between variance and bias is achieved by models that yield small MSEs. This idea motivates the derivation of a model selection criteria called Mellow's C_p, as we shall discuss in the next section.

5.2 All Possible Regressions

In the method of *all possible regressions* (also called *all subset regressions*), one evaluates all model choices by trying out every possible combination of predictors and then compares them via a *model selection criterion*. Suppose there are k predictors in total. Excluding the null model that has the intercept term only, there are a total of $2^k - 1$ different combinations. If $k \geq 20$ for example, which is not uncommon at all in real applications, then $2^k - 1 \geq 1,048,575$ possible distinct combinations need to be considered. It is abundantly clear that this method is feasible only when the number of predictors in the data set is moderately small.

There have been many criteria proposed in the literature for evaluating the performance of competing models. These are often referred to as model selection criteria. As we have seen in section 5.1, a simple model leads to smaller variation, yet with biased or inaccurate estimation, while a complex model yields unbiased estimation or better goodness-of-fit to the current data, yet with inflated variation or imprecise estimation. As we will see, many model selection criteria are intended to balance off between bias and variation by penalizing the goodness-of-fit of the model with its complexity.

5.2.1 *Some Naive Criteria*

A natural yet naive criterion for model evaluation is the sum of square errors,

$$SSE = \sum (y_i - \hat{y}_i)^2,$$

which yields an overall distance between observed responses and their predicted values. A good model is expected to have relatively small SSE. However, it is known that for nested models, a smaller SSE is always associated with a more complex model. As a result, the full model that includes all predictors would have the smallest SSE. Therefore, SSE alone is not a good model selection criterion for direct use, because it measures goodness-of-fit only and fails to take model complexity into consideration.

Similar conclusion applies to the coefficient of determination,

$$R^2 = 1 - \frac{SSE}{SST},$$

which is a monotone increasing function of SSE, noting that the sum of square total SST stays invariant with different model choices. Nevertheless, R^2 is a very popular goodness-of-fit measure due to its easy interpretation.

With a slight modification on R^2, the adjusted R-squared is defined as

$$R_a^2 = 1 - \frac{MSE}{s_y^2},$$

where s_y^2 is the sample variance of Y. After replacing the sum of squares in R^2 by mean squares, R_a^2 is not necessarily increasing with model complexity for nested models. However, R_a^2, as well as MSE, is inadequate as a criterion for variable selection. Besides, it lacks the easy interpretability of R^2 as the proportion of the total variation in observed responses that can be accounted for by the regression model. In fact, R_a^2 turns out to be not very useful for practical purposes.

5.2.2 *PRESS and GCV*

In predictive modeling tasks, candidates models are often evaluated based on their generalization abilities. If the sample size is large, the common approach is to employ an independent test sample. To do so, one partitions the whole data $\mathcal{L}$ into two parts: the training (or learning) set $\mathcal{L}_1$ and the test set $\mathcal{L}_2$. Candidate models are then built or estimated using the training set $\mathcal{L}_1$ and validated with the test set $\mathcal{L}_2$. Models that yield small sum of squared prediction errors for $\mathcal{L}_2$,

$$SSPE = \sum_{i \in \mathcal{L}_2} (y_i - \hat{y}_i)^2, \tag{5.7}$$

are preferable. Note that, in (5.7), $\hat{y}_i$ is the predicted value based on the model fit obtained from training data, would be preferable.

When an independent test data set is unavailable due to insufficient sample size, one has to get around, either analytically or by efficient sample re-use, in order to facilitate validation. The prediction sum of squares

(PRESS) criterion is of the latter type. Related to the *jackknife* technique, PRESS is defined as

$$PRESS = \sum_{i=1}^{n} \left\{ y_i - \hat{y}_{i(-i)} \right\}^2, \tag{5.8}$$

where $\hat{y}_{i(-i)}$ is the predicted value for the ith case based on the regression model that is fitted by excluding the ith case. The jackknife or leave-one-out mechanism utilizes the independence between y_i and $\hat{y}_{i(-i)}$ and facilitates a way of mimicking the prediction errors. A small PRESS value is associated with a good model in the sense that it yields small prediction errors. To compute $PRESS$, one does not have to fit the regression model repeatedly by removing each observation at a time. It can be shown that

$$PRESS = \sum_{i=1}^{n} \left(\frac{e_i}{1 - h_{ii}} \right)^2, \tag{5.9}$$

where $e_i = y_i - \hat{y}_i$ is the ith ordinary residual and h_{ii} is the ith leverage defined as the ith diagonal element of the hat or projection matrix $\mathbf{H} = \mathbf{X}(\mathbf{X}'\mathbf{X})^{-1}\mathbf{X}'$.

PRESS is heavily used in many modern modeling procedures besides linear regression. In scenarios where explicit computation of the components h_{ii}'s is difficult, Craven and Wahba (1979) proposed to replace each of h_i's in $PRESS/n$ by their average $\mathrm{tr}(\mathbf{H})/n = (p+1)/n$, which leads to the well-known generalized cross-validation (GCV) criterion. Nevertheless, it can be easily shown that in linear regression

$$\mathrm{GCV} = \frac{1}{n} \cdot \frac{\sum_{i=1}^{n} e_i^2}{\{1 - (p+1)/n\}^2} = \frac{n \cdot SSE}{\{n - (p+1)\}^2} = \frac{n}{n - (p+1)} \cdot MSE. \tag{5.10}$$

In Section 7.3.4, the use of PRESS and GCV is further discussed in more general settings.

5.2.3 *Mallow's C_P*

Subsequently we shall introduce several criteria that are derived analytically. The first one is the C_p criterion proposed by Mellows (1973). The C_p criterion is concerned with the mean squared errors of the fitted values.

Its derivation starts with assuming that the true model that generates the data is a massive full linear model involving many predictors, perhaps even those on which we do not collect any information. Let $\mathbf{X}$, of dimension

$n \times (k+1)$, denote the design matrix associated with this full model, which can then be stated as

$$\mathbf{y} = \mathbf{X}\boldsymbol{\beta} + \boldsymbol{\varepsilon} \text{ with } \boldsymbol{\varepsilon} \sim \text{MVN}\{\mathbf{0}, \sigma^2 \cdot \mathbf{I}\}. \tag{5.11}$$

Furthermore, it is assumed that the working model is a sub-model nested into the full true model. Let $\mathbf{X}_1$, of dimension $n \times (p+1)$, denote the design matrix associated with the working model:

$$\mathbf{y} = \mathbf{X}_1\boldsymbol{\beta}_1^{\star} + \boldsymbol{\varepsilon}^{\star} \text{ with } \boldsymbol{\varepsilon}^{\star} \sim \text{MVN}\{\mathbf{0}, \sigma_{\star}^2 \cdot \mathbf{I}\}. \tag{5.12}$$

Thus, the setting employed here is exactly the same as that in Section 5.1. The working model (5.12) is assumed to underfit the data.

To evaluate a typical working model (5.12), we consider the performance of its fitted values $\hat{\mathbf{y}} = (\hat{y}_i)$ for $i = 1, \ldots, n$. Let $\mathcal{V}_1 = C(\mathbf{X}_1)$ be the column space of matrix $\mathbf{X}_1$ and $\mathbf{H}_1 = \mathbf{P}_{\mathcal{V}_1} = \mathbf{X}_1(\mathbf{X}_1^t\mathbf{X}_1)^t\mathbf{X}_1^t$ be the projection matrix of $\mathcal{V}_1$. Then the fitted mean vector from the working model (5.12) is

$$\hat{\mathbf{y}} = \mathbf{H}_1\mathbf{y}, \tag{5.13}$$

which is aimed to estimate or predict the true mean response vector $\boldsymbol{\mu} = \mathbf{X}\boldsymbol{\beta}$. Since model (5.12) provides underfitting, $\hat{\mathbf{y}}$ with $E(\hat{\mathbf{y}}) = \mathbf{H}_1\mathbf{X}\boldsymbol{\beta}$ is biased for $\mathbf{X}\boldsymbol{\beta}$. For model selection purpose, however, we seek the best possible working model that provides the smallest mean squared error (MSE), recalling that a small MSE balances the bias-variance tradeoff.

The MSE of $\hat{\mathbf{y}}$, again, can be written as the sum of variance and squared bias:

$$\begin{aligned} MSE(\hat{\mathbf{y}}) &= E\left\{(\hat{\mathbf{y}} - \mathbf{X}\boldsymbol{\beta})^t(\hat{\mathbf{y}} - \mathbf{X}\boldsymbol{\beta})\right\} = \sum_i E(\hat{y}_i - \mathbf{x}_i^t\boldsymbol{\beta})^2 \\ &= \sum_{i=1}^n \left[\{E(\hat{y}_i) - \mathbf{x}_i^t\boldsymbol{\beta}\}^2 + \text{Var}(\hat{y}_i)\right] \\ &= \sum_{i=1}^n \text{bias}^2(\hat{y}_i) + \sum_{i=1}^n \text{Var}(\hat{y}_i). \end{aligned} \tag{5.14}$$

Denote the bias part in (5.14) as B. Using the results established in Section 5.1, it follows that

$$\begin{aligned} B &= \sum_{i=1}^n \{\text{bias}(\hat{y}_i)\}^2 = \{E(\hat{\mathbf{y}}) - \mathbf{X}\boldsymbol{\beta}\}^t \{E(\hat{\mathbf{y}}) - \mathbf{X}\boldsymbol{\beta}\} \\ &= (\mathbf{H}_1\mathbf{X}\boldsymbol{\beta} - \mathbf{X}\boldsymbol{\beta})^t (\mathbf{H}_1\mathbf{X}\boldsymbol{\beta} - \mathbf{X}\boldsymbol{\beta}) = \boldsymbol{\beta}^t\mathbf{X}(\mathbf{I} - \mathbf{H}_1)\mathbf{X}\boldsymbol{\beta} \\ &= (\mathbf{X}_1\boldsymbol{\beta}_1 + \mathbf{X}_2\boldsymbol{\beta}_2)^t(\mathbf{I} - \mathbf{H}_1)(\mathbf{X}_1\boldsymbol{\beta}_1 + \mathbf{X}_2\boldsymbol{\beta}_2) \\ &= (\boldsymbol{\beta}_2^t\mathbf{X}_2)^t(\mathbf{I} - \mathbf{H}_1)(\boldsymbol{\beta}_2^t\mathbf{X}_2), \text{ since } (\mathbf{I} - \mathbf{H}_1)\mathbf{X}_1 = \mathbf{P}_{\mathcal{V}_1^{\perp}}\mathbf{X}_1 = \mathbf{0} \\ &= E(SSE_1) - (n - p - 1)\sigma^2 \text{ using (5.4)}, \end{aligned} \tag{5.15}$$

where SSE_1 is the residual sum of squares from fitting the working model (5.12). Denote the variance part in (5.14) as V, which is

$$\begin{aligned} V &= \sum_{i=1}^{n} \mathrm{Var}(\hat{y}_i) = \mathrm{trace}\,\{\mathrm{Cov}(\hat{\mathbf{y}})\} \\ &= \mathrm{trace}\,\{\mathrm{Cov}(\mathbf{H}_1\mathbf{y})\} = \mathrm{trace}\,\{\sigma^2 \cdot \mathbf{H}_1\} \\ &= (p+1) \cdot \sigma^2. \end{aligned} \tag{5.16}$$

Thus, bringing (5.15) and (5.16) into (5.14) yields that

$$MSE(\hat{\mathbf{y}}) = B + V = E(SSE_1) - \sigma^2 \cdot \{n - 2(p+1)\}. \tag{5.17}$$

The C_p criterion is aimed to provide an estimate of the MSE in (5.17) scaled by σ^2, i.e.,

$$\frac{MSE(\hat{\mathbf{y}})}{\sigma^2} = \frac{E(SSE_1)}{\sigma^2} - \{n - 2(p+1)\}.$$

Thus the C_p criterion is specified as

$$C_p = \frac{SSE_1}{MSE_{\text{full}}} - \{n - 2(p+1)\}, \tag{5.18}$$

where the true error variance σ^2 is estimated by MSE_{full}, the mean squared error from the full model that includes all predictors.

When there is no or little bias involved in the current working model so that $B \approx 0$ in (5.17), the expected value of C_p is about $V/\sigma^2 = (p+1)$. Models with substantial bias tend to yield C_p values much greater than $(p+1)$. Therefore, Mellows (1975) suggests that any model with $C_p < (p+1)$ be a candidate. It is, however, worth noting that the performance of C_p depends heavily on whether MSE_{full} provides a reliable estimate of the true error variance σ^2.

5.2.4 *AIC, AIC_C, and BIC*

Akaike (1973) derived a criterion from information theories, known as the Akaike information criterion (AIC). The AIC can be viewed as a data-based approximation for the Kullback-Leibler discrepancy function between a candidate model and the true model. It has the following general form:

$$AIC = n \times \text{log-likelihood} + 2 \times \text{number of parameters}.$$

When applied to Gaussian or normal models, it becomes, up to a constant,

$$AIC \;\simeq\; n \cdot \log(SSE) + 2\,p.$$

The first part $2 \cdot \log(SSE)$ measures the goodness-of-fit of the model, which is penalized by model complexity in the second part $2p$. The constant 2 in the penalty term is often referred to as complexity or penalty parameter. The smaller AIC results in the better candidate model.

It is noteworthy that C_p is equivalent to GCV in (5.10) in view of approximation $1/(1-x)^2 = 1 + 2x$ (see Problem 5.7), both having similar performance as AIC. All these three criteria are asymptotically equivalent.

Observing that AIC tends to overfit when the sample size is relatively small, Hurvich and Tsai (1989) proposed a bias-corrected version, called AIC_C, which is given by

$$AIC_C \simeq n \cdot \log(SSE) + \frac{n \cdot (n+p+1)}{n-p-3}.$$

Within the Bayesian framework, Schwarz (1978) developed another criterion, labeled as BIC for Bayesian information criterion (or also SIC for Schwarz information criterion and SBC for Schwarz-Bayesian criterion). The BIC, given as

$$BIC \simeq n \cdot \log(SSE) + \log(n)\, p,$$

applies a larger penalty for overfitting. The information-based criteria have received wide popularity in statistical applications mainly because of their easy extension to other regression models. There are many other information-based criteria introduced in the literature.

In large samples, a model selection criterion is said to be asymptotically *efficient* if it is aimed to select the model with minimum mean squared error, and *consistent* if it selects the true model with probability one. No criterion is both consistent and asymptotically efficient. According to this categorization, MSE, R_a^2, GCV, C_p, AIC, and AIC_C are all asymptotically efficient criteria while BIC is a consistent one. Among many other factors, the performance of these criteria depends on the available sample size and the signal-to-noise ratio. Based on the extensive simulation studies, McQuarrie and Tsai (1998) provided some general advice on the use of various model selection criteria, indicating that AIC and C_p work best for moderately-sized sample, AIC_C provides the most effective selection with small samples, while BIC is most suitable for large samples with relatively strong signals.

5.3 Stepwise Selection

As mentioned earlier, the method of all possible regressions is infeasible when the number of predictors is large. A common alternative in this case is to apply a stepwise algorithm. There are three types of stepwise procedures available: backward elimination, forward addition, and stepwise search. In these algorithmic approaches, variables are added into or deleted from the model in an iterative manner, one at a time. Another important feature of stepwise selection is that competitive models evaluated in a particular step have the same complexity level. For this reason, measures such as SSE or R^2 can be directly used for model comparisons with equivalence to other model selection criteria such as AIC. In order to determine a stopping rule for the algorithm, a test statistic is often employment instead, which, nevertheless, inevitably raises the multiplicity concern. A natural choice, as implemented in SAS, is the F statistic that evaluates the reduction in Type III sum of squares (SS) or the main effect of each individual predictor. In the R implementation `stepAIC()`, AIC is used instead, in which case the procedure stops when further modification (i.e., adding or deleting a selected variable) on the model no longer decreases AIC.

5.3.1 *Backward Elimination*

The backward elimination procedure proceeds as follows. Start with fitting the whole model that includes all k predictors. For each predictor X_j, compute the F test statistic that compares the whole model with the reduced model that excludes X_j. Identify the least significant predictor $X^{\star}$ that corresponds to the largest p-value associated with the F test. If this largest p-value is smaller than a threshold significance level α_{stay}, the procedure is stopped and this whole model is claimed as the final model. Otherwise, we remove $X^{\star}$ from the current model and fit the model that contains the remaining $k-1$ predictors. Next, the least significant predictor is identified and may be removed by examining the F test statistics and their p-values. This procedure is repeated till all p-values in the model are less than α_{stay}. The resultant model is then claimed as the final model.

The final model identified by backward elimination is sensitive to the threshold α_{stay}. Common choices for α_{stay} are 0.10 and 0.05. The default value in SAS is 0.10. Backward elimination is the most computationally efficient one among the three stepwise procedures. It takes at most k steps or model fittings to arrive at a final model. This property renders it a popular variable selection or screening method in situations where the data set contains a large number of predictors. Nevertheless, there is one problem with backward elimination. That is, a dropped variable would have no more chance to re-enter the model. However, a variable that has been excluded in an earlier stage may become significant after dropping other predictors.

5.3.2 *Forward Addition*

Forward addition is sort of like a reversed procedure of backward elimination. Starting with the null model that has the intercept only, the procedure fits all k simple linear regression models, each with one predictor included only. Again, the F test statistic that compares each simple linear model with the null is computed, as well as its corresponding p-value. Let $X^\star$ denote the most significant predictor, the one associated with the smallest p-value. This very first step amounts to find the predictor that has the highest correlation with Y. If this smallest p-value is smaller than a threshold significance level α_{entry}, the procedure is stopped and the null model is claimed as the final model. Otherwise, $X^\star$ is added into the null model. Subsequently, with $X^\star$ included, one identifies the next most significant variable to add. This procedure is repeated till no more variable is eligible to add.

The default value for α_{entry} in SAS is 0.05. Computationally, forward addition is slower than backward deletion. It involves at most $k+(k-1)+\cdots+1 = k(k+1)/2$ model fittings to finally stop. The problem associated with forward addition is analogous to the one with backward elimination. Once added, a variable would always stay in the final model. However, a variable that was added in an earlier stage may become insignificant after including other predictors into the model. As a matter of fact, forward addition is not very often used in applications.

5.3.3 *Stepwise Search*

The stepwise search method is intended to avoid the problems with both backward elimination and forward addition so that variables already in the model may be removed due to insignificance and variables excluded may be added later on when it becomes significant. The procedure itself is more similar to the forward addition algorithm. As in forward addition, the most significant variable is added to the model at each step, if its corresponding

F test is significant at the level of α_{entry}. Before the next variable is added, however, the stepwise search method takes an additional look-back step to check all variables included in the current model and deletes any variable that has a p-value greater than α_{stay}. Only after the necessary deletions are accomplished can the procedure move to the next step of adding another variable into the model. The stepwise search continues till every variable in the model is significant at the α_{stay} level and every variables not in the model is insignificant at the α_{entry} level if added.

Among all three stepwise selection procedures, the stepwise search algorithm performs best, although computationally the backward elimination algorithm is the fastest one.

One should be very careful with these automatic selection procedures. Since a largest number of F tests have been conducted, there is a very high probability of making Type I error (including unimportant predictors) and Type II error (excluding important predictors). As demonstrated via simulation in Cook and Weisberg (1999), the algorithmic stepwise model selection can considerably overstate the significance of predictors. The predictors left in the final model may appear much more important than they really are. Furthermore, there are arguments indicating that, same as other data-adaptive algorithms, results from stepwise procedures are sample-specific and unlikely to replicate. See, e.g., Thompson (2001). This is because a small change or variation that is specific to one sample may give a predictor an advantage over another that it would not have been seen in other samples, making stepwise selection results unlikely to generalize. Besides, in order to arrive at a successful model fit, one needs consider transformations such as high-order terms and interactions among predictors. Thus it is sometimes advised to use stepwise procedures for variable screening purposes only, instead of for final model determination. Furthermore, even if a 'best' model is obtained, there are still model diagnostic issues to follow. Diagnostic results may suggest further modifications of the model. Thus it is also advised to conduct the model building process in an iterative manner by alternating between model selection and model diagnostics.

5.4 Examples

To illustrate, we first consider an astronomical data set taken from Ex. 4.9 in Mendenhall and Sinich (2003).

Table 5.1 The Quasar Data from Ex. 4.9 in Mendenhall and Sinich (2003)

Quasar	Redshift (X_1)	Line Flux (X_2)	Line Luminosity (X_3)	AB_{1450} (X_4)	Absolute Magnitude (X_5)	Rest Frame Equivalent Width (Y)
1	2.81	-13.48	45.29	19.5	-26.27	117
2	3.07	-13.73	45.13	19.65	-26.26	82
3	3.45	-13.87	45.11	18.93	-27.17	33
4	3.19	-13.27	45.63	18.59	-27.39	92
5	3.07	-13.56	45.3	19.59	-26.32	114
6	4.15	-13.95	45.2	19.42	-26.97	50
7	3.26	-13.83	45.08	19.18	-26.83	43
8	2.81	-13.5	45.27	20.41	-25.36	259
9	3.83	-13.66	45.41	18.93	-27.34	58
10	3.32	-13.71	45.23	20	-26.04	126
11	2.81	-13.5	45.27	18.45	-27.32	42
12	4.4	-13.96	45.25	20.55	-25.94	146
13	3.45	-13.91	45.07	20.45	-25.65	124
14	3.7	-13.85	45.19	19.7	-26.51	75
15	3.07	-13.67	45.19	19.54	-26.37	85
16	4.34	-13.93	45.27	20.17	-26.29	109
17	3	-13.75	45.08	19.3	-26.58	55
18	3.88	-14.17	44.92	20.68	-25.61	91
19	3.07	-13.92	44.94	20.51	-25.41	116
20	4.08	-14.28	44.86	20.7	-25.67	75
21	3.62	-13.82	45.2	19.45	-26.73	63
22	3.07	-14.08	44.78	19.9	-26.02	46
23	2.94	-13.82	44.99	19.49	-26.35	55
24	3.2	-14.15	44.75	20.89	-25.09	99
25	3.24	-13.74	45.17	19.17	-26.83	53

[a] *Source*: Schmidt, M., Schneider, D. P., and Gunn, J. E. (1995) Spectroscopic CCD surveys for quasars at large redshift. *The Astronomical Journal*, **110**, No. 1, p. 70 (Table 1).

Example 5.1 A *quasar* is a distant celestial object that is at least four billion light-years away from earth. The *Astronomical Journal* (Schmidt, M., Schneider, D. P., and Gunn, J. E., 1995) reported a study of 90 quasars detected by a deep space survey. Based on the radiations provided by each quasar, astronomers were able to measure several of its quantitative characteristics, including redshift range (X_1), line flux in erg/cm^2 (X_2), line lunminosity in erg/s (X_3), AB_{1450} magnitude (X_4), absolute magnitude (X_5), and rest frame equivalent width (Y). One objective of the study is to model the rest frame equivalent width (Y) using other characteristics. The data for a sample of 25 large quasars, as given in Table 5.1, are used in this example.

```
data quasar;
input obs x1 x2 x3 x4 x5 y;
y = log(y);
cards;
1 2.81 -13.48 45.29 19.50 -26.27 117
2 3.07 -13.73 45.13 19.65 -26.26  82
......
25 3.24 -13.74 45.17 19.17 -26.83  53
;
```

Table 5.2 All Possible Regression Selection for the Quasar Data.

# of Variables in Model	R^2	Adjusted R^2	C_p	AIC	BIC	Variables in the Model
1	0.4383	0.4139	63741.50	−48.4771	−52.4740	X_5
	0.3938	0.3674	68798.74	−46.5690	−50.5661	X_4
	0.0317	−.0104	109901.7	−34.8618	−38.8600	X_3
	0.0219	−.0206	111008.5	−34.6114	−38.6096	X_2
	0.0001	−.0433	113483.1	−34.0603	−38.0585	X_1
2	0.9997	0.9997	16.6433	−233.7111	−233.6811	X_3 X_5
	0.9637	0.9604	4100.948	−114.9603	−120.8996	X_2 X_4
	0.8468	0.8328	17377.12	−78.9501	−84.9357	X_3 X_4
	0.7560	0.7339	27674.90	−67.3260	−73.3169	X_2 X_5
	0.4387	0.3876	63703.35	−46.4929	−52.4890	X_1 X_5
	0.4383	0.3873	63743.21	−46.4772	−52.4733	X_4 X_5
	0.4308	0.3790	64598.78	−46.1440	−52.1402	X_1 X_4
	0.0342	−.0536	109621.4	−32.9261	−38.9239	X_1 X_2
	0.0317	−.0563	109897.7	−32.8632	−38.8609	X_1 X_3
	0.0317	−.0563	109902.7	−32.8621	−38.8598	X_2 X_3
3	0.9998	0.9998	2.4574	−246.8445	−242.7279	X_2 X_3 X_4
	0.9998	0.9997	8.2375	−240.3419	−238.4174	X_2 X_4 X_5
	0.9997	0.9996	18.2860	−231.9630	−232.4650	X_2 X_3 X_5
	0.9997	0.9996	18.3879	−231.8909	−232.4116	X_1 X_3 X_5
	0.9997	0.9996	18.4107	−231.8748	−232.3996	X_3 X_4 X_5
	0.9995	0.9994	41.6504	−219.2603	−222.5086	X_1 X_2 X_4
	0.9990	0.9988	101.2419	−201.7318	−207.2841	X_1 X_3 X_4
	0.9978	0.9975	232.8692	−183.0267	−189.8461	X_1 X_2 X_5
	0.4993	0.4278	56821.44	−47.3509	−55.3457	X_1 X_4 X_5
	0.0507	−.0849	107747.7	−31.3575	−39.3547	X_1 X_2 X_3
4	0.9998	0.9998	4.0256	−245.4055	−240.4626	X_1 X_2 X_3 X_4
	0.9998	0.9998	4.4329	−244.8759	−240.1753	X_2 X_3 X_4 X_5
	0.9998	0.9997	9.5825	−238.9992	−236.8300	X_1 X_2 X_4 X_5
	0.9997	0.9996	18.7720	−231.0593	−231.7917	X_1 X_2 X_3 X_5
	0.9997	0.9996	20.2756	−229.9703	−231.0530	X_1 X_3 X_4 X_5
5	0.9998	0.9998	6.0000	−243.4392	−237.8491	X_1 X_2 X_3 X_4 X_5

To proceed, a logarithm transformation is first applied to the response

(Y). In RPOC REG, the SELECTION= option in the MODEL statement is designed for variable selection purposes. The following program shows how to do the all possible regressions selection with the quasar data. Note that, with SELECTION=RSQUARE, all the possible models are sorted in order of model complexity. The option BEST=m outputs the best m model choices only. Table 5.2 presents the SAS output for all possible regression selection. Since the SAS output are self-explanatory and easy to follow, we shall explain all the results very briefly. It can be seen that all three criteria C_p, AIC, and BIC yield the same selection: $\{X_2, X_3, X_4\}$.

```
proc reg data=quasar;
ods output SubsetSelSummary = dat;
ods select SubsetSelSummary;
model y = x1-x5/ selection=RSQUARE CP AIC BIC ADJRSQ;
model y = x1-x5/ selection=CP RSQUARE AIC BIC ADJRSQ BEST=1;
run;

ods rtf file="C:\...\SAS-5-1.rtf" bodytitle startpage=no
        keepn notoc_data;
ods ps FILE="C:\...\SAS-5-1.EPS";
title "ALL POSSIBLE REGRESSIONS SELECTION FOR QUASAR DATA";
title2 "The Last Line is the best model selected by Cp.";
data dat; set dat;
drop Model Dependent Control;
proc print data=dat;
run;
ods _all_ close;
```

The SELECTION=STEPWISE option requests the stepwise search algorithm. Other choices would be BACKWARD and FORWARD, as shown in the next program. The threshold significance levels are set by two parameters, SLENTRY and SLSTAY.

```
ods rtf file="C:\...\SAS-5-2.rtf";
proc reg data=quasar;
ods select SelParmEst SelectionSummary RemovalStatistics
           EntryStatistics;
model y = x1-x5 / selection=backward slstay=0.05 details;
model y = x1-x5 / selection=forward slentry=0.05 details=steps;
model y = x1-x5 / selection=stepwise slentry=0.05 slstay=0.05
                  details=summary;
run;
ods rtf close;
```

The DETAILS and DETAILS= options help control the level of details produced in these stepwise procedures. Table 5.3 displays the SAS output for backward elimination. Penal (a) shows the detailed steps. Starting with the full model that included all five covariates, the procedure seeks the one showing least significance with the F test. One column lists the hierarchical Type II sum of squares in SAS, which is reduction in SSE due to adding a term or variable to the model that contains all terms other than those containing that variable (e.g., a cross-product interaction). Since there is no interaction or polynomial term considered in the model, Type II SSE is essentially the same as the Type III SSE, which is the reduction in error SS due to adding the term after all other terms have been added to the model. One is referred to *SAS/STAT User's Guide* (SAS Institute Inc., 2004) for more details on different types of sum of squares in analysis of variance. As a result, X_5, with a p-value of 0.8746 greater than SLSTAY=0.05, is dropped. Next, the resultant model with $X_1 - X_4$ is checked and X_1 is dropped. The procedure stops at the model containing $\{X_2, X_3, X_4\}$ as none of them has a p-value greater than SLSTAY. Panel (b) summarizes all the elimination steps taken in the procedure. The partial R^2 presented in the summary is calculated as the ratio of the Type II SSE versus total variation.

Table 5.4 presents the results from forward addition. It starts with adding X_5 which has the highest correlation with the response, followed by adding X_3. At this step, no other covariate makes it to the list because none of their associated p-values is lower than SLENTRY. The stepwise selection procedure results in the same results as those provided by forward addition in this example. We have omitted the presentation. Basically, it performs an additional check, e.g., to see whether X_5 becomes insignificant after adding X_3 in Step 2.

It is interesting to notice that these selection methods did not pick up the same model even with this simple example, which somehow demonstrates the inherent instability and great variability in general model selection. This motivates the Bayesian model averaging (BMA) method in Section 9.2. □

Another noteworthy issue is that PROC REG cannot handle categorical variables directly. One needs to define dummy variables manually. For each categorical predictor, we want to evaluate all its related dummy variables together with the F test. In this case, a stepwise regression method that works on sets of variables would be useful, as illustrated by the following example. With this approach, the dummy variables defined for the same categorical predictor are either all included into or completely excluded from the model.

Example 5.2 We consider another data set used in Ex. 6.4 from Mendenhall and Sinich (2003). In any production process where one or more workers are engaged in a variety of tasks, the total time spent in production varies as a function of the size of the work pool and the level of output of the various activities. The data in Table 5.5 were collected from a large metropolitan department store, in which the number of hours worked (Y) per day by the clerical staff may depend on the following variables: number of pieces of mail processed (X_1), number of money orders and gift certificates sold (X_2), number of window payments transacted (X_3), number of change order transactions processed (X_4), number of checks cashed (X_5), number of pieces of miscellaneous mail processed on an "as available" basis (X_6), number of bus tickets sold (X_7) and the day of work (X_8). Here, X_8 is categorical taking 6 values $\{M, T, W, Th, F, S\}$. To account for it, five 0-1 binary dummy variables $\{Zt, Zw, Zth, Zf, Zs\}$ are defined with the reference cell coding scheme, leaving the level of 'M' (Monday) as baseline.

```
data clerical;
input weekday $ y x1-X7;
Zt =0; Zw=0; Zth=0; Zf=0; Zs=0;
if weekday="T" then Zt= 1;
if weekday="W" then Zw = 1;
if weekday="Th" then Zth = 1;
if weekday="F" then Zf = 1;
if weekday="S" then Zs = 1;
datalines;
M  128.5  7781  100  886  235  644  56   737
T  113.6  7004  110  962  388  589  57  1029
......
Th  86.6  6847   14  810  230  547  40   614
;

title "ALL POSSIBLE REGRESSIONS SELECTION FOR CLERICAL DATA";
ods rtf file="C:\...\SAS-5-2.rtf";
proc reg data=clerical;
ods select SelParmEst SelectionSummary RemovalStatistics
      EntryStatistics;
model y = x1-x7 {Zt Zw Zth Zf Zs}/
      selection=backward slstay=0.05 details;
model Y = x1-x7 {Zt Zw Zth Zf Zs}/
```

```
      selection=forward slentry=0.05 details=steps;
model Y = x1-x7 {Zt Zw Zth Zf Zs}/selection=stepwise
      slentry=0.05 slstay=0.05 details=summary;
run;
ods rtf close;
```

The above SAS program shows how to perform stepwise selection while keeping the five dummy variables processed in a groupwise manner. Note that this groupwise method, unfortunately, is not available for the all possible regressions selection method.

Table 5.6 presents the results from backward elimination. The variables dropped from the model are X_7, X_1, X_2, X_3, and X_4 in the order of elimination. In every step, the dummy variables $\{Z_t, Z_w, Z_{th}, Z_f, Z_s\}$ are evaluated altogether with an F test. The final model in penal (b) selects X_5, X_6, and the categorical predictor. □

Another alternative SAS procedure for linear regression is PROC GLM. The CLASS statement in PROC GLM automatically define dummy variables for categorical predictors. However, it does not have the facility for automatic model selection. One has to get around by writing macros or applying a number of MODEL statements interactively.

5.5 Other Related Issues

Model selection is crucial and fundamental in statistical analysis. Whenever a statistical model, simple or complicated, is attempted, one has to face the issue of model identification. Its general scope, much broader than selecting the best subset of variables, ranges from variable screening, throughout bandwidth selection in kernel smoothing, tree size selection in decision trees, weight decay in neural networks and etc., to model ensemble methods such as boosting. On the other hand, model selection is tricky and sometimes controversial. It is common to see different selection methods result in quite different model choices.

In the analysis of large data sets, *variable screening* is an important preliminary step. Very often it has been overlooked from classical statistical textbooks. But it is critical for practical data analysis. Mainly motivated by fear of information loss and desire to optimize the profit of the experiment, investigators always try to collect as much information as they could in a study. The resultant data set is often of enormous size, both in number of observations and number of variables, so that sometimes a simple computation of sample means may take seconds or minutes. This is not entirely unrealistic when, for example, it comes to data mining practices. In these scenarios, applying a modeling package directly could be overwhelmingly expensive in computation. Variable screening helps identify and remove predictors that are not of much predictive values to the response.

Nevertheless, one should, in actuality, be cautious about dropping variables. This is because data collection on these variables has been costly. Moreover, the relationships with the response can be of complicated forms and hard to decipher. Thus we suggest to try out a number of different variable screening techniques in order to facilitate a comprehensive evaluation on each predictor. Only these making it to the top drop list for all or most screening methods can then be considered for exclusion.

In this section, we discuss several commonly used variable screening methods. There have also been quite a few recent advances in model selection. Among others, two fruitful developments are particularly remarkable: least absolute shrinkage and selection operator (LASSO) and Bayesian model averaging (BMA). We shall postpone their coverage to later chapters. The LASSO technique, covered in Chapter 7, makes available a continuous variable selecting process. The BMA, discussed in Section 9.2, is intended to integrate the uncertainties revolving around all candidate models.

5.5.1 *Variance Importance or Relevance*

The correlation of coefficient between each predictor and the response provides a numerical summary of how they are linearly related. Variables with a very small absolute correlation can potentially be screened out into the removal list. One might also compute correlations or the variance inflation factors (VIF) among predictors to identify those redundant ones. Recall that the VIF technique helps find out variables that can be mostly accounted for by linear combinations of other variables.

The concept of *variable importance or relevance* initially emerges from the machine learning literatures. It ranks all variables in order of their importance or relevance by fully examining the predictive power furnished by each variable. The variable importance ranking feature provided in random forests (RF) (Breiman, 2001), which is among the newest and most promising developments in this regards, has been increasingly applied as a tool for variable selection in various fields. The method takes a cost-of-exclusion approach, in which the relevance of a variable is assessed by

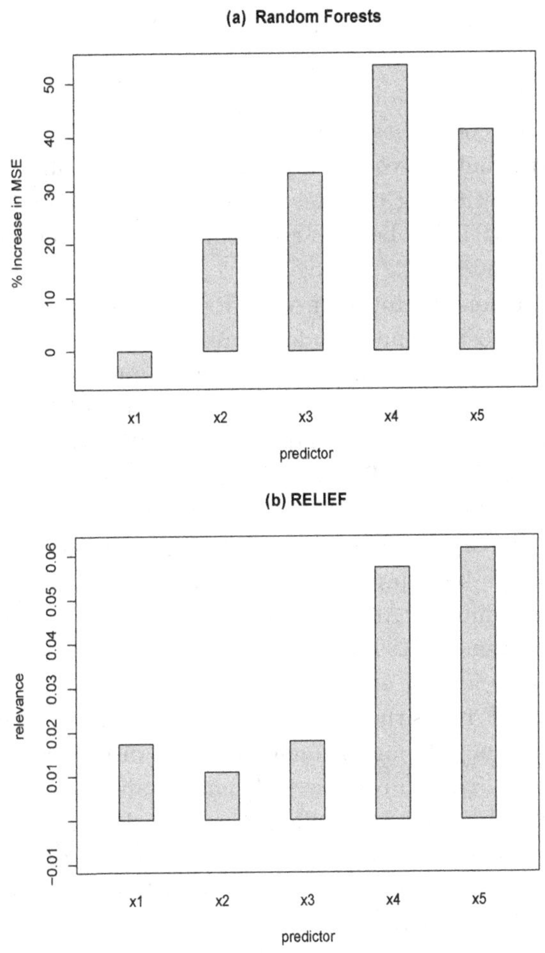

Fig. 5.2 Variable Importance for the Quasar Data: Random Forests vs. RELIEF.

comparing some model performance measure with and without the given feature included in a modeling process. To approximate the underlying regression function, random forests averages piecewise constants furnished by a number of tree models.

The tree method, also called recursive partitioning, was first proposed by Morgan and Sonquist (1963). By recursively bisecting the predictor space, the hierarchical tree structure partitions the data into meaningful groups and makes available a piecewise approximation of the underlying relationship between the response and its associated predictors. The applications of tree models have been greatly advanced in various fields especially since the development of classification and regression trees (CART) by Breiman et al. (1984). Tree construction in CART consists of three major steps: (1) growing a large initial tree; (2) a pruning algorithm; and (3) a validation method for determining the best tree size. The pruning idea for tree size selection in CART has become and remains the current standard in constructing tree models.

Recursive partitioning, together with its various extensions such as MARS, bagging, boosting, and random forests, has come into play very actively in modern statistical modeling and data mining practice. A detailed coverage of this techniques is beyond the scope. We refer interested readers to Breiman et al. (1984) and Breiman (2001) for a full account of trees and random forests. The variable importance ranking in random forests (RF) is available from the R package `randomForest`.

With the advent of modern computing facilities, various feature selection algorithms have been proposed and studied in the field of machine learning and data mining. Among others, the RELIEF algorithm developed by Kira and Rendel (1992), Kononenko, Simec, and Robnik-Sikonja (1997), and Robnik-Sikonja and Kononenko (1997) is particularly worth mentioning. RELIEF ranks the relevance of each feature or variable by accumulatively collecting information on its ability to distinguish responses of different values and agree on responses of similar values or patterns across sets of neighborhoods. Here, we shall only outline the RELIEF algorithm initially proposed for classification problems where the response variable Y is binary. A handy way of how to apply it to data with continuous responses is described in the example.

Suppose that the data set available consists of $\{(y_i, \mathbf{x}_i) : i = 1, \ldots, n\}$, where y_i is the binary outcome with value being 0 or 1 and $\mathbf{x}_i \in \mathcal{R}^p$ is a p-dimensional predictor vector. The RELIEF algorithm is illustrated in Algorithm 5.1. At each step of an iterative process, an instance is chosen at random from the data set and the importance measure for each variable is updated according to how well it recognizes distanced and similar response values.

Algorithm 5.1: RELIEF for Data with Binary Responses.

- Initialize all importance measures $W_j = W(X_j) = 0$ for $j = 1, 2, \ldots, p$;
- Do $k = 1, 2, \ldots, K$
  - Randomly select an observation $\{k : (y_k, \mathbf{x}_k)\}$, call it "Obs-$k$";
  - Find the observation m, called 'the nearest miss' , which is closest, in terms of the distance in $\mathbf{x}$, to "Obs-k" and has response y_m equal to y_k;
  - Find the observation h, called 'the nearest hit' , which is closest, in terms of the distance in $\mathbf{x}$, to "Obs-k" and has response y_h unequal to y_k;
  - Do $j = 1, 2, \ldots, p$,

    Update $W_j \longleftarrow W_j - \text{diff}(x_{kj}, x_{mj})/m + \text{diff}(x_{kj}, x_{hj})/m$;
  - End do;
- End do.

In the outlined algorithm, function $\text{diff}(x_{kj}, x_{k'j})$ computes the difference between x_{kj} and $x_{k'j}$, the values of predictor X_j for observations k and k'. While various distance definitions can apply to measure the difference, it is defined in the original proposal, for categorical predictors, as

$$\text{diff}(x_{kj}, x_{k'j}) = \begin{cases} 0 & \text{if } x_{kj} = x_{k'j} \\ 1 & \text{otherwise} \end{cases}$$

and for continuous predictors as

$$\text{diff}(X_{kj}, X_{k'j}) = \frac{|x_{kj} - x_{k'j}|}{\max(X_j) - \min(X_j)},$$

where $\max(X_j)$ and $\min(X_j)$ are the maximum and minimum of X_j, respectively. With this formulation, it is no need to normalize or standardize variables that are measured in different scales. An implementation of RELIEF is available in R package: `dprep`.

Example We use the quasar data again to illustrate the two variable importance ranking techniques. Figure 5.2 (a) plots the importance, measured by percentage of increase in MSE, for each predictor from a 2,000 random trees. Although it has been extended to continuous responses, the initial rationale in RELIEF works best in classification problems with binary responses. One simple alternative way is to dichotomize the continuous response y into two classes by some randomly selected cutoff point, then proceed with RELIEF as if it is a classification problem. Repeat the procedure for multiple times, each time with different cutoff points, and finally integrate the results by averaging the importance measures.

Figure 5.2 (b) plots the averaged relevance measure of each variable obtained from RELIEF in the manner just described. Since the sample size $n = 25$ is small, two classes are used with 500 repetitions and, at each repetition, five randomly selected observations are used for computing the relevance. It can be seen that the two methods match somehow in their findings, both ranking X_4 and X_5 higher than others, except for that RF found X_4 as the most important while X_5 is deemed most important by RELIEF. □

While the variable importance techniques in RF packages are becoming very popular due to their novelty and nonparametric nature, we would like to call the reader's attention to a selection bias problem brought up by, e.g., Strobl et al. (2007). It has been found that continuous predictors and categorical predictors are treated differently. In particular, a variable that has more distinct values or levels tends to be more weighted than a variable that has much fewer distinct values or levels. This is an inherent problem in recursive partitioning, as a predictor with more distinct values generates a much larger number of permissible splits. To put it in another way, if one would like to apply some kind of critical values to signal the statistical significance for the importance measures, the critical value can be quite different for different predictors. Thus great caution should be exercised when comparing the importance measures cross predictors.

To illustrate, we consider a simple simulation study on classification. In this study, each data set contains four covariates (X_1, X_2, X_3, X_4), none of which is really related to a binary response Y. Here, X_1 is binary taking values 0 and 1; X_2 is categorical with 10 levels; X_3 is continuous with values being randomly taken from $\{0.0, 0.1, \ldots, 1.0\}$; X_4 is continuous with values in $\{0.00, 0.01, \ldots, 1.00\}$; and the response Y is generated independently as $(+1, -1)$ with $\text{pr}(Y = +1) = \text{pr}(Y = -1) = 0.5$. Two sample sizes are considered: $n = 100$ and $n = 1000$. For each generated data sets, both RELIEF and RF are applied to extract variable importance.

Figure 5.3 gives boxplots of the resultant importance measures obtained in 200 simulation runs. For random forests, the Gini importance, which describes the improvement in the "Gini gain" splitting criterion, is used. We can see that X_2 is systematically weighted more than X_1 merely due to more levels it has. Similar behaviors can be observed for the comparison between X_4 over X_3. This selection bias problem remains even with increased sample size. On the other hand, RELIEF does not seem to suffer much from this

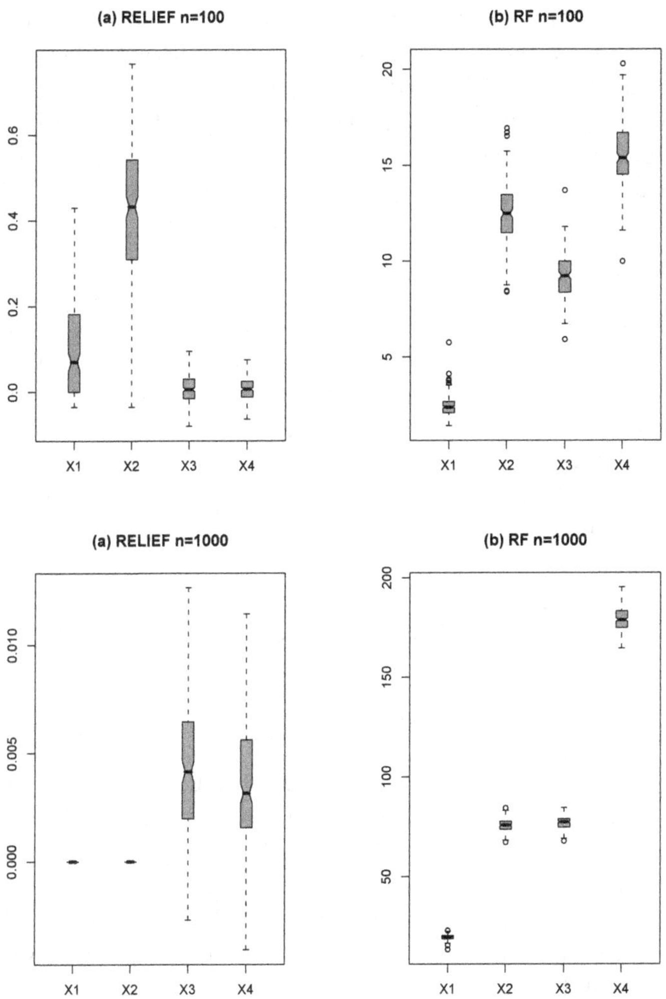

Fig. 5.3 A Simulation Example on Variable Importance: Random Forests vs. RELIEF.

bias problem. The relevance measures it provided stay around 0 for all four predictors, although variations of the measures for categorical and continuous variables are inherently different.

To amend the selection bias in random forests, Strobl et al. (2007), following the series work of Loh (2002) in tree construction, suggests first selecting the most important variable at every split and then partitioning data according to the most important variable selected. Furthermore, they found that subsampling without replacement, instead of bootstrapping, helps a great deal in producing unbiased variable importance measures. This is because subsampling without replacement tends to yield less correlated importance measures. Alternatively, one might consider obtaining a critical value for each variable to refer to by efficient use of permutation.

5.5.2 *PCA and SIR*

Dimension reduction techniques can be useful for variable selection in some circumstances. The classical principal component analysis (PCA) finds directions or linear combinations of predictors, $\mathbf{a}^t\mathbf{x}$, that explain most of the variations in predictors. Note that PCA is executed with no account of the response. When applying PCA in regression problems, one is hoping that the response also varies most in the directions that predictors vary most. A slightly more elaborated explanation on PCA can be found in Section 7.3.2.

Another technique, the sliced inverse regression (SIR), addresses dimension reduction in regression directly. SIR assumes a multiple index model given as

$$y_i = f(\mathbf{a}_1^t\mathbf{x}_i, \ldots, \mathbf{a}_K^t\mathbf{x}_i, \varepsilon_i). \tag{5.19}$$

Namely, y depends on $\mathbf{x}$ only through a set of indices. The dimension reduction is achieved when K is much smaller than p, the dimension of $\mathbf{x}$. The space spanned by $\mathbf{a}_k$'s, $k = 1, \ldots, K$ is called *effective dimension reduction* (EDR) space. Under fairly general conditions specified by Li (1991), it can be shown that the standardized inverse regression curve $E(\Sigma^{-1/2}\mathbf{x}|y)$ is contained in the subspace spanned by $\Sigma^{1/2}\mathbf{a}_k$'s, where $\Sigma = \text{Cov}(\mathbf{x})$. Apply PCA or spectral decomposition on matrix

$$\Sigma^{-1/2}\text{Cov}\{E(\mathbf{x}|y)\}\,\Sigma^{-1/2} \tag{5.20}$$

and let $\mathbf{b}_k$, for $k = 1\ldots, K$, denotes the nonzero eigenvalues. The $\mathbf{b}_k$'s are the EDR directions spanning the EDR space. To compute $\mathbf{b}_k$, Σ is replaced by the sample variance-covariance matrix of $\mathbf{x}$ and $\text{Cov}\{E(\mathbf{x}|y)\}$ can be estimated by slicing the data according to y values. Alternatively, Li (1992) suggested to replace $\text{Cov}\{E(\mathbf{x}|y)\}$ in (5.20) by

$$E\left[\{\mathbf{z} - E(\mathbf{z})\}\{\mathbf{x} - E(\mathbf{x})\}\{\mathbf{x} - E(\mathbf{x})\}^t\right],$$

where the random vector $\mathbf{z}$ can be either the response vector $\mathbf{y}$ or the residual vector obtained from a linear regression of y on $\mathbf{x}$.

One remaining practical question is how the identified PCA or SIR directions can be used for regression or prediction. Plugging the PCA or SIR directions directly into linear models is not a very appealing approach due to the blurred interpretation. Two better model choices are projection pursuit regression (see, e.g., Friedman and Stuetzle, 1981) and neural networks (see, e.g., Bishop, 1995) when, in particular, the multi-layer perceptron (MLP) with one hidden layer is used. Both facilitate a *ridge approximation* to the underlying regression function through models of the form

$$y_i = \sum_{k=1}^{K} g_k(\mathbf{a}_k^t \mathbf{x}_i) + \varepsilon_i. \tag{5.21}$$

A *ridge function* is referred to a function of some linear combination of multivariate components. While projection pursuit achieves nonparametric ridge approximation with unspecified $g_k(\cdot)$'s, the neural networks model can be viewed as a parametric version assuming that the functions $g_k(\cdot)$ have a known form. The directions identified by PCA or SIR fit naturally into these two modeling processes.

Problems

1. Verify the following properties of partitioned matrices. Given a symmetric p.d. matrix $\mathbf{A}$ is partitioned in the form
$$\mathbf{A} = \begin{pmatrix} \mathbf{A}_{11} & \mathbf{A}_{12} \\ \mathbf{A}_{21} & \mathbf{A}_{22} \end{pmatrix},$$
where $\mathbf{A}_{11}$ and $\mathbf{A}_{22}$ are square matrices, both $\mathbf{A}_{11}$ and $\mathbf{A}_{22}$ are p.d. and the inverse of $\mathbf{A}$ is given by
$$\mathbf{A}^{-1} = \begin{pmatrix} \mathbf{A}_{11} + \mathbf{A}_{11}^{-1}\mathbf{A}_{12}\mathbf{B}^{-1}\mathbf{A}_{21}\mathbf{A}_{11}^{-1} & -\mathbf{A}_{11}^{-1}\mathbf{A}_{12}\mathbf{B}^{-1} \\ -\mathbf{B}\mathbf{A}_{21}\mathbf{A}_{11}^{-1} & \mathbf{B}^{-1} \end{pmatrix},$$
with $\mathbf{B} = \mathbf{A}_{22} - \mathbf{A}_{21}\mathbf{A}_{11}^{-1}\mathbf{A}_{12}$. Besides, by straightforward algebra, one can verify that
$$(\mathbf{x}_1' \ \mathbf{x}_2')\ \mathbf{A}^{-1} \begin{pmatrix} \mathbf{x}_1 \\ \mathbf{x}_2 \end{pmatrix} = \mathbf{x}_1'\mathbf{A}_{11}^{-1}\mathbf{x}_1 + \mathbf{b}'\mathbf{B}^{-1}\mathbf{b},$$
where $\mathbf{b} = \mathbf{x}_2 - \mathbf{A}_{21}\mathbf{A}_{11}^{-1}\mathbf{x}_1$.

2. Complete the remaining steps in the proof of Theorem 5.2.

3. Use the following simulation experiment (Cook and Weisberg, 1999, pp. 280-282) to inspect how stepwise selection overstates significance. Generate a data set of $n = 100$ cases with a response and 50 predictors, all from the standard normal distribution independently. Thus none of the predictors is really related to the response. Fit the following three models: a, Including all 50 predictors, fit the whole model without selection; b, obtain the final model by performing a forward addition procedure with $\alpha_{\text{entry}} = 0.25$; c, Obtain the final model by performing a forward addition procedure with $\alpha_{\text{entry}} = 0.05$. Repeat the experiment with $n = 1,000$. Use Table 5.7 to summarize your results and make comments.

4. The variance of the regression prediction and variance of the least squares estimators tend to be larger when the regression model is over-fitted. Show the following results for an over-fitted regression model:

   (a) Let $x_1, x_2, \cdots, x_k$ be the regressors and $b_0, b_1, \cdots, b_k$ be the estimates of the regression model, if $(k+1)$-th regressor x_{k+1} is added into the model and the estimates are denoted by $b_0^*, b_1^*, \cdots, b_k^*, b_{k+1}^*$ then
$$\text{Var}(b_i^*) \geq \text{Var}(b_i) \quad \text{for } \ i = 0, 1, 2, \cdots, k.$$

(b) The variance of fitted value based on an over-fitted model is larger. i.e., denote $\hat{y}_1 = \sum_{i=0}^{k} b_i x_i$ and $\hat{y}_2 = \sum_{i=0}^{k+1} b_i^* x_i$, then

$$\text{Var}(\hat{y}_1) \leq \text{Var}(\hat{y}_2).$$

5. Consider the quasar data used in Example 5.1. Define $Y' = log(Y)$. Perform variable selection via both the all possible regressions method and the three stepwise procedures. Examine the results and make comments.

6. All-subset-regression approach is an exhausted search and it is feasible only in situation where the number of the regressors is not too large. Table 5.8 is the female teacher effectiveness data that has 7 independent variables. In the data set, the response variable y is the quantitative evaluation made by the cooperating female teachers and the regressors are the scores of the seven standard tests. Then the all-subset-regression approach would search for $2^7 - 1 = 127$ regression models if the regression model with no regressors is excluded. Using the SAS procedure REG and PRESS criteria to verify among all possible 127 regression models that the best regression models containing one, two, $\cdots$, seven regressors are the models in Table 5.9.

7. As a criterion for comparing models, The C_p measure in (5.18) can alternatively be given by

$$C_p \approx MSE + \frac{2 \cdot (p+1)}{n} \hat{\sigma}^2.$$

Suppose that MSE from the current model is used to estimate the true noise variance, i.e., $\hat{\sigma}^2 = MSE$. Show that C_p is similar to the GCV criterion given in (5.10), using approximation

$$\frac{1}{1-x^2} \approx 1 + 2x.$$

Table 5.3 Backward Elimination Procedure for the Quasar Data.

(a) Steps in backward elimination.

Step 0: The Full Model

Variable	Parameter Estimate	Standard Error	Type II SS	F Test Value	P-Value
Intercept	−33.9611	11.64951	0.00040842	8.50	0.0089
X_1	−0.0295	0.04481	0.00002081	0.43	0.5184
X_2	1.4690	0.36414	0.00078215	16.28	0.0007
X_3	0.8549	0.36184	0.00026828	5.58	0.0290
X_4	0.9662	0.25139	0.00070989	14.77	0.0011
X_5	−0.0400	0.25021	0.00000123	0.03	0.8746

Step 1: After Removing X_5

Variable	Parameter Estimate	Standard Error	Type II SS	F Test Value	P-Value
Intercept	−35.2616	8.1375	0.00085839	18.78	0.0003
X_1	−0.0262	0.0389	0.00002075	0.45	0.5082
X_2	1.4154	0.1386	0.00477	104.30	$< .0001$
X_3	0.9080	0.1407	0.00190	41.65	$< .0001$
X_4	0.9260	0.0028	5.17789	113264	$< .0001$

Step 2: After Removing X_1

Variable	Parameter Estimate	Standard Error	Type II SS	F Test Value	P-Value
Intercept	−29.79820	0.67081	0.08786	1973.23	$< .0001$
X_2	1.50849	0.01102	0.83506	18754.3	$< .0001$
X_3	0.81358	0.01223	0.19706	4425.58	$< .0001$
X_4	0.92575	0.00269	5.28153	118615	$< .0001$

(b) Summary of Backward Elimination.

Step	Variable Removed	Num of Vars in Model	Partial R-Square	Model R^2	C_p	F Test Value	P-Value
1	X_5	4	0.0000	0.9998	4.0256	0.03	0.8746
2	X_1	3	0.0000	0.9998	2.4574	0.45	0.5082

Table 5.4 Forward Addition Procedure for the Quasar Data.

(a) Steps in forward addition.

Step 1: X_5 Added

Variable	Parameter Estimate	Standard Error	Type II SS	F Test Value	P-Value
Intercept	17.19298	3.02830	4.29432	32.23	$< .0001$
X_5	0.48707	0.11497	2.39122	17.95	0.0003

Step 2: X_3 Added

Variable	Parameter Estimate	Standard Error	Type II SS	F Test Value	P-Value
Intercept	−75.76681	0.47440	1.98599	25507.5	$< .0001$
X_3	2.31295	0.01166	3.06250	39333.9	$< .0001$
X_5	0.92205	0.00354	5.28081	67825.2	$< .0001$

(b) Summary of Forward Addition.

Step	Variable Entered	Num of Vars in Model	Partial R-Square	Model R^2	C_p	F Test Value	P-Value
1	X_5	1	0.4383	0.4383	63741.5	17.95	0.0003
2	X_3	2	0.5614	0.9997	16.6433	39333.9	$< .0001$

Table 5.5 The Clerical Data from Ex. 6.4 in Mendenhall and Sinich (2003)

Obs.	Day of Week	Y	X_1	X_2	X_3	X_4	X_5	X_6	X_7
1	M	128.5	7781	100	886	235	644	56	737
2	T	113.6	7004	110	962	388	589	57	1029
3	W	146.6	7267	61	1342	398	1081	59	830
4	Th	124.3	2129	102	1153	457	891	57	1468
5	F	100.4	4878	45	803	577	537	49	335
6	S	119.2	3999	144	1127	345	563	64	918
7	M	109.5	11777	123	627	326	402	60	335
8	T	128.5	5764	78	748	161	495	57	962
9	W	131.2	7392	172	876	219	823	62	665
10	Th	112.2	8100	126	685	287	555	86	577
11	F	95.4	4736	115	436	235	456	38	214
12	S	124.6	4337	110	899	127	573	73	484
13	M	103.7	3079	96	570	180	428	59	456
14	T	103.6	7273	51	826	118	463	53	907
15	W	133.2	4091	116	1060	206	961	67	951
16	Th	111.4	3390	70	957	284	745	77	1446
17	F	97.7	6319	58	559	220	539	41	440
18	S	132.1	7447	83	1050	174	553	63	1133
19	M	135.9	7100	80	568	124	428	55	456
20	T	131.3	8035	115	709	174	498	78	968
21	W	150.4	5579	83	568	223	683	79	660
22	Th	124.9	4338	78	900	115	556	84	555
23	F	97	6895	18	442	118	479	41	203
24	S	114.1	3629	133	644	155	505	57	781
25	M	88.3	5149	92	389	124	405	59	236
26	T	117.6	5241	110	612	222	477	55	616
27	W	128.2	2917	69	1057	378	970	80	1210
28	Th	138.8	4390	70	974	195	1027	81	1452
29	F	109.5	4957	24	783	358	893	51	616
30	S	118.9	7099	130	1419	374	609	62	957
31	M	122.2	7337	128	1137	238	461	51	968
32	T	142.8	8301	115	946	191	771	74	719
33	W	133.9	4889	86	750	214	513	69	489
34	Th	100.2	6308	81	461	132	430	49	341
35	F	116.8	6908	145	864	164	549	57	902
36	S	97.3	5345	116	604	127	360	48	126
37	M	98	6994	59	714	107	473	53	726
38	T	136.5	6781	78	917	171	805	74	1100
39	W	111.7	3142	106	809	335	702	70	1721
40	Th	98.6	5738	27	546	126	455	52	502
41	F	116.2	4931	174	891	129	481	71	737
42	S	108.9	6501	69	643	129	334	47	473
43	M	120.6	5678	94	828	107	384	52	1083
44	T	131.8	4619	100	777	164	834	67	841
45	W	112.4	1832	124	626	158	571	71	627
46	Th	92.5	5445	52	432	121	458	42	313
47	F	120	4123	84	432	153	544	42	654
48	S	112.2	5884	89	1061	100	391	31	280
49	M	113	5505	45	562	84	444	36	814
50	T	138.7	2882	94	601	139	799	44	907
51	W	122.1	2395	89	637	201	747	30	1666
52	Th	86.6	6847	14	810	230	547	40	614

[a] *Source*: Adapted from Work Measurement, by G. L. Smith, Grid Publishing Co., Columbus, Ohio, 1978 (Table 3.1).

Table 5.6 Backward Elimination Results for the Clerical Data.

(a) Summary of the Steps in Backward Elimination.

Step	Variable Removed	Num of Vars in Model	Partial R-Square	Model R^2	C_p	F Test Value	P-Value
1	X_7	11	0.0024	0.6479	11.2717	0.27	0.6051
2	X_1	10	0.0066	0.6413	10.0045	0.75	0.3928
3	X_2	9	0.0060	0.6354	8.6689	0.68	0.4141
4	X_3	8	0.0187	0.6166	8.7571	2.16	0.1494
5	X_4	7	0.0232	0.5934	9.3472	2.60	0.1139

(b) The Final Model Selected.

Variable		Parameter Estimate	Standard Error	Type II SS	F Test Value	P-Value
intercept		77.1920	7.85734	10981	96.51	< .0001
Group				1972.6617	3.47	0.0100
	Zt	3.43653	5.3905	46.2429	0.41	0.5271
	Zw	−0.7867	6.0585	1.9184	0.02	0.8973
	Zth	−13.7984	5.3820	747.8711	6.57	0.0138
	Zf	−9.5425	5.3646	359.9975	3.16	0.0822
	Zs	0.5064	5.1973	1.0801	0.01	0.9228
X_5		0.0410	0.0104	1763.8950	15.50	0.0003
X_6		0.3289	0.1223	822.7474	7.23	0.0101

Table 5.7 Results of a Simulated Example for Inspecting Overall Significance in Stepwise Selection.

Sample Size (n)	α_{entry}	# of Terms in the Model	R^2	Overall F p-Value	# of Terms with p-Value ≤ 0.25	# of Terms with p-Value ≤ 0.05
100	1.00					
	0.25					
	0.05					
1,000	1.00					
	0.25					
	0.05					

Table 5.8 Female Teachers Effectiveness Data

y Effectiveness Score	x1 Standard Test 1	x2 Standard Test 2	x3 Standard Test 3	x4 Standard Test 4	x5 Standard Test 5	x6 Standard Test 6	x7 Standard Test 7
410	69	125	93	3.70	59.00	52.5	55.66
569	57	131	95	3.64	31.75	56.0	63.97
425	77	141	99	3.12	80.50	44.0	45.32
344	81	122	98	2.22	75.00	37.3	46.67
324	0	141	106	1.57	49.00	40.8	41.21
505	53	152	110	3.33	49.35	40.2	43.83
235	77	141	97	2.48	60.75	44.0	41.61
501	76	132	98	3.10	41.25	66.3	64.57
400	65	157	95	3.07	50.75	37.3	42.41
584	97	166	120	3.61	32.25	62.4	57.95
434	76	141	106	3.51	54.50	61.9	57.90

Data Source: Raymond H. Myers, *Classical and Modern Regression with Applications.* Duxbury, p. 191.

Table 5.9 All-subset-regression for Female Teacher Effectiveness Data

Vars	Intercept	x1	x2	x3	x4	x5	x6	x7	CP	PRESS
x7	14.72	.	.	.	.	.	.	8.14	23.71	82293
x2x7	-577.84	.	3.73	.	.	.	.	9.47	12.42	51498
x2x6x7	-911.68	.	5.48	.	.	.	-10.10	20.97	5.64	35640.82
x1x2x6x7	-920.41	-0.32	5.57	.	.	.	-10.00	21.22	7.35	40279.35
x1x2x5x6x7	-1654.82	-1.32	8.16	.	.	3.24	-11.64	27.96	6.81	34607.30
x1x2x4x5x6x7	-1510.98	-1.42	7.44	.	22.80	2.89	-10.69	25.37	8.38	42295
x1x2x3x4x5x6x7	-1499.47	-1.22	5.05	4.10	39.44	2.24	-11.54	23.79	8.00	115019

Chapter 6

Model Diagnostics: Heteroscedasticity and Linearity

Model diagnostics involve two aspects: outlier detection and model assumption checking. In Chapter 4, we have discussed various criteria on how to detect regression outliers that have unusual response or predictor values and influential observations whose inclusion in the data set would have substantial impact on model estimation and prediction. In this chapter, we continue to study the second aspect of model diagnostics, i.e., how to evaluate the validity of model assumptions and how to make remedies if any of the model assumptions is found violated.

There are four major assumptions involved in the classical linear regression model:

$$\mathbf{y} = \mathbf{X}\boldsymbol{\beta} + \boldsymbol{\varepsilon} \text{ with } \boldsymbol{\varepsilon} \sim \mathcal{N}(\mathbf{0}, \sigma^2\mathbf{I}), \tag{6.1}$$

where $\mathbf{X}$ is $n \times (k+1)$ of rank $(k+1)$. These are listed below.

- *Linearity*: The linearity assumption is referred to as the assumed linear relationship between the mean response $E(y_i|\mathbf{x}_i)$ and the predictors $\mathbf{x}_i$. It is this assumption that leads to $E(\varepsilon_i) = 0$.
- *Independence*: The independence assumption states that the observations in the data set are randomly selected, which corresponds to the independence of random errors.
- *Homoscedasticity*: The homoscedasticity assumption requires that all random errors have the same constant variance.
- *Normality*: Normally distributed random errors are assumed.

All the four assumptions can be summarized in short by $\varepsilon_i \overset{iid}{\sim} \mathcal{N}(0, \sigma^2)$ for $i = 1, \ldots, n$.

We shall briefly discuss how to check the independence and normality assumptions. Among these four, the independence assumption is the hardest to check. There are only a couple of limited tests available for testing independence. For example, the Durbin-Watson test, which tests for autocorrelation, is discussed on page 235 and illustrated on page 266. Nevertheless, the plausibility of independence usually can be inspected from data collection schemes. If data are collected by simple random sampling, it is reasonable to assume independence. If, on the other hand, the data are collected over time in a longitudinal study or several measures are taken repeatedly from the same unit in a cluster study, then the dependence among observations becomes manifest and should not be ignored. Random and/or mixed effect models (McCulloch, Searle, and Neuhaus, 2008) are often used to model the intra-cluster correlation or time-related dependence structure.

The normality assumption can be checked (see Section 3.17) by graphically or numerically examining the studentized jackknife residuals,

$$r_{(-i)} = \frac{y_i - \hat{y}_i}{\sqrt{MSE_{(-i)} \cdot (1 - h_{ii})}} = r_i \sqrt{\frac{(n-1)-(k+1)}{\{n-(k+1)\} - r_i^2}}, \tag{6.2}$$

where $MSE_{(-i)}$ is the resultant MSE from fitting the linear model with i-th observation excluded; h_{ii} is the i-th leverage; and

$$r_i = \frac{y_i - \hat{y}_i}{\sqrt{MSE \cdot (1 - h_{ii})}}$$

is the i-th studentized residual. Under the model assumptions, it can be shown that $r_{(-i)}$ follows the t distribution with $\{(n-1)-(k+1)\}$ degrees of freedom, which be approximated by $\mathcal{N}(0, 1)$ for large dfs. Graphically, the quantile-quantile (or Q-Q) normal probability plot of the studentized jackknife residuals can be used to examine normality. Numerically, various goodness-of-fit tests can be used to formally test for normality. The most commonly used one, for example, is the Shapiro-Wilk test, which is available in PROC UNIVARIATE with the NORMAL option in SAS and in the R function `shapiro.test()`. Nevertheless, we would like to comment that the normality assumption is of somewhat lesser concern for large data in the spirit of the central limit theorem. The inferences in linear regression show considerable robustness to violation of normality as long as the sample size is reasonably large.

However, the two other assumptions, linearity and homoscedasticity, are closely related to the bias and variance in model estimation. Violation of either assumption may result in severely misleading results. The errors induced by misspecified homoscedasticity or linearity will not be lessened with increased sample size and the problems remain for large samples. In this chapter, we shall consider tests for assessing homoscedasticity and strategies for exploring the appropriate functional form.

6.1 Test Heteroscedasticity

In classical regression, the equal variance assumption simply states that, given X, the conditional variance of the error terms are constant for all observations. In other words, whenever X varies, the corresponding response Y has the same variance around the regression line. This is so-called homoscedasticity. Referring to the weighted least square estimation in Section 7.2.4, this assumption implies that all Y values corresponding to various X's are equally important and should be evenly weighted. In contrast, the condition of the error variance not being constant over all observations is called heteroscedasticity.

6.1.1 *Heteroscedasticity*

In statistics, a sequence or a vector of random variables is said to be heteroscedastic, if the random variables have different variances. When statistical techniques, such as the ordinary least squares (OLS), is applied to the regression model, a number of assumptions are typically made. One of these is that the error term in the model has a constant variance. This will be true if the observations of the error term are assumed to be drawn from identical distributions. Heteroscedasticity is the violation of this assumption. What is the validity of this assumption and what happens if this assumption is not fulfilled? Specifically, the following questions can be asked:

1. What is the nature of heteroscedasticity?
2. What are the consequences of heteroscedasticity?
3. How do we detect heteroscedasticity?
4. How do we control for heteroscedasticity?

In general, there are several reasons why stochastic disturbance terms in the model may be heteroscedastic:

1. Response variable may change its magnitude according to the values of one or more independent variables in the model. Therefore, it may induce heteroscedastic error.
2. As data collecting techniques improve, it is likely to decrease the variation and to commit fewer errors.
3. Outlier observations are much different in relation to the other observations in the sample. The inclusion or exclusion of such observations,

especially if the sample size is small, can substantially alter the results of regression analysis.
4. Very often the heteroscedasticity may be due to the misspecified regression model such that some important variables are omitted from the model.

The consequences of heteroscedasticity are as follows:

1. The ordinary least squares (OLS) estimators and regression predictions based on them remain unbiased and consistent.
2. The OLS estimators are no longer the BLUE because they are no longer efficient. As a result, regression predictions will be inefficient as well.
3. Because of the inconsistency of the covariance matrix of the estimated regression coefficients, the tests of hypotheses, that is, t-tests or F-tests, are no longer valid.

In short, if we persist in using the usual testing procedures despite the fact of heteroscedasticity in the regression model, whatever conclusions we draw or inferences we make may be very misleading.

An informal method for detecting heteroscedasticity is the graphical method. If there is no a priori or empirical information about the nature of heteroscedasticity, in practice one can do the regression analysis on the assumption that there is no heteroscedasticity and then do a postmortem examination of the residual squared e_i^2 to see if they exhibit any systematic pattern. Although e_i^2's are not the same thing as ε^2, they can be used as proxies especially if the sample size is sufficiently large. To carry out this informal method, one can simply plot e_i^2 against either $\hat{y}_i$ or any of the explanatory variables.

6.1.2 *Likelihood Ratio Test, Wald, and Lagrange Multiplier Test*

Since the tests we will discuss in the subsequent sections will be related to the Langrange multiplier test we first briefly discuss the standard likelihood ratio test (LR), Wald (W) test, and Lagrange multiplier test (LM). Denote the log likelihood to be $L(x, \theta) = \log l(x, \theta)$ and the second derivative of the likelihood to be $C(\theta) = \dfrac{\partial^2 L(x, \theta)}{\partial \theta^2}$. The standard likelihood ratio test is defined as

$$LR = 2[\log l(x, \hat{\theta}) - \log l(x, \tilde{\theta})], \tag{6.3}$$

where $\hat{\theta}$ is the likelihood estimate without restrictions, $\tilde{\theta}$ is the likelihood estimate with restrictions, and g is the number of the restrictions imposed. The asymptotic distribution of the LR test is the χ^2 with g degrees of freedom. The Wald test is defined as

$$W = (\hat{\theta} - \theta_0)' I(\hat{\theta})(\hat{\theta} - \theta_0), \tag{6.4}$$

where $I(\theta) = E_\theta\left[\frac{\partial^2 \log l(x,\theta)}{\partial \theta^2}\right]$, the information matrix.

The basic idea of the Lagrange multiplier test focuses on the characteristic of the log-likelihood function when the restrictions are imposed on the null hypothesis. Suppose that the null hypothesis is $H_0 : \boldsymbol{\beta} = \boldsymbol{\beta}_0$, we consider a maximization problem of log-likelihood function when the restriction of the null hypothesis which we believe to be true is imposed. That is, we try to solve for maximization problem under the constrain $\boldsymbol{\beta} = \boldsymbol{\beta}_0$. This restricted maximization problem can be solved via the Lagrange multiplier method. i.e., we can solve for the unconstrained maximization problem:

$$\max_{\boldsymbol{\beta}} \left[L(\boldsymbol{\beta}) + \lambda(\boldsymbol{\beta} - \boldsymbol{\beta}_0)\right].$$

Differentiating with respect to $\boldsymbol{\beta}$ and λ and setting the results equal to zero yield the restricted maximum likelihood estimation, $\boldsymbol{\beta}^* = \boldsymbol{\beta}_0$, and the estimate of the Lagrange multiplier, $\lambda^* = S(\boldsymbol{\beta}^*) = S(\boldsymbol{\beta}_0)$, where $S(\cdot)$ is the slope of the log-likelihood function, $S(\boldsymbol{\beta}) = \frac{dL(\boldsymbol{\beta})}{d\boldsymbol{\beta}}$, evaluated at the restricted value $\boldsymbol{\beta}_0$. The greater the agreement between the data and the null hypothesis, i.e., $\hat{\boldsymbol{\beta}} \approx \boldsymbol{\beta}_0$, the closer the slope will be to zero. Hence, the Lagrange multiplier can be used to measure the distance between $\hat{\boldsymbol{\beta}}$ and $\boldsymbol{\beta}_0$. The standard form for the LM test is defined as

$$LM = [S(\theta_0)]^2 I(\theta_0)^{-1} \sim \chi_1^2 \tag{6.5}$$

The generalization to the multivariate version is a straightforward extension of (6.5) and can be written as

$$LM = S(\tilde{\theta})' I(\tilde{\theta})^{-1} S(\tilde{\theta}) \sim \chi_g^2, \tag{6.6}$$

where g is the number of linear restrictions imposed.

Example We now give an example to compute the LR, Wald, and LM tests for the elementary problem of the null hypothesis $H_0 : \mu = \mu_0$ against $H_a : \mu \neq \mu_0$ from a sample size n drawn from a normal distribution with variance of unity. i.e., $X \sim N(\mu, 1)$ and the log-likelihood function is

$$L(\mu) = -\frac{n}{2}\log(2\pi) - \frac{1}{2}\sum_{i=1}^{n}(X_i - \mu)^2,$$

which is a quadratic form in μ. The first derivative of the log-likelihood is

$$\frac{dL(\mu)}{d\mu} = \sum_{i=1}^{n}(X_i - \mu) = n(\bar{X} - \mu),$$

and the first derivative of the log-likelihood is a quadratic form in μ. The second derivative of the log-likelihood is a constant:

$$\frac{d^2L(\mu)}{d\mu^2} = -n.$$

The maximum likelihood estimate of μ is $\hat{\mu} = \bar{X}$ and LR test is given by

$$\begin{aligned} LR &= 2[L(\hat{\mu}) - L(\mu_0)] \\ &= \sum_{i=1}^{n}(X_i - \mu_0)^2 - \sum_{i=1}^{n}(X_i - \bar{X})^2 \\ &= n(\bar{X} - \mu_0)^2. \end{aligned} \tag{6.7}$$

The Wald test is given by

$$W = (\mu - \mu_0)^2 I(\theta_0) = n(\bar{X} - \mu_0)^2. \tag{6.8}$$

Since $\dfrac{dL(\mu_0)}{d\mu} = n(\bar{X} - \mu_0)$, the LM test is given by

$$W = S(\mu_0)^2 C(\theta_0)^{-1} = n(\bar{X} - \mu_0)^2. \tag{6.9}$$

Note that $\bar{X} \sim N(\mu_0, n^{-1})$, each statistic is the square of a standard normal variable and hence is distributed as χ^2 with one degree of freedom. Thus, in this particular example the test statistics are χ^2 for all sample sizes and therefore are also asymptotically χ^2_1. □

6.1.3 *Tests for Heteroscedasticity*

There are several formal tests that can be used to test for the assumption that the residuals are homoscedastic in a regression model. White's (1980) test is general and does not presume a particular form of heteroscedasticity. Unfortunately, little can be said about its power and it has poor small sample properties unless there are very few number of regressors. If we have prior knowledge that the variance σ_i^2 is a linear (in parameters) function of explanatory variables, the Breusch-Pagan (1979) test is more powerful. In addition, Koenker (1981) proposes a variant of the Breusch-Pagan test that does not assume normally distributed errors. We explain these methods as follows.

6.1.3.1 *White's Test*

In statistics, White's test (1980), named after Halbert White, is a test which establishes whether the variance in a regression model is constant (homoscedasticity). To test for constant variance we regresses the squared residuals from a regression model onto the regressors, the cross-products of the regressors, and the squared regressors. Then the White's test statistic will be used to perform the test. The White's test does not require any prior knowledge about the source of heteroscedasticity. It is actually a large sample Lagrange Multiplier (LM) test, and it does not depend on the normality of population errors. We use the regression model $y_i = \beta_0 + \beta_1 x_{1i} + \beta_2 x_{2i} + \varepsilon_i$ to illustrate the White's test.

(1) Given the data, estimate the regression model and obtain the residuals $e_i = y_i - \hat{y}_i$.
(2) Next, estimate the following auxiliary regression model and obtain its R^2:

$$e_i = \beta_0 + \beta_1 x_{1i} + \beta_2 x_{2i} + \beta_3 x_{1i}^2 + \beta_4 x_{2i}^2 + \beta_5 x_{1i} x_{2i} + \varepsilon_i.$$

(3) Compute White's test statistic: Under the null hypothesis that there is no heteroscedasticity, it can be shown that the sample size n times the R^2 obtained from the auxiliary regression asymptotically follows the chi-square distribution with degrees of freedom equal to the number of regressors (not including the constant term) in the auxiliary regression. That is,

$$nR^2 \sim \chi_5^2.$$

(4) Perform the test by comparing nR^2 to the chi-square critical value. If $nR^2 > \chi^2_{\alpha,\,5}$, the conclusion is that there is heteroscedasticity. If $nR2 < \chi^2_{\alpha,\,5}$, there is no heteroscedasticity, which implies that

$$\beta_1 = \beta_2 = \beta_3 = \beta_4 = \beta_5 = 0.$$

6.1.3.2 *Park, Glesjer, and Breusch-Pagan-Godfrey Tests*

All three of these tests are similar. Like White's test, each of these tests is the Lagrange Multiplier (LM) test and thus follows the same general procedure. Given the regression model,

$$y_i = \beta_0 + \beta_1 x_{1i} + \beta_2 x_{2i} + \cdots + \beta_k x_{ki} + \varepsilon_i.$$

The Park, Glesjer, and Breusch-Pagan-Godfrey Tests carry out the following steps:

(1) Given the data, estimate the regression model and obtain the residuals, $e_i = y_i - \hat{y}_i$.

(2) Estimate the following auxiliary regression models and obtain their R^2's.

(a) For Park Test the auxiliary regression model is:

$$\log e_i^2 = \alpha_0 + \alpha_1 \log Z_{1i} + \alpha_2 \log Z_{2i} + \cdots + \alpha_p \log Z_{pi} + \varepsilon_i$$

(b) For Glesjer test the auxiliary regression is:

$$e_i^2 = \alpha_0 + \alpha_1 Z_{1i} + \alpha_2 Z_{2i} + \cdots + \alpha_p Z_{pi} + \varepsilon_i$$

(c) For Breusch-Pagan-Godfrey test the auxiliary regression is:

$$\tilde{e}_i^2 = \alpha_0 + \alpha_1 Z_{1i} + \alpha_2 Z_{2i} + \cdots + \alpha_p Z_{pi} + \varepsilon_i,$$

where $\tilde{e}_i^2 = e_i^2 / \left(\sum_{i=1}^n e_i^2/n\right)$. In each auxiliary regression, the Z_i's may be some or all of the X_i's.

(3) Compute the LM test statistic: Under the null hypothesis that there is no heteroscedasticity, it can be shown that the sample size n times the R^2 obtained from the auxiliary regressions asymptotically follows the chi-square distribution with degrees of freedom equal to the number of regressors (not including the constant term) in the auxiliary regression model. That is,

$$nR^2 \sim \chi^2_p.$$

It is important to note that the test statistics originally proposed by Park and Glesjer are Wald test statistics, and the test statistic originally suggested in the Breusch-Pagan-Godfrey Test is one-half of the auxiliary regressions explained sum of squares, distributed chi-square with p degrees of freedom. However, as pointed out by Engle (1984), since all of these tests are simply large-sample tests, they are all operationally equivalent to the LM test.

(4) Perform the LM test by comparing nR^2 to the chi-square critical value. If $nR^2 > \chi^2_{\alpha,\, p}$, the conclusion is that there is heteroscedasticity. If $nR^2 < \chi^2_{\alpha,\, p}$, there is no heteroscedasticity, which implies that

$$\alpha_1 = \alpha_2 = \cdots = \alpha_p = 0.$$

The Park, Glesjer, and Breusch-Pagan-Godfrey tests all require knowledge about the source of heteroscedasticity. i.e., the Z variables are known to be responsible for the heteroscedasticity. These tests are all, in essence, LM tests. In the Park test, the error term in the auxiliary regression may not satisfy the classical nonlinear regression model (CNLRM) assumptions and may be heteroscedastic itself. In the Glejser test, the error term e_i is nonzero, is serially correlated, and is ironically heteroscedastic. In the Breusch-Pagan-Godfrey test, the error term is quite sensitive to the normality assumption in small samples.

6.1.3.3 *Goldfeld-Quandt test*

If population errors are homoscedastic, and thus share the same variance over all observations, then the variance of residuals from a part of the sample observations should be equal to the variance of residuals from another part of the sample observations. Thus, a "natural" approach to test for heteroscedasticity would be to perform the F-test for the equality of residual variances, where the F-statistic is simply the ratio of two sample variances. Consider the following regression model:

$$y_i = \beta_0 + \beta_1 x_{1i} + \beta_2 x_{2i} + \cdots + \beta_k x_{ki} + \varepsilon_i.$$

(1) Identify a variable to which the population error variance is related. For illustrative purpose, we assume that X_1 is related to $\mathrm{Var}(\varepsilon_i)$ positively.
(2) Order or rank the observations according to the values of X_1, beginning with the lowest X_1 value.
(3) Omit c central observations, where c is specified a priori, and divide the remaining $n - c$ observations into two groups each of $(n - c)/2$

observations. The choice of c, for the most part, is arbitrary; however, as a rule of thumb, it will usually lie between one-sixth and one-third of the total observations.

(4) Run separate regressions on the first $(n - c)/2$ observations and the last $(n - c)/2$ observations, and obtain the respective residual sum of squares: ESS_1 representing the residual sum of squares from the regression corresponding to the smaller X_1 values (the small variance group) and ESS_2 from the larger X_1 values (the large variance group).

(5) Compute the F-statistic

$$F = \frac{ESS_1/df}{ESS_2/df},$$

where $df = \dfrac{n - c - 2(k+1)}{2}$ and k is the number of estimated slope coefficients.

(6) Perform the F-test. If ε_i's are normally distributed, and if the homoscedasticity assumption is valid, then it can be shown that F follows the F distribution with degrees of freedom in both the numerator and denominator. If $F > F_{\alpha,\, df,\, df}$, then we can reject the hypothesis of homoscedasticity, otherwise we cannot reject the hypothesis of homoscedasticity.

The Goldfeld-Quandt test depends importantly on the value of c and on identifying the correct X variable with which to order the observations. This test cannot accommodate situations where the combination of several variables is the source of heteroscedasticity. In this case, because no single variable is the cause of the problem, the Goldfeld-Quandt test will likely conclude that no heteroscedasticity exists when in fact it does.

6.2 Detection of Regression Functional Form

Despite all the favorable properties of linear regression, the effects of predictors might be curvilinear in reality and the straight line can only offer a rather poor approximation. The partial residual plot or the partial regression plot graphically helps depict the linear or nonlinear relationship between the response and an individual predictor after adjusting for other predictors. However, eventually it would be preferable to have available an explicit functional form that characterizes their relationship.

6.2.1 *Box-Cox Power Transformation*

Transformation on response and/or predictors may help deal with nonlinearity. In the early development, Box and Cox (1964) considered the power family of transformations on the response:

$$f_\lambda(y) = \begin{cases} (y^\lambda - 1)/\lambda & \text{when } \lambda \neq 0, \\ \log(y) & \text{when } \lambda = 0. \end{cases} \tag{6.10}$$

Note that

$$\lim_{\lambda \to 0} (y^\lambda - 1)/\lambda = \log(y),$$

which motivates the form in (6.10). The aim of the Box-Cox transformations is to ensure the usual assumptions for linear model are more likely to hold after the transformation. That is, $f_\lambda(\boldsymbol{y}) \sim N(\boldsymbol{X\beta}, \sigma^2 I)$, where $f_\lambda(\boldsymbol{y})$ is the transformed response. The Box-Cox transformation provides the option to simultaneously estimate transformation parameter λ and the regression parameters $\beta_0, \beta_1, \cdots, \beta_k$.

When the Box-Cox transformation is applied to the response variable Y in the multiple regression model

$$f_\lambda(y_i) = \beta_0 + \beta_1 x_{1i} + \beta_2 x_{2i} + \cdots + \beta_k x_{ki} + \varepsilon_i \ \text{ for } i = 1, 2, \cdots, n, \tag{6.11}$$

we can jointly estimate λ and all regression parameters, we hope that for the transformed data the classical linear model assumptions are satisfied so that model (6.11) gives a better fit to the data.

One main convenience in the Box-Cox transformation is that statistical inference on the transformation parameter λ is available via the maximum likelihood (ML) approach. This ML method is commonly used since it is conceptually easy to understand and a profile likelihood function is easy to compute in this case. Also it is convenient to obtain an approximate CI for λ based on the asymptotic property of MLE. This allows us to evaluate the necessity of transformation or adequacy of linearity.

In order to jointly estimate the transformation parameter λ and regression parameters $\boldsymbol{\beta}$ in model (6.11), we start with the density of $f_\lambda(\boldsymbol{y})$, which is given by

$$\frac{1}{(2\pi\sigma^2)^{n/2}} \exp\left\{ -\frac{\| f_\lambda(\boldsymbol{y}) - \boldsymbol{X\beta} \|^2}{2\sigma^2} \right\}.$$

Thus, the likelihood function $l(\boldsymbol{y}|\boldsymbol{\beta}, \lambda, \sigma^2)$ is

$$l(\boldsymbol{y}|\boldsymbol{\beta}, \lambda, \sigma^2) = \frac{1}{(2\pi\sigma^2)^{n/2}} \exp\left\{ -\frac{\| f_\lambda(\boldsymbol{y}) - \boldsymbol{X\beta} \|^2}{2\sigma^2} \right\} J(\lambda, \boldsymbol{y}), \tag{6.12}$$

where $J(\lambda, \boldsymbol{y})$ is the Jocobian of the transformation

$$J(\lambda, \boldsymbol{y}) = \frac{\partial f_\lambda(\boldsymbol{y})}{\partial \boldsymbol{y}} = \prod_{i=1}^{n} y_i^{\lambda-1}.$$

To obtain the MLE from the likelihood equation, observe that for each fixed λ, the likelihood equation is proportional to the likelihood equation for estimating $(\boldsymbol{\beta}, \sigma^2)$ for observed $f_\lambda(\boldsymbol{y})$. Thus the MLEs for regression parameter $\boldsymbol{\beta}$ and error variance σ^2 are given by

$$\widehat{\boldsymbol{\beta}} = (\boldsymbol{X}^t\boldsymbol{X})^{-1}\boldsymbol{X} f_\lambda(\boldsymbol{y}) \tag{6.13}$$

$$\hat{\sigma}^2 = \frac{f_\lambda(\boldsymbol{y})^t(I - \boldsymbol{X}(\boldsymbol{X}^t\boldsymbol{X})^{-1}\boldsymbol{X}^t)f_\lambda(\boldsymbol{y})}{n} = \frac{SSE}{n}. \tag{6.14}$$

Substituting $\widehat{\boldsymbol{\beta}}$ and $\hat{\sigma}^2$ into the likelihood equation, we could obtain the so-called profile log likelihood for λ, given by

$$\begin{aligned} L_p(\lambda) &= L\left\{\lambda|\boldsymbol{y}, \boldsymbol{X}, \hat{\boldsymbol{\beta}}(\lambda), \hat{\sigma}^2(\lambda)\right\} \\ &\propto -\frac{n}{2}\log\{\hat{\sigma}^2(\lambda)\} + (\lambda - 1)\sum_{i=1}^{n}\log y_i, \end{aligned}$$

where the notation $\propto$ means "up to a constant, is equal to." Note that the profile likelihood is the likelihood function maximized over $(\boldsymbol{\beta},\ \sigma^2)$.

Let

$$g = \left(\prod_{i=1}^{n} y_i\right)^{1/n}$$

be the geometric mean of the responses and define

$$f_\lambda(\boldsymbol{y}, g) = \frac{f_\lambda(\boldsymbol{y})}{g^{\lambda-1}}.$$

Then the profile log likelihood function can be rewritten as follows

$$L_p(\lambda) \propto -\frac{n}{2}\log\left\{\frac{f_\lambda(\boldsymbol{y}, g)^t(I - \boldsymbol{X}(\boldsymbol{X}^t\boldsymbol{X})^{-1}\boldsymbol{X}^t)f_\lambda(\boldsymbol{y}, g)}{n}\right\}. \tag{6.15}$$

To maximize (6.15), we only need to find the λ that minimizes

$$\begin{aligned} SSE(\lambda) &= f_\lambda(\boldsymbol{y}, g)^t(I - \boldsymbol{X}(\boldsymbol{X}^t\boldsymbol{X})^{-1}\boldsymbol{X}^t)f_\lambda(\boldsymbol{y}, g) \\ &= SSE \cdot g^{2-2\lambda}. \end{aligned} \tag{6.16}$$

We can apply likelihood ratio test to test for $H_0 : \lambda = \lambda_0$. The test statistic is

$$\begin{aligned} LRT &= 2\big[L_P(\hat{\lambda}) - L_P(\lambda_0)\big] \\ &= n\ \log\left(\frac{SSE(\lambda_0)}{SSE(\hat{\lambda})}\right) \xrightarrow{H_0} \chi_1^2. \end{aligned} \tag{6.17}$$

Asymptotically LRT is distributed as χ_1^2 under the null. An approximate $(1-\alpha)\times 100\%$ CI for λ is given by

$$\left\{\lambda : \log\left(\frac{SSE(\lambda)}{SSE(\hat{\lambda})}\right) \leq \frac{\chi^2_{1-\alpha}(1)}{n}\right\}. \tag{6.18}$$

To apply the Box-Cox transformation $f_\lambda(\cdot)$, the response Y has to be positive. In the case of having negative Y values, we can always subtract every observation by a constant that is less than the minimum, e.g., $y_i' = y_i - b$ with $b < \min(y_1, y_2, \cdots, y_n)$. The Box-Cox transformation is reported to be successful in transform unimodal skewed distributions into normal distributions, but is not quite helpful for bimodal or U-shaped distributions. There are also various extensions of it, e.g., transformations on both the response and the predictors, for which we shall not go into further details. Implementation of the Box-Cox transformation on the response is available in SAS PROC TRANSREG and the R `MASS` library.

6.2.2 *Additive Models*

Stone (1985) proposed the additive models, which provide a nonparametric way of exploring and assessing the functional forms of predictors. These models approximate the unknown underlying regression function through additive functions of the predictors. The general form of an additive model is given by

$$y_i = \beta_0 + \sum_{j=1}^{k} f_j(x_{ij}) + \varepsilon_i, \tag{6.19}$$

where $f_j(\cdot)$'s are some smooth functions for x_j. This model specification is nonparametric in nature with $f_j(\cdot)$'s left as totally unspecified smooth functions and, at the same time, can be flexible by mixing in linear and other parametric forms. For example, a semi-parametric model of the following form is a special case:

$$y_i = \beta_0 + \sum_{j=1}^{m} \beta_j x_{ij} + \sum_{j=m+1}^{k} f_i(x_{ij}) + \varepsilon_i. \tag{6.20}$$

This semi-parametric additive model is of practical importance and is frequently used in detecting functional forms. Furthermore, it allows for nonlinear interaction components in two independent variables. For example, the following model is interaction additive model:

$$y_i = \beta_0 + f_{12}(x_{i1}, x_{i2}) + \sum_{j=3}^{k} f_j(x_{ij}) + \varepsilon_i. \tag{6.21}$$

An iterative algorithm, known as *backfitting*, is used to estimate the model. In this approach, each function $f_j(\cdot)$ is estimated using a *scatterplot smoother* and, by iteration, all p functions can be simultaneously estimated. The scatterplot smoother is referred to as a two-dimensional (i.e., one y versus one x) smoothing technique. Any commonly used smoothing methods, i.e., the kernel method (including local polynomial regression) or splines (including regression splines or smoothing splines), can be employed to do the smoothing. The nonparametric nonlinear interaction terms $f_{12}(x_{i1}, x_{i2})$ in (6.21) are often fit with the so-called thin-plate splines. See, e.g., Chapters 5 and 6 of Hastie, Tibshirani, and Friedman (2002) for a coverage of smoothing techniques. If, in particular, a linear smoother such as smoothing spline is used, the fitted additive model remains linear, with nonlinearity achieved by predefined spline basis functions.

Algorithm 6.1: Backfitting Algorithm for Fitting Additive Models.

Initialize $\hat{\beta}_0 = \bar{y}$ and $\hat{f}_j = 0$ for $j = 1, 2, \ldots, k$.
Do $j = 1, 2, \ldots, k;\ 1, 2, \ldots, k;\ \ldots,$

- Compute the partial residuals $e_{ij} = y_i - \left\{\hat{\beta}_0 + \sum_{l \neq j} \hat{f}_l(x_{il})\right\}$.
- Update $\hat{f}_j$ as the sactterplot smoother regressing e_{ij} on x_{ij}.
- Numerical zero adjustment

$$\hat{f}_j = \hat{f}_j - (1/n) \cdot \sum_{i=1}^{n} \hat{f}_j(x_{ij})$$

  to prevent slippage due to machine rounding.

End do till convergence.

The backfitting algorithm is outlined in Algorithm 6.1. Set $\hat{\beta}_0 = \bar{y}$ and initialize all $\hat{f}_j$ to zeroes. Start with finding the partial residuals e_{1i} for the first predictor x_{i1}. Smoothing e_{i1} versus x_{i1} yields an estimator of $\hat{f}_1$. Next, obtain the partial residual e_{2i} for the second predictor x_{2i}, with updated $\hat{f}_1$. Smoothing e_{i2} versus x_{i2} yields an estimator of $\hat{f}_2$. Continue this to the p-th predictor x_{ip}. And then repeat the whole procedure for multiple rounds till all estimates of $\hat{f}_j$ stabilize.

The main idea of backfitting is about partial residuals. The effect of each predictor is well preserved under the additivity assumption and can be isolated by its associated partial residuals. Another way of understanding the algorithm is that, if a linear smoother is used, backfitting is equivalent to a Gauss-Seidel algorithm for solving a linear system of equations. In order to solve for $\mathbf{z}$ in a linear system of equations $\mathbf{Az} = \mathbf{b}$, one may proceed successively to solve each z_j in the j-th equation with the current guess of other z's and repeat the process until convergence.

The estimated $\hat{f}_j$'s help identify and characterize nonlinear covariate effects. The smoothness of $\hat{f}_j$ is controlled by some smoothing parameters, e.g., the bandwidth in kernel methods and the penalty in smoothing splines. In practice, selection of the smoothing parameter is made either by specifying the number of degrees of freedom or via minimum generalized cross-validation (GCV). Recall that GCV is yet another model selection criterion obtained as an approximation to PRESS in Chapter 5. We shall further explore its use in Section 7.3.4.

The additive models provide a useful extension of linear models and make them more flexible. With retained additivity, the resultant model fit can be interpreted in the same way as in linear models. This technique is generalized by Hastie and Tibshirani (1986) and Hastie and Tibshirani (1990) to handle other typed of responses, in which case it is called a *Generalized Additive Model.* A GAM assumes that the mean of the dependent variable depends on an additive form of functions of predictors through a link function. GAM permits the probability distribution of the response to be any member from the exponential family. Many widely used statistical models belong to this general family class, including Gaussian models for continuous data, logistic models for binary data, and Poisson models for count data. We shall study these models in Chapter 8.

Algorithm 6.2: alternating conditional expectation (ACE)

(i) Initialization: set $v_i = \hat{g}(y_i) = (y_i - \bar{y})/s_y$.

(ii) Regressing v_i on x_{ij}, apply the backfitting algorithm for additive models to estimate $\hat{f}_j(\cdot)$'s.

(iii) Compute $u_i = \sum_j \hat{f}_j(x_{ij})$. Regressing u_i on y_i, update the estimate $\hat{g}(\cdot)$ as a scatterplot smoother.

(iv) Obtain the updated $v_i = \hat{g}(y_i)$ values and standardize them such that $v_i \leftarrow (v_i - \bar{v})/s_v$.

(v) Go back to Step (ii) and iterate till convergence.

In terms of implementation, additive models and generalized additive models can be fit with PROC GAM in SAS. Two R packages, `gam` and `mgcv`, are also available.

6.2.3 *ACE and AVAS*

In the same vein as additive models, Breiman and Friedman (1985) put forward the ACE (alternating conditional expectation) algorithm, which seeks optimal nonparametric transformations of both the response and predictors. The working model of ACE is

$$g(y_i) = \beta_0 + \sum_{j=1}^{k} f_j(x_{i1}) + \varepsilon_i. \tag{6.22}$$

Consider the simple regression setting when $k = 1$. ACE approaches the problem by minimizing the squared-error loss

$$E\{g(Y) - f(X)\}^2.$$

Note that for fixed g, the function f that minimizes the squared loss is $f(X) = E\{g(Y)|X\}$, and conversely, for fixed f the function g that minimizes the squared loss is $g(Y) = E\{f(X)|Y\}$. This is the key idea of the ACE algorithm: it alternates between computations of these two conditional expectations via smoothing until convergence. When both g and f are zero functions, the squared error loss is perfectly minimized. To prevent the procedure from convergence to this trivial case, ACE standardizes $g(Y)$ so that $\text{var}\{g(Y)\} = 1$ at each step. The ACE procedure is outlined in Algorithm 6.2.

Algorithm 6.3: AVAS Modification to Step (3) in the ACE Algorithm

---

- Compute $\hat{f}_j(x_{ij})$'s and obtain $u_i = \sum_j \hat{f}_j(x_{ij})$;
- Compute the squared residuals $r_i^2 = (v_i - u_i)^2$.
- Estimate the variance function of $g(Y)$, $\hat{V}(\cdot)$, as a scatterplot smoother that regresses r_i^2 against u_i.
- Update the estimate $\hat{g}(\cdot)$ as the variance stabilizing transformation (see Ex. 6.2.4)

$$\hat{g}(y) = \int_0^y \frac{1}{\sqrt{V(u)}} du.$$

ACE has a close connection to canonical correlation analysis. ACE basically searches for transformations to maximize the correlation between the transformed response and the sum of the transformed predictors. Thus, ACE is a closer kin to correlation analysis than to regression. One undesirable feature of ACE is that it treats the dependent and independent variables symmetrically. Therefore, small changes can lead to radically different solutions. See Hastie and Tibshirani (1990) for a discussion of other anomalies of ACE.

To overcome the shortcomings of the ACE, Tibshirani (1988) proposed the AVAS (additivity and variance stabilization) algorithm. The AVAS estimates the same working model (6.22) as ACE with a similar algorithm. However, instead of fitting a scatterplot smoother to obtain $\hat{g}(\cdot)$ directly in step (3), it first estimates the variance of $g(y_i)$ as a function of v_i and then estimates $g(\cdot)$ as an asymptotic variance stabilizing transformation, as detailed in Algorithm 6.3. With this alteration, the role of the response and predictors become quite distinct and functions g and f_j become fixed points of the optimization problem in various model configurations. Besides, AVAS is invariant under monotone transformations. Both ACE and AVAS are available in the R package `acepack`.

In general, transformations not only lead to enhanced linearity in a regression model but also help improve normality and stabilize variances. On the other hand, one should keep in mind that transforming variables would complicate the model interpretation, especially when a nonparametric smoothing transformation is used. Consider, for example, a regression problem where prediction is the main purpose and the ACE is used. One would have to find the inverse transformation of $g(\cdot)$ in order to compute the predicted values, which may not always work well. So the ACE or GAM findings, often showing a rough form of thresholds or change-points, is frequently used to suggest parametric transformations.

6.2.4 *Example*

We use the navy manpower data for illustration. First, we consider the Box-Cox power transformation for the response Y only. Figure 6.1 plots the resultant log-likelihood score versus a number of λ values, together with 95% confidence interval about the best choice of λ. It can be seen that the maximum log-likelihood score occurs around $\hat{\lambda} = 1$, suggesting that no transformation on Y is necessary.

Next, we fit an additive model with the data. Due to the very limited sample size ($n = 10$), smoothing splines with three degrees of freedom are used to model both the effect of x_1 and x_2, leaving $10 - (1 + 3 + 3) = 3$

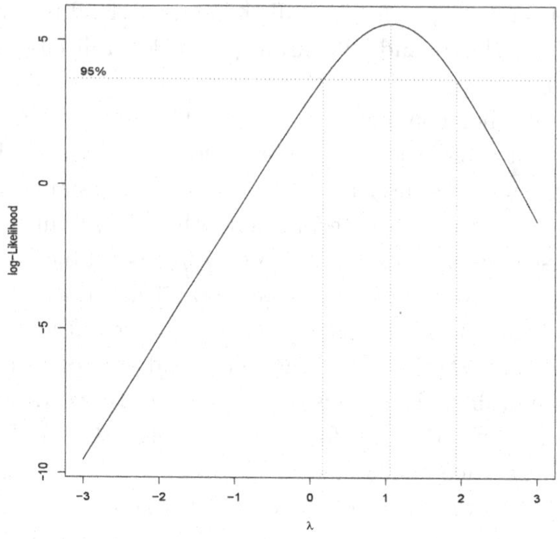

Fig. 6.1 Box-Cox Power Transformation of Y in the Navy Manpower Data: Plot of the Log-likelihood Score vs. λ.

degrees of freedom for the additive model.

Figure 6.2 depicts the resultant partial prediction for both of them. Also superimposed are their associated confidence bands, a result due to Wahba (1983). It can be seen that the three degrees of freedom correspond to two thresholds or change points on each plot. Although both curves show some nonlinear pattern, neither is striking. A linear fit might be adequate.

Treating the additive model as the full model and the model that replaces the nonparametric term by a linear one as the reduced model, the F test can be used here as a formal test for nonlinearity versus linearity. It yields values of 3.3176 and 1.2493 for x_1 and x_2, respectively. Referring to the $F(2,3)$ distribution, the p-values are 0.1737 and 0.4030, both suggesting negligible nonlinearity.

We finally try out the ACE and AVAS procedures. Since they give very similar outputs in this example, only the results from AVAS are presented. Figure 6.3 shows the transformations for Y, X_1 and X_2, respectively. It can

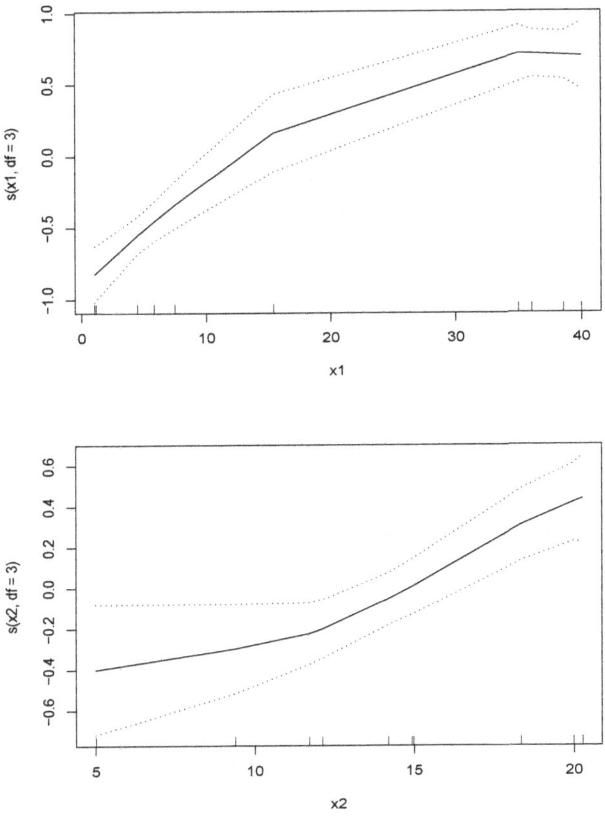

Fig. 6.2 Detecting the Nonlinear Effect of x_1 and x_2 via Additive Models in the Navy Manpower Data.

be seen that all of them are roughly linear, although $f_2(X_2)$ shows a slight threshold effect. Figure 6.4 gives the 3-D scatter plots before and after the transformation. The plot after transformation, presented in Fig. 6.4(b), clearly shows some enhanced linearity.

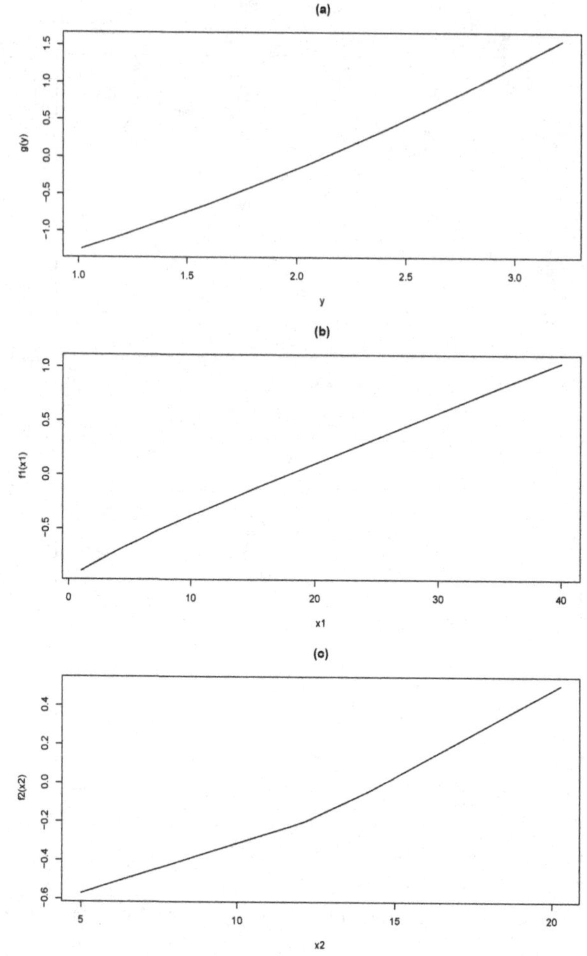

Fig. 6.3 AVAS Results for the Navy Manpower Data: (a) Plot of $g(Y)$; (b) Plot of $f_1(X_1)$; (c) Plot of $f_2(X_2)$.

Problems

1. Let $\mathbf{y}^{(\lambda)}$ be the response vector after applying a Box-Cox power transformation with λ. Assuming that $y^{(\lambda)}$ is normally distributed with mean $\mathbf{X}\boldsymbol{\beta}$ and variance-covariance $\sigma^2\mathbf{I}$, show that the maximized log-likelihood score is given as in (6.15).

2. Given that variable W has mean u and variance $V(u)$, show, using the

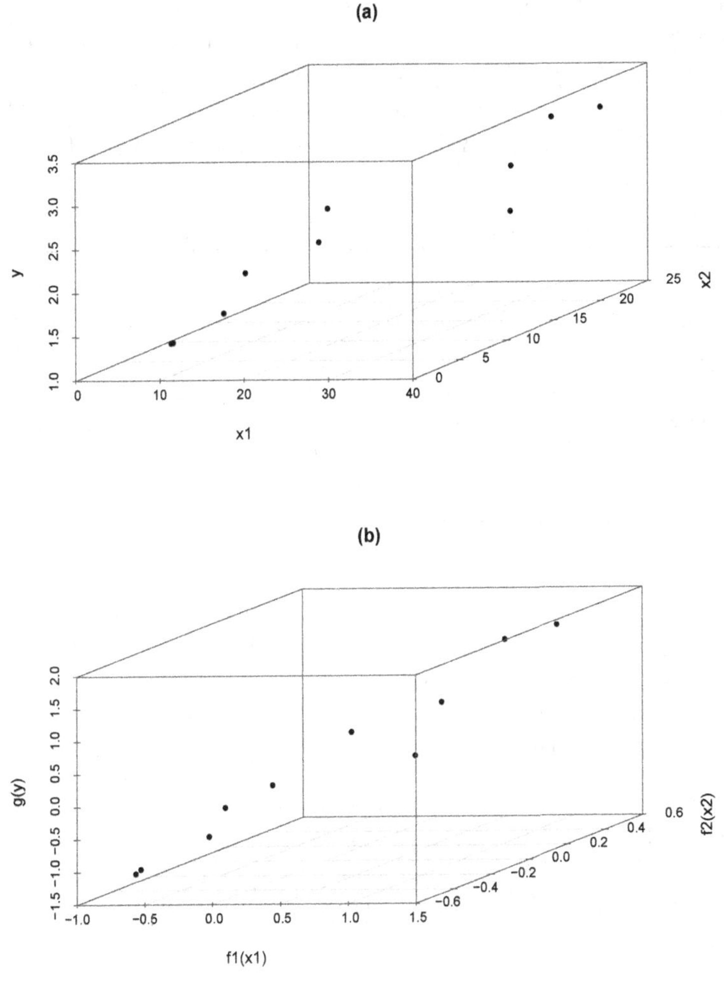

Fig. 6.4 AVAS Results for the Navy Manpower Data: (a) 3-D Scatter Plot of the Original Data; (b) 3-D Scatter Plot of $g(Y)$ vs. $f_1(X_1)$ and $f_2(X_2)$ after Transformation.

Delta method, that the asymptotic variance-stabilizing transformation for W is given by $g(W)$

$$g(w) = \int_0^w \frac{1}{\sqrt{V(u)}} du$$

and find the stabilized variance.

3. Suppose an oil company get it crude oil from 4 different sources, refines it in 3 different refineries, using the same 2 processes in each refinery. In one part if the refining process a measurement of efficiency is taken as a percentage and recorded as an integer between 0 and 100. The following table shows the available measurements of efficiency for different samples of oil.

Table 6.1 Results of Efficiency Tests

		Source			
Refinery	Peocess	Texas	Oklahoma	Gulf of Mexico	Iran
Galveston	1	31, 33, 44, 36	38	26	-
	2	37, 59	42	-	-
Newark	1	-	-	42	34, 42, 28
	2	39	36	32, 38	-
Savannah	1	42	36	-	22
	2	-	42, 46	26	37, 43

(a). For the 8 observations on Texas oil write out the equations for a regression on dummy variables for considering the effect of refinery and process on efficiency.

(b). Rewrite the equations in terms of a linear model.

(c). Write down the equations in terms of a linear model.

4. Repeat Exercise 1 for the Oklahoma data.
5. Repeat Exercise 1 for the Gulf of Mexico data.
6. Repeat Exercise 1 for the Iran data.
7. Repeat Exercises 1-4 with interactions between refinery and processes included.

8(a). For all 25 observations in Table 6.1 write down the equations of the linear model for considering the effect of source, refinery and process in efficiency. Do not include interactions.

(b). Write down the equation of the general model for this situation.

(c). Write down the normal equations.

9. Repeat Exercise 6 with interactions between source and refinery and between refinery and process.
10. Repeat Exercise 6 with all possible interactions included.
11. In any of the above exercises derive two solutions of the normal equations and investigate functions of the elements of the solutions that

might be invariant to whatever solution is used.

12. Repeat the above exercises assuming that processes are nested within refineries, suitably modifying the interaction where necessary.

Chapter 7

Extensions of Least Squares

In this chapter, we study four extended uses of least squares for dealing with a variety of regression problems often encountered in practice. The first section deals with non-full rank linear models in which the design matrix is not of full column rank. The generalized least squares (GLS) method is then introduced in Section 2. The GLS is useful for handling correlated observations or nonconstant variances. In Section 3, we discuss the ridge regression and the least absolute shrinkage and selection operator (LASSO) proposed by Tibshirani (1996). Both are shrinkage estimators. The ridge regression is a remedial measure taken to handle severe multicollinearity problems while the LASSO is particularly useful for variable selection. Finally, parametric nonlinear regression, where numerical methods are necessary for solving the optimization with least squares, is briefly covered in Section 4.

7.1 Non-Full-Rank Linear Regression Models

We first consider linear models with a non-full-rank design matrix $\mathbf{X}$. The model specification is given as $\mathbf{y} = \mathbf{X}\boldsymbol{\beta} + \boldsymbol{\varepsilon}$, where $\mathbf{X}$ is $n \times (k+1)$ of rank $p < (k+1)$ and $\boldsymbol{\varepsilon} \sim (\mathbf{0}, \sigma^2\mathbf{I})$. Non-full-rank linear models often emerge in dealing with data from analysis-of-variance (ANOVA) designs.

Example Consider, for instance, the *One-Way ANOVA* setting, which compares several population means due to different treatments received. This is essentially a regression problem that involves one continuous response Y and a categorial predictor Z. Suppose Z has c levels. With the regression approach, $(c-1)$ dummy or indicator variables are defined in order to adequately and necessarily account for the c levels of Z. Inclusion

Table 7.1 Layouts for Data Collected in One-way ANOVA Experiments.

(a) ANOVA Data Layout

category			
1	2	$\cdots$	c
Y_{11}	Y_{21}	$\cdots$	Y_{c1}
Y_{12}	Y_{22}	$\cdots$	Y_{c2}
$\vdots$	$\vdots$		$\vdots$
$\vdots$	Y_{2n_2}	$\cdots$	$\vdots$
$\vdots$			Y_{cn_c}
Y_{1n_1}			

(b) Regression Data Layout

				Dummy Variables			
subject	Y		Z	Z_1	Z_1	$\cdots$	Z_c
1	Y_{11}	$\longrightarrow Y_1$	1	1	0	$\cdots$	0
$\vdots$	$\vdots$	$\vdots$	$\vdots$	$\vdots$	$\vdots$		$\vdots$
n_1	Y_{1n_1}	$\longrightarrow Y_{n_1}$	1	1	0	$\cdots$	0
n_1+1	Y_{21}	$\longrightarrow Y_{n_1+1}$	2	0	1	$\cdots$	0
$\vdots$	$\vdots$	$\vdots$	$\vdots$	$\vdots$	$\vdots$		$\vdots$
n_1+n_2	Y_{2n_2}	$\longrightarrow Y_{n_1+n_2}$	2	0	1	$\cdots$	0
$\vdots$	$\vdots$	$\vdots$	$\vdots$	$\vdots$	$\vdots$		$\vdots$
n	Y_{cn_c}	$\longrightarrow Y_n$	c	0	0	$\cdots$	1

of the $(c-1)$ dummy variables into the linear model results in a full-rank design matrix $\mathbf{X}$ and the estimation can be proceeded with ordinary least squares as usual.

On the other hand, one may alternatively define c indicator variables $\{Z_1, Z_2, \ldots, Z_c\}$, one for each level, such that

$$Z_{ij} = \begin{cases} 1, & \text{if the } i\text{-th subject is in the } j\text{-th catgory;} \\ 0, & \text{otherwise} \end{cases} \tag{7.1}$$

for $j = 1, \ldots, c$ and $i = 1, \ldots, n$ and then consider model

$$\begin{aligned} Y_i &= \beta_0 + \beta_1 Z_{i1} + \cdots + \beta_c Z_{ic} + \varepsilon_i, \\ &= \beta_0 + \beta_j + \varepsilon_i, \quad \text{if the } i\text{-th subject falls into the } j\text{-th category,} \end{aligned} \tag{7.2}$$

for $i = 1, \ldots, n$. This model has the equivalent form to the classical one-way ANOVA model given by

$$Y_{ij} = \mu + \alpha_j + \varepsilon_{ij} \tag{7.3}$$

for $j = 1, \ldots, n_i$ and $i = 1, \ldots, c$ with correspondences $\mu = \beta_0$, $\alpha_j = \beta_j$, and $n = n_1 + n_2 + \ldots + n_c$.

To better see the correspondences, it is instructive to take a look at the different data layouts for ANOVA and regression analysis, as illustrated in Table 7.1. Data for one-way ANOVA involve several independent random samples. The data originally collected arrive in with the layout shown in Table 7.1(a). In the study, n_i individuals are randomly assigned to the i-th treatment category for $i = 1, \ldots, c$. When n_i are all the same, the design is said to be *balanced*; otherwise, it is an *unbalanced* design. Let Y_{ij} denote the response value on j-th observation assigned to the i-th treatment or category. To prepare data for the regression approach, we put all response values Y_{ij}'s into one column and relabel the data as shown in Table 7.1(b). Also displayed in the table is the categorical variable Z which indicates the assignment of treatment, as well as the c dummy variables induced by Z.

Models (7.3) and (7.2) have very easy interpretation. Consider the balanced design for example. With side condition $\sum_j \alpha_j = 0$, $\mu = \sum \mu_i / c$ corresponds to the overall mean and $\alpha_j = \mu_j - \mu$ is the difference between the j-th treatment mean and the overall mean, measuring the effect of the j-th treatment. With model (7.2), the $n \times (c+1)$ design matrix $\mathbf{X}$ is given by

$$\mathbf{X} = \begin{pmatrix} 1 & 1 & 0 & \cdots & 0 \\ \vdots & \vdots & \vdots & & \vdots \\ 1 & 1 & 0 & \cdots & 0 \\ 1 & 0 & 1 & \cdots & 0 \\ \vdots & \vdots & \vdots & & \vdots \\ 1 & 0 & 1 & \cdots & 0 \\ \vdots & \vdots & \vdots & & \vdots \\ 1 & 0 & 0 & \cdots & 1 \\ \vdots & \vdots & \vdots & & \vdots \\ 1 & 0 & 0 & \cdots & 1 \end{pmatrix}, \tag{7.4}$$

which is of rank c as adding the last c columns gives the first column $\mathbf{j}$. ¶

For non-full rank models, the matrix $\mathbf{X}^t\mathbf{X}$ is no longer non-singular or invertible. Thus the ordinary LSE $\widehat{\boldsymbol{\beta}} = (\mathbf{X}^t\mathbf{X})^{-1}\mathbf{X}^t\mathbf{y}$ is not directly applicable. There are a few ways to get around this problem. We shall focus on the one that utilizes the generalized inverse of $\mathbf{X}^t\mathbf{X}$.

7.1.1 *Generalized Inverse*

We first give the definition of the generalized inverse and outline some of its general properties.

Definition 7.1. Given an $n \times m$ matrix $\mathbf{X}$, a *generalized inverse* of $\mathbf{X}$, denoted by $\mathbf{X}^-$, is an $m \times n$ matrix such that $\mathbf{X}\mathbf{X}^-\mathbf{X} = \mathbf{X}$.

The following theorem establishes the existence of generalized inverses.

Theorem 7.1. *Any matrix has a generalized inverse.*

Proof. Consider an $n \times m$ matrix $\mathbf{X}$ of rank p. Let $\mathbf{A}$ be an $n \times p$ matrix whose columns are a maximal set of linearly independent columns of $\mathbf{X}$. Then every column of $\mathbf{X}$ can be expressed as a linear combination of $\mathbf{A}$. Thus $\mathbf{X} = \mathbf{AB}$ for some $p \times m$ matrix $\mathbf{B}$ which consists of the coefficients for these linear combinations. Since $\text{rank}(AB) \leq \min\{\text{rank}(A), \text{rank}(B)\}$, it follows that both $\mathbf{A}$ and $\mathbf{B}$ are of rank p.

Since $\mathbf{A}$ is of full column rank, $\mathbf{A^tA}$ must be positive definite. This can be easily verified by considering quadratic form $\mathbf{a}^t\mathbf{A^tAa} = (\mathbf{Aa})^t\mathbf{Aa} \geq 0$ for any nonzero vector $\mathbf{a}$. Moreover, $\mathbf{Aa} = 0$ implies $\mathbf{a} = 0$ as the columns of $\mathbf{A}$ are linear independent. So $\mathbf{a}^t\mathbf{A^tAa} = (\mathbf{Aa})^t\mathbf{Aa} > 0$. Similarly, since $\mathbf{B}$ is of full row rank, $\mathbf{BB^t}$ must be positive definite.

Rewrite

$$\mathbf{X} = \mathbf{AB} = \mathbf{ABB^t(BB^t)^{-1}(A^tA)^{-1}A^tAB} = \mathbf{XB^t(BB^t)^{-1}(A^tA)^{-1}A^tX}.$$

In other words, matrix $\mathbf{A}^- = \mathbf{B^t(BB^t)^{-1}(A^tA)^{-1}A^t}$ is a generalized inverse of $\mathbf{X}$. □

A generalized inverse is not unique unless $\mathbf{X}$ is square and nonsingular, in which case $\mathbf{X}^- = \mathbf{X}^{-1}$. On the other hand, it can be made unique by imposing additional constraints. For example, the unique Moore-Penrose generalized inverse matrix of $\mathbf{X}$ satisfies that $\mathbf{XX^-X} = \mathbf{X}$, $\mathbf{X^-XX^-} = \mathbf{X}^-$, $(\mathbf{XX^-})^t = \mathbf{X^-X}$, and $(\mathbf{X^-X})^t = \mathbf{XX^-}$.

The following theorem provides important basis for inferences in non-full-rank models. We start with two useful lemmas.

Lemma 7.1. *Given an $n \times m$ matrix $\mathbf{X}$ of rank p,*

(i) $\mathbf{XP} = \mathbf{O}$ *for some matrix* $\mathbf{P}$ *if and only if* $\mathbf{X^tXP} = \mathbf{O}$.
(ii) $\mathbf{X} = \mathbf{X(X^tX)^-X^tX}$ *and* $\mathbf{X}^t = (\mathbf{X^tX})^-(\mathbf{X^tX})\mathbf{X^t}$.

Proof.

(i) Let $\mathbf{X} = \mathbf{AB}$ where $\mathbf{A}$ is $n \times p$ of rank p and $\mathbf{B}$ is $p \times m$ of rank p. Matrix $\mathbf{A}$ has linearly independent columns while matrix $\mathbf{B}$ has linearly independent rows. So both $\mathbf{A^A}$ and $\mathbf{BB^t}$ are invertible. It follows that

$$\mathbf{X^tXP} = \mathbf{O} \Longleftrightarrow \mathbf{(AB)^t(AB)P} = \mathbf{B^tA^tABP} = \mathbf{O}$$
$$\Longleftrightarrow \mathbf{(BB^t)(A^tA)BP} = \mathbf{O} \Longleftrightarrow \mathbf{BP} = \mathbf{O}$$
$$\Longleftrightarrow \mathbf{ABP} = \mathbf{XP} = \mathbf{O}.$$

(ii) Consider $(\mathbf{X}^t\mathbf{X})(\mathbf{X}^t\mathbf{X})^-(\mathbf{X}^t\mathbf{X}) = \mathbf{X}^t\mathbf{X}$ by the definition of $(\mathbf{X}^t\mathbf{X})^-$, which can be rewritten as $(\mathbf{X}^t\mathbf{X})\{(\mathbf{X}^t\mathbf{X})^-(\mathbf{X}^t\mathbf{X}) - \mathbf{I}\} = \mathbf{O}$. Thus, by part (1) of the lemma, $\mathbf{X}\{(\mathbf{X}^t\mathbf{X})^-(\mathbf{X}^t\mathbf{X}) - \mathbf{I}\} = \mathbf{O}$. The desired results follow.

□

Theorem 7.2. *Given an $n \times m$ matrix $\mathbf{X}$ of rank p, let*

$$\mathcal{V} = C(\mathbf{X}) = \{\boldsymbol{\mu} \in \mathcal{R}^n : \boldsymbol{\mu} = X\mathbf{b} \text{ for some vector } \mathbf{b} \text{ in } \mathcal{R}^m\} \tag{7.5}$$

denote the column space of $\mathbf{X}$. The projection matrix of $\mathcal{V}$ is given by

$$\mathbf{P}_{\mathcal{V}} = \mathbf{X}(\mathbf{X}^t\mathbf{X})^-\mathbf{X}^t \tag{7.6}$$

for any generalized inverse of $\mathbf{X}^t\mathbf{X}$. It follows that the matrix $\mathbf{X}(\mathbf{X}^t\mathbf{X})^-\mathbf{X}^t$ remains invariant with any choice of $(\mathbf{X}^t\mathbf{X})^-$.

Proof. To show $\mathbf{X}(\mathbf{X}^t\mathbf{X})^-\mathbf{X}^t\mathbf{y}$ is the projection of $\mathbf{y}$ on $\mathcal{V}$ for any given $\mathbf{y} \in \mathcal{R}^n$, one needs to verify two conditions:

(i) $\mathbf{X}(\mathbf{X}^t\mathbf{X})^-\mathbf{X}^t\mathbf{y} \in \mathcal{V}$, which is obvious as

$$\mathbf{X}(\mathbf{X}^t\mathbf{X})^-\mathbf{X}^t\mathbf{y} = \mathbf{X}\{(\mathbf{X}^t\mathbf{X})^-\mathbf{X}^t\mathbf{y}\} = \mathbf{X}\mathbf{b}.$$

(ii) $\{\mathbf{y} - \mathbf{X}(\mathbf{X}^t\mathbf{X})^-\mathbf{X}^t\mathbf{y}\} \perp \mathcal{V}$. This is because, for any $\mathbf{v} \in \mathcal{V}$, $\mathbf{v} = \mathbf{X}\mathbf{b}$ for some $\mathbf{b}$ by the definition of $\mathcal{V}$. We have

$$\{\mathbf{y} - \mathbf{X}(\mathbf{X}^t\mathbf{X})^-\mathbf{X}^t\mathbf{y}\}^t\mathbf{X}\mathbf{b} = 0$$

using (ii) of Lemma 7.1.

The invariance of $\mathbf{X}(\mathbf{X}^t\mathbf{X})^-\mathbf{X}^t$ with respect to the choices of $(\mathbf{X}^t\mathbf{X})^-$ is due to the uniqueness of projection matrix. □

7.1.2 *Statistical Inference on Null-Full-Rank Regression Models*

With the least squares criterion, we seek $\widehat{\boldsymbol{\beta}}$ that minimizes

$$Q(\widehat{\boldsymbol{\beta}}) = \| \mathbf{y} - \mathbf{X}\boldsymbol{\beta} \|^2 .$$

Put in another way, we seek $\widehat{\boldsymbol{\mu}} = \mathbf{X}\boldsymbol{\beta}$ that minimizes

$$Q(\widehat{\boldsymbol{\beta}}) = \| \mathbf{y} - \boldsymbol{\mu} \|^2, \text{for all } \boldsymbol{\mu} \in \mathcal{V}.$$

The mean vector $\boldsymbol{\mu}$ is of form $\mathbf{X}\boldsymbol{\beta}$ and hence must be in $\mathcal{V}$. However,

$$\begin{aligned} Q(\widehat{\boldsymbol{\beta}}) &= \| \mathbf{y} - \boldsymbol{\mu} \|^2 \\ &= \| (\mathbf{y} - \mathbf{P}_{\mathcal{V}}\mathbf{y}) + (\mathbf{P}_{\mathcal{V}}\mathbf{y} - \boldsymbol{\mu}) \|^2 \\ &= \| \mathbf{y} - \mathbf{P}_{\mathcal{V}}\mathbf{y} \|^2 + \| \mathbf{P}_{\mathcal{V}}\mathbf{y} - \boldsymbol{\mu} \|^2, \quad \text{since } (\mathbf{y} - \mathbf{P}_{\mathcal{V}}\mathbf{y}) \perp \boldsymbol{\mu} \in \mathcal{V} \\ &\geq \| \mathbf{y} - \mathbf{P}_{\mathcal{V}}\mathbf{y} \|^2 . \end{aligned}$$

Note that $\mathbf{P}_{\mathcal{V}}\mathbf{y} \in \mathcal{V}$. Therefore, we have

$$\widehat{\mu} = \mathbf{P}_{\mathcal{V}}\mathbf{y} = \mathbf{X}(\mathbf{X}^t\mathbf{X})^-\mathbf{X}^t\mathbf{y} = \mathbf{X}\widehat{\boldsymbol{\beta}}$$

for any generalized inverse $(\mathbf{X}^t\mathbf{X})^-$. A solution of $\widehat{\boldsymbol{\beta}}$ is given by

$$\widehat{\boldsymbol{\beta}} = (\mathbf{X}^t\mathbf{X})^-\mathbf{X}^t\mathbf{y}. \tag{7.7}$$

Unfortunately, the form $\widehat{\boldsymbol{\beta}} = (\mathbf{X}^t\mathbf{X})^-\mathbf{X}^t\mathbf{y}$ is not invariant with different choices of $(\mathbf{X}^t\mathbf{X})^-$. Furthermore, the expected value of $\widehat{\boldsymbol{\beta}}$ is

$$E(\widehat{\boldsymbol{\beta}}) = (\mathbf{X}^t\mathbf{X})^-\mathbf{X}^t E(\mathbf{y}) = (\mathbf{X}^t\mathbf{X})^-\mathbf{X}^t\mathbf{X}\boldsymbol{\beta} \neq \boldsymbol{\beta}.$$

Namely, $\widehat{\boldsymbol{\beta}}$ in (7.7) is not an unbiased estimator of $\boldsymbol{\beta}$. In fact, $\boldsymbol{\beta}$ does not have a linear unbiased estimator (LUE) at all. Suppose that there exists a $(k+1) \times n$ matrix $\mathbf{A}$ such that $E(\mathbf{A}\mathbf{y}) = \boldsymbol{\beta}$. Then

$$\boldsymbol{\beta} = E(\mathbf{A}\mathbf{y}) = \mathbf{A}E(\mathbf{y}) = \mathbf{A}\mathbf{X}\boldsymbol{\beta},$$

which holds for any vector $\boldsymbol{\beta} \in \mathcal{R}^{k+1}$. So we must have $\mathbf{A}\mathbf{X} = \mathbf{I}_{k+1}$. This is however impossible because, otherwise,

$$k + 1 = \text{rank}(\mathbf{I}_{k+1}) = \text{rank}(\mathbf{A}\mathbf{X}) \leq \text{rank}(\mathbf{X}) = p < (k+1).$$

In fact, we can say that $\boldsymbol{\beta}$ is not *estimable* according to the following definition.

Definition 7.2. A linear combination of $\boldsymbol{\beta}$, $\boldsymbol{\lambda}^t\boldsymbol{\beta}$, is said to be *estimable* if it has an LUE. That is, there exists a vector $\mathbf{a} \in \mathcal{R}^n$ such that $E(\mathbf{a}^t\mathbf{y}) = \boldsymbol{\lambda}^t\boldsymbol{\beta}$.

The following theorem characterizes estimable functions by providing a necessary and sufficient condition.

Theorem 7.3. *In the non-full-rank linear model, $\boldsymbol{\lambda}^t\boldsymbol{\beta}$ is estimable if and only if $\boldsymbol{\lambda}^t = \mathbf{a}^t\mathbf{X}$, for some $\mathbf{a} \in \mathcal{R}^n$. In this case, $\boldsymbol{\lambda}^t\boldsymbol{\beta} = \mathbf{a}^t\mathbf{X}\boldsymbol{\beta} = \mathbf{a}^t\boldsymbol{\mu}$. In other words, functions of form $\mathbf{a}^t\boldsymbol{\mu}$ for any $\mathbf{a} \in \mathcal{R}^n$ are estimable and this form takes in ALL estimable functions available.*

Proof. If $\boldsymbol{\lambda}^t\boldsymbol{\beta}$ is estimable, then there must exist a vector $\mathbf{a}$ such that $E(\mathbf{a}^t\mathbf{y}) = \boldsymbol{\lambda}^t\boldsymbol{\beta}$. Consider $\boldsymbol{\lambda} \in R(\mathbf{X})$, consider

$$E(\mathbf{a}^t\mathbf{y}) = \mathbf{a}^t\mathbf{X}\boldsymbol{\beta} = \boldsymbol{\lambda}\boldsymbol{\beta}.$$

Since this holds for any $\boldsymbol{\beta}$, we have $\mathbf{a}^t\mathbf{X} = \boldsymbol{\lambda}^t$, which implies that $\boldsymbol{\lambda} \in R(\mathbf{X})$, the row space of $\mathbf{X}$ defined as

$$\mathcal{R}(\mathbf{X}) = \left\{\mathbf{r} \in \mathcal{R}^{k+1} : \mathbf{r} = \mathbf{a}^t\mathbf{X} \text{ for some } \mathbf{a} \in \mathcal{R}^n\right\}.$$

On the other hand, if $\mathbf{a^t X} = \boldsymbol{\lambda}^t$ for some vector $\mathbf{a}$, then

$$\begin{aligned} E\{\mathbf{a^t X}(\mathbf{X^t X})^- \mathbf{X}^t \mathbf{y}\} &= \mathbf{a^t X}(\mathbf{X^t X})^- \mathbf{X}^t E(\mathbf{y}) \\ &= \mathbf{a^t}\left\{\mathbf{X}(\mathbf{X^t X})^- \mathbf{X^t X}\right\}\boldsymbol{\beta} \\ &= \mathbf{a^t X}\boldsymbol{\beta} \quad \text{using Lemma 7.1(ii)}, \\ &= \boldsymbol{\lambda}^t \boldsymbol{\beta} \end{aligned}$$

Namely, $\boldsymbol{\lambda}^t\boldsymbol{\beta}$ has an LUE. □

A set of estimable functions $\boldsymbol{\lambda}_1^t\boldsymbol{\beta}$, $\boldsymbol{\lambda}_2^t\boldsymbol{\beta}$, ..., $\boldsymbol{\lambda}_m^t\boldsymbol{\beta}$, is said to be linearly independent if the vectors $\boldsymbol{\lambda}_1, \ldots, \boldsymbol{\lambda}_m$ are linearly independent. From the above theorem, $\boldsymbol{\lambda}_i \in R(\mathbf{X})$. In addition, every row of $\mathbf{X}\boldsymbol{\beta}$ must be estimable. Thus the following corollary becomes obvious.

Corollary 7.1. *The maximum number of linearly independent estimable functions of $\boldsymbol{\beta}$ is p, the rank of $\mathbf{X}$.*

The next question is how to come up with a best linear unbiased estimator (BLUE) for an estimable function $\boldsymbol{\lambda}^t\boldsymbol{\beta} = \mathbf{a}^t\boldsymbol{\mu}$. This question is nicely addressed by the following Guass-Markov theorem. It turns out that the solution can simply be obtained by plugging $\widehat{\boldsymbol{\beta}}$ or $\widehat{\boldsymbol{\mu}} = \mathbf{X}\widehat{\boldsymbol{\beta}}$.

Theorem 7.4. *(Gauss-Markov)* $\mathbf{a}^t\widehat{\boldsymbol{\mu}}$ *is the BLUE of* $\mathbf{a}^t\boldsymbol{\mu}$.

Proof. First

$$\mathbf{a}^t\widehat{\boldsymbol{\mu}} = \mathbf{a}^t\mathbf{P}_{\mathcal{V}}\mathbf{y} = \mathbf{a}^t\mathbf{X}(\mathbf{X^t X})^-\mathbf{X^t y}$$

is linearly in $\mathbf{y}$. Also, since $\boldsymbol{\mu} \in \mathcal{V}$ and hence $\mathbf{P}_{\mathcal{V}}\boldsymbol{\mu} = \boldsymbol{\mu}$, it follows that

$$E(\mathbf{a}^t\widehat{\boldsymbol{\mu}}) = \mathbf{a}^t\mathbf{P}_{\mathcal{V}}E(\mathbf{y}) = \mathbf{a}^t\mathbf{P}_{\mathcal{V}}\boldsymbol{\mu} = \mathbf{a}^t\boldsymbol{\mu}.$$

Therefore, $\mathbf{a}^t\widehat{\boldsymbol{\mu}}$ is an LUE of $\mathbf{a}^t\boldsymbol{\mu}$.

Furthermore,

$$\begin{aligned} \mathrm{Var}(\mathbf{a}^t\widehat{\boldsymbol{\mu}}) &= \mathrm{Var}(\mathbf{a}^t\mathbf{P}_{\mathcal{V}}\mathbf{y}) = \sigma^2\mathbf{a^t}\mathbf{P}_{\mathcal{V}}\mathbf{P}_{\mathcal{V}}^t\mathbf{a} \\ &= \sigma^2\mathbf{a^t}\mathbf{P}_{\mathcal{V}}\mathbf{a} = \sigma^2 \parallel \mathbf{P}_{\mathcal{V}}\mathbf{a} \parallel^2, \end{aligned}$$

since the projection matrix $\mathbf{P}_{\mathcal{V}}$ is idempotent. Suppose $\mathbf{b^t y}$ is any other LUE of $\mathbf{a}^t\boldsymbol{\mu}$. It follows from its unbiasedness that

$$\begin{aligned} & E(\mathbf{b^t y}) = \mathbf{b^t}\boldsymbol{\mu} = \mathbf{a}^t\boldsymbol{\mu} \\ \iff & \langle \mathbf{a} - \mathbf{b},\ \boldsymbol{\mu}\rangle = 0 \text{ for any } \boldsymbol{\mu} \in \mathcal{V}; \\ \iff & (\mathbf{a} - \mathbf{b}) \in \mathcal{V}^{\perp} \\ \iff & \mathbf{P}_{\mathcal{V}}(\mathbf{a} - \mathbf{b}) = 0 \text{ or } \mathbf{P}_{\mathcal{V}}\mathbf{a} = \mathbf{P}_{\mathcal{V}}\mathbf{b}. \end{aligned}$$

Consider

$$\begin{aligned}\mathrm{Var}(\mathbf{b^t y}) &= \sigma^2 \mathbf{b^t b} = \sigma^2 \parallel \mathbf{b} \parallel^2 = \sigma^2 \parallel \mathbf{b} - \mathbf{P}_{\mathcal{V}}\mathbf{b} \parallel^2 + \sigma^2 \parallel \mathbf{P}_{\mathcal{V}}\mathbf{b} \parallel^2 \\ &= \sigma^2 \parallel \mathbf{b} - \mathbf{P}_{\mathcal{V}}\mathbf{b} \parallel^2 + \sigma^2 \parallel \mathbf{P}_{\mathcal{V}}\mathbf{a} \parallel^2 \\ &\geq \sigma^2 \parallel \mathbf{P}_{\mathcal{V}}\mathbf{a} \parallel^2 = \mathrm{Var}(\mathbf{a}^t\widehat{\boldsymbol{\mu}}),\end{aligned}$$

with "=" held if and only if $\mathbf{b} = \mathbf{P}_{\mathcal{V}}\mathbf{b} = \mathbf{P}_{\mathcal{V}}\mathbf{a}$, in which case $\mathbf{b^t y} = \mathbf{a^t P}_{\mathcal{V}}\mathbf{y} = \mathbf{a}^t\widehat{\boldsymbol{\mu}}$. Therefore, $\mathbf{a}^t\widehat{\boldsymbol{\mu}}$ is the BLUE. □

The following corollary gives the BLUE of $\boldsymbol{\lambda}^t\boldsymbol{\beta}$.

Corollary 7.2. *If $\boldsymbol{\lambda}^t\boldsymbol{\beta}$ is estimable, then $\boldsymbol{\lambda}^t\widehat{\boldsymbol{\beta}} = \boldsymbol{\lambda}(\mathbf{X^t X})^{-}\mathbf{X^t y}$, for any generalized inverse $(\mathbf{X^t X})^{-}$, is the BLUE.*

Most theories in linear models can be justified using either properties of multivariate normal distributions or arguments from subspace and projects. The latter approach offers attractive geometric interpretations. We have followed this route to prove the Gauss-Markov (GM) theorem. It is clear that the arguments in the proof apply with entirety to the full-rank linear models. The key point about the GM theorem is that, in order for $\mathbf{a}^t\widehat{\boldsymbol{\mu}}$ to be the BLUE of $\mathbf{a}^t\boldsymbol{\mu}$, $\widehat{\boldsymbol{\mu}}$ needs to be the projection of $\mathbf{y}$ on $\mathcal{V} = C(\mathbf{X})$. Whether or not $\mathbf{X}$ is of full column rank only affects the way of computing the projection matrix.

The sum of squared error (SSE) in non-full-rank models can be analogously defined:

$$\begin{aligned}SSE &= (\mathbf{y} - \mathbf{X}\widehat{\boldsymbol{\beta}})^t(\mathbf{y} - \mathbf{X}\widehat{\boldsymbol{\beta}}) = \mathbf{y^t y} - \widehat{\boldsymbol{\beta}}^t\mathbf{X^t y} \\ &= \parallel \mathbf{y} - \mathbf{P}_{\mathcal{V}}\mathbf{y} \parallel^2 = \parallel \mathbf{P}_{\mathcal{V}^\perp}\mathbf{y} \parallel^2 = \mathbf{y}^t\mathbf{P}_{\mathcal{V}^\perp}\mathbf{y},\end{aligned} \tag{7.8}$$

which is invariant with different choices of $\widehat{\boldsymbol{\beta}}$. It is easy to check that $\mathrm{trace}(\mathbf{P}_{\mathcal{V}^\perp}) = n - \mathrm{rank}(X) = n - p$ and hence $E(SSE) = (n-p)\cdot\sigma^2$ using the fact that

$$E(\mathbf{y^t A y}) = \mathrm{trace}(\mathbf{A}\boldsymbol{\Sigma}) + \boldsymbol{\mu}^t\mathbf{A}\boldsymbol{\mu}$$

for $\mathbf{y} \sim (\boldsymbol{\mu}, \boldsymbol{\Sigma})$. A natural unbiased estimator of σ^2 is given by

$$\hat{\sigma}^2 = \frac{SSE}{n-p}. \tag{7.9}$$

It is important to note that we have not used the normality assumption in any of the steps so far. In other words, all the above results hold without the normality assumption.

Example We continue with the earlier one-way ANOVA example. It can be found that

$$\mathbf{X}^{\mathbf{t}}\mathbf{X} = \begin{pmatrix} n & n_1 & n_2 & \cdots & n_c \\ n_1 & n_1 & 0 & \cdots & 0 \\ n_2 & 0 & n_2 & \cdots & 0 \\ \vdots & \vdots & \vdots & & \vdots \\ n_c & 0 & 0 & \cdots & n_c \end{pmatrix} \tag{7.10}$$

and

$$\mathbf{X}^{\mathbf{t}}\mathbf{y} = \left(\sum_{i=1}^{c}\sum_{j=1}^{n_i} y_{ij}, \ \sum_{j=1}^{n_1} y_{1j}, \ \ldots, \ \sum_{j=1}^{n_c} y_{cj} \right)^t. \tag{7.11}$$

A generalized inverse of $\mathbf{X}^{\mathbf{t}}\mathbf{X}$ is given by

$$(\mathbf{X}^{\mathbf{t}}\mathbf{X})^{-} = \begin{pmatrix} 0 & 0 & 0 & \cdots & 0 \\ 0 & 1/n_1 & 0 & \cdots & 0 \\ 0 & 0 & 1/n_2 & \cdots & 0 \\ \vdots & \vdots & \vdots & & \vdots \\ 0 & 0 & 0 & \cdots & 1/n_c \end{pmatrix}. \tag{7.12}$$

With this generalized inverse, it can be found that

$$\widehat{\boldsymbol{\beta}} = (\mathbf{X}^{\mathbf{t}}\mathbf{X})^{-}\mathbf{X}^{\mathbf{t}}\mathbf{y} = (0, \ \bar{y}_1, \ \bar{y}_2, \ \ldots, \ \bar{y}_c)^t \tag{7.13}$$

and

$$\widehat{\boldsymbol{\mu}} = \mathbf{X}\widehat{\boldsymbol{\beta}} = (\hat{y}_1, \ldots, \hat{y}_1, \hat{y}_2, \ldots, \bar{y}_2, \ldots, \bar{y}_c, \ldots, \bar{y}_c)^t. \tag{7.14}$$

There are other choices for $(\mathbf{X}^{\mathbf{t}}\mathbf{X})^{-}$, which yield different forms of $\widehat{\boldsymbol{\beta}}$. Nevertheless, the predicted mean vector $\widehat{\boldsymbol{\mu}}$ would remain the same. □

Now we add the normality assumption so that the non-full-rank model is specified as

$$\mathbf{y} \sim \mathcal{N}\left(\mathbf{X}\boldsymbol{\beta}, \sigma^2\mathbf{I}\right), \text{ where } \mathbf{X} \text{ is } n \times (k+1) \text{ of rank } p. \tag{7.15}$$

It follows immediately that

$$\boldsymbol{\lambda}^t\widehat{\boldsymbol{\beta}} \sim \mathcal{N}\left\{\boldsymbol{\lambda}^t\boldsymbol{\beta}, \sigma^2 \cdot \boldsymbol{\lambda}^t(\mathbf{X}^{\mathbf{t}}\mathbf{X})^{-}\boldsymbol{\lambda}\right\} \tag{7.16}$$

for any estimable linear function $\boldsymbol{\lambda}^t\boldsymbol{\beta}$, since

$$\begin{aligned} \text{Cov}(\boldsymbol{\lambda}^t\widehat{\boldsymbol{\beta}}) &= \text{Cov}(\boldsymbol{\lambda}^t(\mathbf{X}^{\mathbf{t}}\mathbf{X})^{-}\mathbf{X}^{\mathbf{t}}\mathbf{y}) = \sigma^2\boldsymbol{\lambda}^t(\mathbf{X}^{\mathbf{t}}\mathbf{X})^{-}\mathbf{X}^{\mathbf{t}}\mathbf{X}(\mathbf{X}^{\mathbf{t}}\mathbf{X})^{-}\boldsymbol{\lambda} \\ &= \sigma^2\mathbf{a}^{\mathbf{t}}\mathbf{X}(\mathbf{X}^{\mathbf{t}}\mathbf{X})^{-}\mathbf{X}^{\mathbf{t}}\mathbf{X}(\mathbf{X}^{\mathbf{t}}\mathbf{X})^{-}\boldsymbol{\lambda}, \text{ plugging in } \boldsymbol{\lambda}^t = \mathbf{a}^{\mathbf{t}}\mathbf{X} \\ &= \sigma^2 \cdot \mathbf{a}^{\mathbf{t}}\mathbf{X}(\mathbf{X}^{\mathbf{t}}\mathbf{X})^{-}\boldsymbol{\lambda}, \text{ using Lemma 7.1(ii)} \\ &= \sigma^2 \cdot \boldsymbol{\lambda}^t(\mathbf{X}^{\mathbf{t}}\mathbf{X})^{-}\boldsymbol{\lambda}. \end{aligned}$$

Also,

$$\frac{(n-p)\cdot\hat{\sigma}^2}{\sigma^2} = \frac{SSE}{\sigma^2} \sim \chi^2(n-p). \tag{7.17}$$

Furthermore, it can be verified that $SSE = \| \mathbf{P}_{\mathcal{V}^\perp}\mathbf{y} \|^2$ is independent of $\widehat{\boldsymbol{\mu}} = \mathbf{P}_{\mathcal{V}}\mathbf{y}$ (and hence $\boldsymbol{\lambda}^t\widehat{\boldsymbol{\beta}}$), using the fact that $\mathbf{P}_{\mathcal{V}}\mathbf{y}$ and $\mathbf{P}_{\mathcal{V}^\perp}\mathbf{y}$ jointly follow a multivariate normal distribution with

$$\mathrm{Cov}(\mathbf{P}_{\mathcal{V}}\mathbf{y}, \mathbf{P}_{\mathcal{V}^\perp}\mathbf{y}) = \sigma^2\mathbf{P}_{\mathcal{V}}\mathbf{P}_{\mathcal{V}^\perp}^t = \mathbf{O}.$$

Putting (7.16), (7.17), and their independence together yields

$$t = \frac{\dfrac{(\boldsymbol{\lambda}^t\widehat{\boldsymbol{\beta}} - \boldsymbol{\lambda}^t\boldsymbol{\beta})}{\sqrt{\sigma^2\cdot\boldsymbol{\lambda}^t(\mathbf{X}^t\mathbf{X})^-\boldsymbol{\lambda}}}}{\sqrt{\dfrac{SSE/\sigma^2}{n-p}}} = \frac{(\boldsymbol{\lambda}^t\widehat{\boldsymbol{\beta}} - \boldsymbol{\lambda}^t\boldsymbol{\beta})}{\sqrt{\hat{\sigma}^2\cdot\boldsymbol{\lambda}^t(\mathbf{X}^t\mathbf{X})^-\boldsymbol{\lambda}}} \sim t(n-p) \tag{7.18}$$

according to the definition of the central t distribution. This fact can be used to construct confidence intervals or conduct hypothesis testings for $\boldsymbol{\lambda}^t\boldsymbol{\beta}$. For example, a $(1-\alpha)\times 100\%$ CI for an estimable $\boldsymbol{\lambda}^t\boldsymbol{\beta}$ is given by

$$\boldsymbol{\lambda}^t\widehat{\boldsymbol{\beta}} \pm t_{1-\alpha/2}^{(n-p)}\sqrt{\hat{\sigma}^2\cdot\boldsymbol{\lambda}^t(\mathbf{X}^t\mathbf{X})^-\boldsymbol{\lambda}}. \tag{7.19}$$

More generally, let $\boldsymbol{\Lambda}$ be an $m\times(k+1)$ matrix with rows λ_i^t for $i=1,\ldots,m$. $\boldsymbol{\Lambda}\boldsymbol{\beta}$ is said to be estimable if every of its components, $\lambda_i^t\boldsymbol{\beta}$, is estimable. Furthermore, we assume that $\boldsymbol{\Lambda}\boldsymbol{\beta}$ is a set of m linearly independent estimable functions, in which case $\boldsymbol{\Lambda}$ is of full row rank m. Under this assumption, it can be verified that matrix $\Lambda(\mathbf{X}^t\mathbf{X})^-\boldsymbol{\Lambda}^t$ is invertible (See Problem 0c). Following the same arguments as above, we have

$$\boldsymbol{\Lambda}\widehat{\boldsymbol{\beta}} \sim \mathcal{N}\left\{\boldsymbol{\Lambda}\boldsymbol{\beta}, \sigma^2\cdot\boldsymbol{\Lambda}(\mathbf{X}^t\mathbf{X})^-\boldsymbol{\Lambda}^t\right\}$$

and hence

$$\frac{(\boldsymbol{\Lambda}\widehat{\boldsymbol{\beta}} - \boldsymbol{\Lambda}\boldsymbol{\beta})^t\left[\Lambda(\mathbf{X}^t\mathbf{X})^-\boldsymbol{\Lambda}^t\right]^{-1}(\boldsymbol{\Lambda}\widehat{\boldsymbol{\beta}} - \boldsymbol{\Lambda}\boldsymbol{\beta})}{\sigma^2} \sim \chi^2(m), \tag{7.20}$$

which is independent of $SSE/\sigma^2 \sim \chi^2(n-p)$. Therefore, according to the definition of F distribution,

$$\begin{aligned} F &= \frac{(\boldsymbol{\Lambda}\widehat{\boldsymbol{\beta}} - \boldsymbol{\Lambda}\boldsymbol{\beta})^t\left[\Lambda(\mathbf{X}^t\mathbf{X})^-\boldsymbol{\Lambda}^t\right]^{-1}(\boldsymbol{\Lambda}\widehat{\boldsymbol{\beta}} - \boldsymbol{\Lambda}\boldsymbol{\beta})/(\sigma^2\cdot m)}{SSE/(\sigma^2\cdot(n-p))} \\ &= \frac{(\boldsymbol{\Lambda}\widehat{\boldsymbol{\beta}} - \boldsymbol{\Lambda}\boldsymbol{\beta})^t\left[\Lambda(\mathbf{X}^t\mathbf{X})^-\boldsymbol{\Lambda}^t\right]^{-1}(\boldsymbol{\Lambda}\widehat{\boldsymbol{\beta}} - \boldsymbol{\Lambda}\boldsymbol{\beta})/m}{SSE/(n-p)} \\ &\sim F(m, n-p), \end{aligned} \tag{7.21}$$

a fact being useful for testing general linear hypothesis $\mathrm{H}_0 : \boldsymbol{\Lambda}\boldsymbol{\beta} = \mathbf{b}$.

7.2 Generalized Least Squares

In an ordinary linear regression model, the assumption $\text{Cov}(\boldsymbol{\varepsilon}) = \sigma^2\mathbf{I}$ implies that y_i's are independent sharing the same constant variance σ^2. We now consider data in which the continuous responses are correlated or have differing variances. The model can be stated in a more general form given by

$$\mathbf{y} = \mathbf{X}\boldsymbol{\beta} + \boldsymbol{\varepsilon} \text{ with } \boldsymbol{\varepsilon} \sim \left(\mathbf{0}, \sigma^2\mathbf{V}\right), \tag{7.22}$$

where $\mathbf{X}$ is $n \times (p+1)$ of full rank $(p+1)$ and $\mathbf{V}$ is a *known* positive definite matrix. The notation $\boldsymbol{\varepsilon} \sim \left(\mathbf{0}, \sigma^2\mathbf{V}\right)$ means that $E(\boldsymbol{\varepsilon}) = \mathbf{0}$ and $\text{Cov}(\boldsymbol{\varepsilon}) = \sigma^2\mathbf{V}$.

Estimation of Model (7.22) can be generally proceeded in the following way. By applying a transformation, we first establish a connection between model (7.22) and an ordinary linear regression model where random errors are independent with constant variance. All the statistical inference associated with model (7.22) can then be made through available results with the ordinary linear model.

Using spectral decomposition, it can be established that $\mathbf{V}$ is positive definite if and only if there exists a nonsingular matrix $\mathbf{P}$ such that $\mathbf{V} = \mathbf{P}\mathbf{P}^t$. In particular, $V^{1/2}$ satisfies this condition. Multiplying Model (7.22) by $\mathbf{P}^{-1}$ on both sides yields that

$$\mathbf{P}^{-1}\mathbf{y} = \mathbf{P}^{-1}\mathbf{X}\boldsymbol{\beta} + \mathbf{P}^{-1}\boldsymbol{\varepsilon},$$

where

$$\begin{aligned}
\text{Cov}(\mathbf{P}^{-1}\boldsymbol{\varepsilon}) &= \mathbf{P}^{-1}\text{Cov}(\boldsymbol{\varepsilon})(\mathbf{P}^{-1})^t \\
&= \mathbf{P}^{-1}\sigma^2\mathbf{V}(\mathbf{P}^{-1})^t = \sigma^2 \cdot \mathbf{P}^{-1}\mathbf{P}\mathbf{P}^{\mathbf{t}}(\mathbf{P}^{-1})^t \\
&= \sigma^2\mathbf{I}.
\end{aligned}$$

Define $\mathbf{u} = \mathbf{P}^{-1}\mathbf{y}$, $\mathbf{Z} = \mathbf{P}^{-1}\mathbf{X}$, and $\mathbf{e} = \mathbf{P}^{-1}\boldsymbol{\varepsilon}$. Then Model (7.22) can be written as an ordinary linear model form

$$\mathbf{u} = \mathbf{Z}\boldsymbol{\beta} + \mathbf{e} \text{ with } \mathbf{e} \sim \left(\mathbf{0}, \sigma^2\mathbf{I}\right), \tag{7.23}$$

where the design matrix $\mathbf{Z}$ is $n \times (p+1)$ also of full rank $(p+1)$. All the inferences with Model (7.22) can be first processed through Model (7.23) and then re-expressed in terms of quantities in (7.22).

7.2.1 *Estimation of* $(\boldsymbol{\beta}, \sigma^2)$

First of all, the least squares criterion is

$$\begin{aligned} Q(\boldsymbol{\beta}) &= (\mathbf{u} - \mathbf{Z}\boldsymbol{\beta})^t(\mathbf{u} - \mathbf{Z}\boldsymbol{\beta}) \\ &= (\mathbf{y} - \mathbf{X}\boldsymbol{\beta})^t(\mathbf{P}^{-1})^t\mathbf{P}^{-1}(\mathbf{y} - \mathbf{X}\boldsymbol{\beta}) \\ &= (\mathbf{y} - \mathbf{X}\boldsymbol{\beta})^t\mathbf{V}^{-1}(\mathbf{y} - \mathbf{X}\boldsymbol{\beta}) \end{aligned} \tag{7.24}$$

According to Model (7.23), the least squares estimator (LSE)

$$\widehat{\boldsymbol{\beta}} = \left(\mathbf{Z}^t\mathbf{Z}\right)^{-1}\mathbf{Z}^t\mathbf{u}$$

is the BLUE of $\boldsymbol{\beta}$. Re-expressing it gives

$$\begin{aligned} \widehat{\boldsymbol{\beta}} &= \left\{(\mathbf{P}^{-1}\mathbf{X})^t(\mathbf{P}^{-1}\mathbf{X})\right\}^{-1}(\mathbf{P}^{-1}\mathbf{X})^t\mathbf{P}^{-1}\mathbf{y} \\ &= \left\{\mathbf{X}^t(\mathbf{P}^t)^{-1}\mathbf{P}^{-1}\mathbf{X}\right\}^{-1}\mathbf{X}^t(\mathbf{P}^t)^{-1}\mathbf{P}^{-1}\mathbf{y} \\ &= \left\{\mathbf{X}^t(\mathbf{P}\mathbf{P}^t)^{-1}\mathbf{X}\right\}^{-1}\mathbf{X}^t(\mathbf{P}\mathbf{P}^t)^{-1}\mathbf{y} \\ &= \left\{\mathbf{X}^t\mathbf{V}^{-1}\mathbf{X}\right\}^{-1}\mathbf{X}^t\mathbf{V}^{-1}\mathbf{y}. \end{aligned} \tag{7.25}$$

The estimator of this form $\widehat{\boldsymbol{\beta}} = \left\{\mathbf{X}^t\mathbf{V}^{-1}\mathbf{X}\right\}^{-1}\mathbf{X}^t\mathbf{V}^{-1}\mathbf{y}$ is often termed as the *generalized least squares* estimator (GLSE). At the same time, the variance-covariance matrix of $\widehat{\boldsymbol{\beta}}$ can also be obtained via Model (7.23):

$$\begin{aligned} \text{Cov}(\widehat{\boldsymbol{\beta}}) &= \sigma^2\left(\mathbf{Z}^t\mathbf{Z}\right)^{-1} = \sigma^2\left\{(\mathbf{P}^{-1}\mathbf{X})^t\mathbf{P}^{-1}\mathbf{X}\right\}^{-1} \\ &= \sigma^2\left\{\mathbf{X}^t\left(\mathbf{P}\mathbf{P}^t\right)^{-1}\mathbf{X}\right\}^{-1} \\ &= \sigma^2\left(\mathbf{X}^t\mathbf{V}^{-1}\mathbf{X}\right)^{-1}. \end{aligned} \tag{7.26}$$

Similarly, an unbiased estimator of σ^2 is

$$\begin{aligned} \hat{\sigma}^2 &= \frac{(\mathbf{u} - \mathbf{Z}\widehat{\boldsymbol{\beta}})^t(\mathbf{u} - \mathbf{Z}\widehat{\boldsymbol{\beta}})}{n - (p+1)} \\ &= \frac{(\mathbf{V}^{-1}\mathbf{y} - \mathbf{V}^{-1}\mathbf{X}\widehat{\boldsymbol{\beta}})^t(\mathbf{V}^{-1}\mathbf{y} - \mathbf{V}^{-1}\mathbf{X}\widehat{\boldsymbol{\beta}})}{n - (p+1)} \\ &= \frac{(\mathbf{y} - \mathbf{X}\widehat{\boldsymbol{\beta}})^t\mathbf{V}^{-1}(\mathbf{y} - \mathbf{X}\widehat{\boldsymbol{\beta}})}{n - (p+1)} \end{aligned} \tag{7.27}$$

Expanding the numerator and plugging (7.25) give

$$\widehat{\sigma}^2 = \frac{\mathbf{y}^t\left\{\mathbf{V}^{-1} - \mathbf{V}^{-1}\mathbf{X}(\mathbf{X}^t\mathbf{V}^{-1}\mathbf{X})^{-1}\mathbf{X}^t\mathbf{V}^{-1}\right\}\mathbf{y}}{n - (p+1)}. \tag{7.28}$$

Many other measures associated with the GLS model (7.22) can be extracted in a similar manner. For example, the project or hat matrix $\mathbf{H}$ in GLS is

$$\begin{aligned} \mathbf{H} &= \mathbf{Z}(\mathbf{Z}^t\mathbf{Z})^{-1}\mathbf{Z}^t \\ &= \mathbf{P}^{-1}\mathbf{X}(\mathbf{X}^t\mathbf{V}^{-1}\mathbf{X})^{-1}\mathbf{X}^t\mathbf{P}^{-1} \\ &= \mathbf{V}^{-1/2}\mathbf{X}(\mathbf{X}^t\mathbf{V}^{-1}\mathbf{X})^{-1}\mathbf{X}^t\mathbf{V}^{-1/2}, \text{ if } \mathbf{P} = \mathbf{V}^{1/2} \text{ is used.} \end{aligned} \tag{7.29}$$

7.2.2 Statistical Inference

In order to facilitate statistical inference, we add the normality assumption so that $\boldsymbol{\varepsilon} \sim \mathcal{N}\left(\mathbf{0}, \sigma^2\mathbf{V}\right)$. Hence Model (7.22) becomes

$$\mathbf{u} = \mathbf{Z}\boldsymbol{\beta} + \mathbf{e} \text{ with } \mathbf{e} \sim \mathcal{N}\left(\mathbf{0},\, \sigma^2\mathbf{I}\right). \tag{7.30}$$

Again this allows us to apply the established results directly and then convert back to the GLS model (7.22).

First of all, $\widehat{\boldsymbol{\beta}}$ and $\widehat{\sigma}^2$ are independent with

$$\widehat{\boldsymbol{\beta}} \sim \mathcal{N}\left\{\boldsymbol{\beta}, \sigma^2(\mathbf{X}^t\mathbf{V}^{-1}\mathbf{X})^{-1}\right\}, \tag{7.31}$$

$$\frac{\{n-(p+1)\}\cdot\widehat{\sigma}^2}{\sigma^2} \sim \chi^2\{n-(p+1)\}. \tag{7.32}$$

These distributional properties make available the statistical inference on an individual regression parameter β_j and the variance parameter σ^2. For example, one may make the following probability statement according to (7.32)

$$\mathrm{P}\left\{\chi^2_{\alpha/2}(n-p-1) < \frac{\{n-(p+1)\}\cdot\widehat{\sigma}^2}{\sigma^2} < \chi^2_{1-\alpha/2}(n-p-1)\right\} = 1-\alpha,$$

where $\chi^2_{\alpha/2}(n-p-1)$ denotes the $(\alpha/2)$-th percentile of the χ^2 with $(n-p-1)$ df. It follows that a $(1-\alpha)\times 100\%$ CI for σ^2 can be given by

$$\left\{\frac{\{n-(p+1)\}\cdot\widehat{\sigma}^2}{\chi^2_{1-\alpha/2}(n-p-1)}, \frac{\{n-(p+1)\}\cdot\widehat{\sigma}^2}{\chi^2_{\alpha/2}(n-p-1)}\right\}.$$

Consider testing a general linear hypothesis about $\boldsymbol{\beta}$ of form $H_0 : \boldsymbol{\Lambda}\boldsymbol{\beta} = \mathbf{b}$, where $\boldsymbol{\Lambda}$ is $m \times (p+1)$ of rank $m < (p+1)$. Working with Model (7.30), the F test applies with test statistic given by

$$\begin{aligned} F &= \frac{(\boldsymbol{\Lambda}\widehat{\boldsymbol{\beta}} - \mathbf{b})^t \left\{\boldsymbol{\Lambda}(\mathbf{Z}^t\mathbf{Z})^{-1}\boldsymbol{\Lambda}^t\right\}^{-1} (\boldsymbol{\Lambda}\widehat{\boldsymbol{\beta}} - \mathbf{b})\,/\,m}{\widehat{\sigma}^2} \\ &= \frac{(\boldsymbol{\Lambda}\widehat{\boldsymbol{\beta}} - \mathbf{b})^t \left\{\boldsymbol{\Lambda}(\mathbf{X}^t\mathbf{V}^{-1}\mathbf{X})^{-1}\boldsymbol{\Lambda}^t\right\}^{-1} (\boldsymbol{\Lambda}\widehat{\boldsymbol{\beta}} - \mathbf{b})\,/\,m}{\widehat{\sigma}^2}, \end{aligned}$$

using $\mathbf{Z}^t\mathbf{Z} = \mathbf{X}^t\mathbf{V}^{-1}\mathbf{X}$. The test statistic F is distributed as $F(m, n-p-1, \lambda)$ with noncentrality parameter

$$\lambda = \frac{(\boldsymbol{\Lambda}\boldsymbol{\beta} - \mathbf{b})^t \left\{\boldsymbol{\Lambda}(\mathbf{X}^t\mathbf{V}^{-1}\mathbf{X})^{-1}\boldsymbol{\Lambda}^t\right\}^{-1} (\boldsymbol{\Lambda}\boldsymbol{\beta} - \mathbf{b})}{\sigma^2},$$

which reduces to central $F(m, n-p-1)$ under H_0.

Under the normality assumption, estimation of $(\boldsymbol{\beta}, \sigma^2)$ can alternatively be done via maximum likelihood. The likelihood function would be

$$\begin{aligned} l(\boldsymbol{\beta}, \sigma^2) &= \frac{1}{(2\pi)^{n/2} |\sigma^2 \mathbf{V}|^{1/2}} \exp\left\{ -\frac{(\mathbf{y} - \mathbf{X}\boldsymbol{\beta})^t (\sigma^2 \mathbf{V})^{-1} (\mathbf{y} - \mathbf{X}\boldsymbol{\beta})}{2} \right\} \\ &= \frac{1}{(2\pi\sigma^2)^{n/2} |\mathbf{V}|^{1/2}} \exp\left\{ -\frac{(\mathbf{y} - \mathbf{X}\boldsymbol{\beta})^t \mathbf{V}^{-1} (\mathbf{y} - \mathbf{X}\boldsymbol{\beta})}{2\sigma^2} \right\}, \end{aligned}$$

since $|\sigma^2 \mathbf{V}| = (\sigma^2)^n |\mathbf{V}|$. Differentiation of the log-likelihood $\log l(\boldsymbol{\beta}, \sigma^2)$ with respect to $\boldsymbol{\beta}$ and σ^2 leads to their MLEs

$$\widehat{\boldsymbol{\beta}} = (\mathbf{X}^t \mathbf{V}^{-1} \mathbf{X})^{-1} \mathbf{X}^t \mathbf{V}^{-1} \mathbf{y}, \tag{7.33}$$

$$\widehat{\sigma}^2 = \frac{(\mathbf{y} - \mathbf{X}\widehat{\boldsymbol{\beta}})^t \mathbf{V}^{-1} (\mathbf{y} - \mathbf{X}\widehat{\boldsymbol{\beta}})}{n}. \tag{7.34}$$

7.2.3 *Misspecification of the Error Variance Structure*

Suppose that the true model underlying the data is $\mathbf{y} = \mathbf{X}\boldsymbol{\beta} + \boldsymbol{\varepsilon}$ with $\boldsymbol{\varepsilon} \sim (\mathbf{0}, \sigma^2 \mathbf{V})$ in (7.22), and one mistakenly fits ordinary linear model with constant variance for the error terms. Let $\widehat{\boldsymbol{\beta}}^{\star} = (\mathbf{X}^t \mathbf{X})^{-1} \mathbf{X}^t \mathbf{y}$ denote the resultant least squares estimator. We will compare it to the BLUE $\widehat{\boldsymbol{\beta}} = (\mathbf{X}^t \mathbf{V}^{-1} \mathbf{X})^{-1} \mathbf{X}' \mathbf{V}^{-1} \mathbf{y}$ in (7.25) to investigate the effects of mis-specification of the error variance structure.

First consider

$$\begin{aligned} E(\widehat{\boldsymbol{\beta}}^{\star}) &= E\left\{ (\mathbf{X}^t \mathbf{X})^{-1} \mathbf{X}^t \mathbf{y} \right\} = (\mathbf{X}^t \mathbf{X})^{-1} \mathbf{X}^t E(\mathbf{y}) \\ &= (\mathbf{X}^t \mathbf{X})^{-1} \mathbf{X}^t \mathbf{X}\boldsymbol{\beta} \\ &= \boldsymbol{\beta}. \end{aligned}$$

Namely, $\widehat{\boldsymbol{\beta}}^{\star}$ remains unbiased for $\boldsymbol{\beta}$ with misspecified error variance structure. Next, consider

$$\begin{aligned} \text{Cov}(\widehat{\boldsymbol{\beta}}^{\star}) &= \text{Cov}\left\{ (\mathbf{X}^t \mathbf{X})^{-1} \mathbf{X}^t \mathbf{y} \right\} = (\mathbf{X}^t \mathbf{X})^{-1} \mathbf{X}^t \text{Cov}(\mathbf{y}) \mathbf{X} (\mathbf{X}^t \mathbf{X})^{-1} \\ &= \sigma^2 \cdot (\mathbf{X}^t \mathbf{X})^{-1} \mathbf{X}^t \mathbf{V} \mathbf{X} (\mathbf{X}^t \mathbf{X})^{-1}. \end{aligned} \tag{7.35}$$

Comparatively, $\text{Cov}(\widehat{\boldsymbol{\beta}}) = \sigma^2 (\mathbf{X}^t \mathbf{V}^{-1} \mathbf{X})^{-1}$ as in (7.26). Since $\widehat{\boldsymbol{\beta}}$ is the BLUE of $\boldsymbol{\beta}$ and $\widehat{\boldsymbol{\beta}}^{\star}$ is an LUE, we must have that matrix $\text{Cov}(\widehat{\boldsymbol{\beta}}^{\star}) - \text{Cov}(\widehat{\boldsymbol{\beta}})$ is semi-positive definite (see Problem 7.4). As a result, the standard errors of the components $\widehat{\beta}_j^{\star}$ in $\widehat{\boldsymbol{\beta}}^{\star}$ are typically larger than the standard errors of the components $\widehat{\beta}_j$ in $\widehat{\boldsymbol{\beta}}$, which may lead to statistical insignificance for important predictors. In model diagnostics, unusually large standard errors are often recognized as signals of misspecified error variance structure such as heteroscedasticity, although other model deficiencies could be the cause.

7.2.4 *Typical Error Variance Structures*

In the section, we list and discuss several common structures for the error variance.

Weighted Least Squares

The first case is when $\mathbf{V} = \text{diag}(v_i)$ is diagonal. Thus $\mathbf{P}^{-1} = \text{diag}(1/\sqrt{v_i})$. The least squares criterion in (7.24) becomes

$$Q(\boldsymbol{\beta}) = \sum_{i=1}^{n} \frac{1}{v_i} \left(y_i - \mathbf{x}_i^t \boldsymbol{\beta}\right)^2.$$

The quantity $1/v_i$ plays the role of weight for the i-th observation. Accordingly, the resultant estimate of $\boldsymbol{\beta}$ is termed as a *weighted least squares* (WLS) estimate. The diagonal structure of $\mathbf{V}$ implies independent errors yet with nonconstant variances, which renders WLS a useful remedial technique for handling the heteroscedasticity problem.

Note that the weights $1/v_i$ are inversely proportional to the variances $\sigma^2 v_i$. Thus, observations y_i with larger variance receive less weight and hence have less influence on the analysis than observations with smaller variance. This weighting strategy intuitively makes sense because the more precise is y_i (i.e., with smaller variance), the more information it provides about $E(y_i)$, and therefore the more weight it should receive in fitting the regression function.

It is important to notice that we have assumed the weights are specified explicitly, or known up to a proportionality constant. This assumption is not entirely unrealistic. In situations when the response variable Y_i itself is the average of a unit containing m_i individuals from a population, the unit size m_i can be naturally used as the weight, recalling that $\text{Var}(Y_i)$ is equal to the population variance divided by m_i. In addition, the error variances sometimes vary with a predictor variable X_j in a systematic fashion. The relation, for instance, could be $\text{Var}(\varepsilon_i) = \sigma^2 x_{ij}$. Hence $\mathbf{V} = \text{diag}(x_{ij})$. The associated weights are $1/x_{ij}$.

On the other hand, the matrix $\mathbf{V}$ is usually unknown in practice. The weights must be estimated by modeling the heteroscedasticity. One common approach formulates the variance as a known function of predictors or the expected response. See, e.g., Carroll and Ruppert (1988) for a detailed treatment. An *iterative weighted least squares* (IWLS) procedure can be employed to get improved estimation. In this procedure, one initially estimates the weights from the data, and then obtain the WLS estimates of $\boldsymbol{\beta}$. Based on the residuals, the weights are re-estimated and the WLS estimation is updated. The process is iterated until the WLS fit gets stable.

AR(1) Variance Structure

Variance structures from time series models can be employed to model data collected over time in longitudinal studies. Consider a multiple linear regression model with random errors that follow a first-order autoregressive process:

$$y_i = \mathbf{x}_i^t \boldsymbol{\beta} + \varepsilon_i \text{ with } \varepsilon_i = \rho \varepsilon_{i-1} + \nu_i, \tag{7.36}$$

where $\nu_i \sim \mathcal{N}(0, \sigma^2)$ independently and $-1 < \rho < 1$.

To find out the matrix $\mathbf{V}$, we first rewrite ε_i

$$\begin{aligned}
\varepsilon_i &= \rho \cdot \varepsilon_{i-1} + \nu_i \\
&= \rho \cdot (\rho \varepsilon_{i-2} + \nu_{i-1}) + \nu_i \\
&= \nu_t + \rho \nu_{i-1} + \rho^2 \nu_{i-2} + \rho^3 \nu_{i-3} + \cdots \\
&= \sum_{s=0}^{\infty} \rho^s \nu_{i-s}.
\end{aligned}$$

It follows that $E(\varepsilon_i) = \sum_{s=0}^{\infty} \rho^s E(\nu_{i-s}) = 0$ and

$$\text{Var}(\varepsilon_i) = \sum_{s=0}^{\infty} \rho^{2s} \text{Var}(\nu_{i-s}) = \sigma^2 \sum_{s=0}^{\infty} \rho^{2s} = \frac{\sigma^2}{1-\rho^2}. \tag{7.37}$$

Furthermore, the general formula for the covariance between two observations can be derived. Consider

$$\begin{aligned}
\text{Cov}(\varepsilon_i, \varepsilon_{i-1}) &= E(\varepsilon_i \varepsilon_{i-1}) = E\left\{(\rho\, \varepsilon_{i-1} + \nu_i) \cdot \varepsilon_{i-1}\right\} \\
&= E\left\{\rho\, \varepsilon_{i-1}^2 + \nu_i \varepsilon_{i-1}\right\} = \rho \cdot E(\varepsilon_{i-1}^2) \\
&= \rho \cdot \text{Var}(\varepsilon_{i-1}) = \rho \cdot \frac{\sigma^2}{1-\rho^2}
\end{aligned}$$

Also, using $\varepsilon_i = \rho \cdot \varepsilon_{i-1} + \nu_i = \rho \cdot (\rho \varepsilon_{i-2} + \nu_{i-1}) + \nu_i$, it can be found that

$$\begin{aligned}
\text{Cov}(\varepsilon_i, \varepsilon_{i-2}) &= E(\varepsilon_i \varepsilon_{i-2}) = E\left[\{\rho \cdot (\rho \varepsilon_{i-2} + \nu_{i-1}) + \nu_i\} \cdot \varepsilon_{i-2}\right] \\
&= \rho^2 \cdot \frac{\sigma^2}{1-\rho^2}.
\end{aligned}$$

In general, the covariance between error terms that are s steps apart is

$$\text{Cov}(\varepsilon_i, \varepsilon_{i-s}) = \rho^s \frac{\sigma^2}{1-\rho^2}. \tag{7.38}$$

Thus, the correlation coefficient is

$$\text{cov}(\varepsilon_i, \varepsilon_{i-s}) = \rho^s,$$

which indicates a smaller correlation for any two observations that are further apart.

From (7.37) and (7.38), we have

$$\mathbf{V} = \frac{\sigma^2}{1-\rho^2} \cdot \begin{pmatrix} 1 & \rho & \rho^2 & \dots \rho^{n-1} \\ \rho & 1 & \rho & \dots \rho^{n-2} \\ \rho^2 & \rho & 1 & \dots \rho^{n-3} \\ \vdots & \vdots & \vdots & \vdots \\ \rho^{n-1} & \rho^{n-2} & \rho^{n-3} & \dots 1 \end{pmatrix},$$

with inverse

$$\mathbf{V}^{-1} = \frac{1}{\sigma^2} \cdot \begin{pmatrix} 1 & -\rho & 0 & \dots 0 & 0 & 0 \\ -\rho & 1+\rho^2 & -\rho & \dots 0 & 0 & 0 \\ 0 & -\rho & 1+\rho^2 & \dots 0 & 0 & 0 \\ \vdots & \vdots & \vdots & \vdots & \vdots & \vdots \\ 0 & 0 & 0 & \dots -\rho & 1+\rho^2 & -\rho \\ 0 & 0 & 0 & \dots 0 & -\rho & 1 \end{pmatrix}.$$

Note that $\rho = 0$ implies independent random errors. The Durbin-Watson test (Durbin and Watson, 1951) for $H_0 : \rho = 0$ is one of the few tests available for checking the independence assumption in linear regression. The test statistic is computed using the residuals from the ordinary least squares fit $\hat{\varepsilon}_i = y_i - \hat{y}_i = y_i - \mathbf{x}_i^t \widehat{\boldsymbol{\beta}}$:

$$DW = \frac{\sum_{i=2}^{n} (\hat{\varepsilon}_i - \hat{\varepsilon}_{i-1})^2}{\sum_{i=1}^{n} \hat{\varepsilon}_i^2}. \tag{7.39}$$

Durbin and Watson (1951) have also obtained the critical values of DW for decision purposes.

Common Correlation

Another variance structure is

$$\mathbf{V} = \begin{pmatrix} 1 & \rho & \dots & \rho \\ \rho & 1 & \dots & \rho \\ \vdots & \vdots & \vdots & \\ \rho & \rho & \dots & 1 \end{pmatrix},$$

in which all observations have the same variance and all pairs of them have the same correlation ρ. This structure is appropriate in certain repeated measures designs. Rewrite

$$\mathbf{V} = (1-\rho)\mathbf{I} + \rho\mathbf{J}, \tag{7.40}$$

where $\mathbf{J}$ is the $n \times n$ matrix with all elements being 1. Its inverse is given by

$$\mathbf{V}^{-1} = \frac{1}{1-\rho}\left(\mathbf{I} - \frac{\rho}{1+(n-1)\rho}\mathbf{J}\right).$$

It can be, very interestingly, shown (see, e.g., Rencher, 2000) that, with the covariance structure in (7.40), the ordinary least squares estimates $\widehat{\boldsymbol{\beta}}^{\star} = (\mathbf{X}^t\mathbf{X})^{-1}\mathbf{X}^t\mathbf{y}$ turns out to be the BLUE of $\boldsymbol{\beta}$.

7.2.5 *Example*

A data set is collected on 54 adult women in a study of the relationship between diastolic blood pressure (DBP) and age. Fig. 7.1(a) gives a scatter plot of the data. There seems to be a very nice linear association, with greater variability are associated with larger DBO values though. Although transformations on the response or predictors may help stabilize the variability, one might prefer the original form of DBP and age for better interpretation.

A simple linear regression model is first fit with ordinary least squares (OLS). The estimated coefficients are given in Panel (a) of Table 7.2. Fig. 7.1(b) plots the resultant residuals from OLS versus age. It can be seen that the variation increases dramatically with age, signaling heterogeneous errors.

In order to apply the weighted least squares method, one must estimate the weights. Common methods formulate the weights or the error variance as a function of the predictors or the expected response. Alternatively, one may group data in a way such that observations within the same group share a common variance.

Fig. 7.1 (c) and (d) plot $|\widehat{\varepsilon}_i|$ and $\hat{\varepsilon}_i^2$ versus age, respectively. In order to estimate weights, we regress the absolute value $|\widehat{\varepsilon}_i|$ of residuals linearly on age. More specifically, the following steps are taken to obtain the WLS fit.

- Fit simple linear regression of $|\widehat{\varepsilon}_i|$ versus age;
- Obtain the fitted values, denoted by $\hat{r}_i$;
- Compute weights as $v_i = 1/\hat{r}_i^2$.
- Refit the linear model of DBP versus age with weights v_i.

Panel (b) of Table 7.2 presents the resulting WLS fit, obtained from PROC REG with the WEIGHT statement. It can be seen that the LSE and WLSE are very close to each other. This is not surprising as both are unbiased

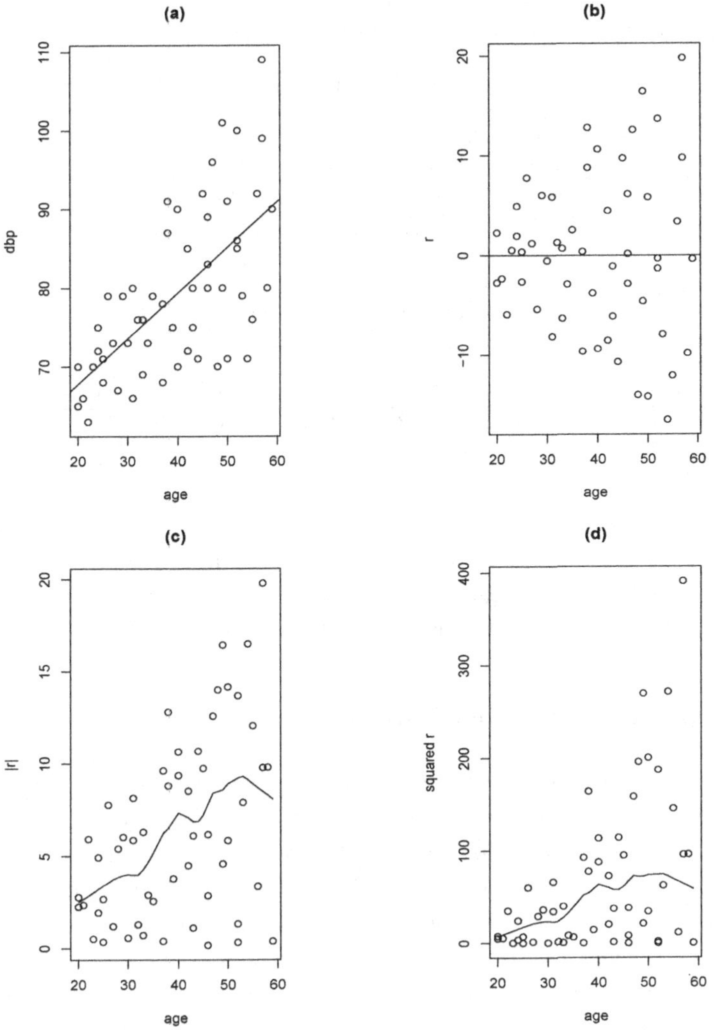

Fig. 7.1 (a) The Scatter Plot of DBP Versus Age; (b) Residual versus Age; (c) Absolute Values of Residuals $|\hat{\varepsilon}_i|$ Versus Age; (d) Squared Residuals $\hat{\varepsilon}_i^2$ Versus Age.

estimates. However, the standard errors from WLS are considerably smaller than those from OLS. As a result, estimates from WLS are statistically more significant.

One may repeat the above steps by iteratively updating the weights and

Table 7.2 Analysis of the Diastolic Blood Pressure Data: (a) Ordinary Least Squares (OLS) Estimates; (b) Weighted Least Squares (WLS) Fitting with Weights Derived from Regressing Absolute Values $|\hat{\varepsilon}_i|$ of the Residuals On Age.

(a) Parameter Estimates from OLS

Variable	DF	Parameter Estimate	Standard Error	t Value	Pr > $\|t\|$	95% Confidence Limits	
Intercept	1	56.15693	3.99367	14.06	<.0001	48.14304	64.17082
age	1	0.58003	0.09695	5.98	<.0001	0.38548	0.77458

(b) WLS Estimates with Weights Derived from $|\hat{\varepsilon}_i|$.

Variable	DF	Parameter Estimate	Standard Error	t Value	Pr > $\|t\|$	95% Confidence Limits	
Intercept	1	55.56577	2.52092	22.04	<.0001	50.50718	60.62436
age	1	0.59634	0.07924	7.53	<.0001	0.43734	0.75534

the WLS fitting. Usually, convergence occurs after a couple of steps. This is because the updated weights are not very different from the initial ones due to the unbiasedness or consistency of OLS estimates.

7.3 Ridge Regression and LASSO

Ridge regression is one of the remedial measures for handling severe multicollinearity in least squares estimation. Multicollinearity occurs when the predictors included in the linear model are highly correlated with each other. When this is the case, the matrix $\mathbf{X}^t\mathbf{X}$ tends to be singular or ill-conditioned and hence identifying the least squares estimates will encounter numerical problems.

To motivate the ridge estimator, we first take a look at the mean squared error, $MSE(\mathbf{b}) = E \parallel \mathbf{b} - \boldsymbol{\beta} \parallel^2$, of least squares estimator of $\boldsymbol{\beta}$. MSE is a commonly-used measure for assessing quality of estimation, which can break into two parts: the squared bias plus the variance:

$$E \parallel \mathbf{b} - \boldsymbol{\beta} \parallel^2 = \sum_j E(b_j - \beta_j)^2 = \sum_j \{E(b_j) - \beta_j\}^2 + \sum_j \text{Var}(b_j). \quad (7.41)$$

According to Gauss-Markov theorem, the least sqaures approach achieves the smallest variance among all unbiased linear estimates. This however does not necessarily guarantee the minimum MSE. To better distinguish different types of estimators, let $\widehat{\boldsymbol{\beta}}^{\mathrm{LS}}$ denote the ordinary least squares estimator of $\boldsymbol{\beta}$. Recall that $E(\widehat{\boldsymbol{\beta}}^{\mathrm{LS}}) = \boldsymbol{\beta}$ and $\mathrm{Cov}(\widehat{\boldsymbol{\beta}}^{\mathrm{LS}}) = \sigma^2 \cdot (\mathbf{X}^t\mathbf{X})^{-1}$. We have

$$\begin{aligned} MSE(\widehat{\boldsymbol{\beta}}^{\mathrm{LS}}) &= E \parallel \widehat{\boldsymbol{\beta}}^{\mathrm{LS}} \parallel^2 - \parallel \boldsymbol{\beta} \parallel^2 \\ &= \mathrm{tr}\left\{\sigma^2(\mathbf{X}^t\mathbf{X})^{-1}\right\} = \sigma^2 \cdot \mathrm{tr}\left\{(\mathbf{X}^t\mathbf{X})^{-1}\right\}. \end{aligned} \tag{7.42}$$

Thus

$$E(\parallel \widehat{\boldsymbol{\beta}}^{\mathrm{LS}} \parallel^2) = \parallel \boldsymbol{\beta} \parallel^2 + \sigma^2 \cdot \mathrm{tr}\left\{(\mathbf{X}^t\mathbf{X})^{-1}\right\}. \tag{7.43}$$

It can be seen that, with ill-conditioned $\mathbf{X}^t\mathbf{X}$, the resultant LSE $\widehat{\boldsymbol{\beta}}^{\mathrm{LS}}$ would be large in length $\parallel \widehat{\boldsymbol{\beta}}^{\mathrm{LS}} \parallel$ and associated with inflated standard errors. This inflated variation would lead to poor model prediction as well.

The ridge regression is a constrained version of least squares. It tackles the estimation problem by producing biased estimator yet with small variances.

7.3.1 *Ridge Shrinkage Estimator*

For any estimator $\mathbf{b}$, the least squares criterion can be rewritten as its minimum, reached at $\widehat{\boldsymbol{\beta}}^{\mathrm{LS}}$, plus a quadratic form in $\mathbf{b}$:

$$\begin{aligned} Q(\mathbf{b}) &= \parallel \mathbf{y} - \mathbf{X}\widehat{\boldsymbol{\beta}}^{\mathrm{LS}} + \mathbf{X}\widehat{\boldsymbol{\beta}}^{\mathrm{LS}} - \mathbf{X}\mathbf{b} \parallel^2 \\ &= (\mathbf{y} - \mathbf{X}\widehat{\boldsymbol{\beta}}^{\mathrm{LS}})^t(\mathbf{y} - \mathbf{X}\widehat{\boldsymbol{\beta}}^{\mathrm{LS}}) + (\mathbf{b} - \widehat{\boldsymbol{\beta}}^{\mathrm{LS}})^t\mathbf{X}^t\mathbf{X}(\mathbf{b} - \widehat{\boldsymbol{\beta}}^{\mathrm{LS}}) \\ &= Q_{\min} + \phi(\mathbf{b}). \end{aligned} \tag{7.44}$$

Contours for each constant of the quadratic form $\phi(\mathbf{b})$ are hyper-ellipsoids centered at the ordinary LSE $\widehat{\boldsymbol{\beta}}^{\mathrm{LS}}$. In view of (7.43), it is reasonable to expect that, if one moves away from $Q_{\min}$, the movement is in a direction which shortens the length of $\mathbf{b}$.

The optimization problem in ridge regression can be stated as

$$\text{minimizing } \parallel \boldsymbol{\beta} \parallel^2 \text{ subject to } (\boldsymbol{\beta} - \widehat{\boldsymbol{\beta}}^{\mathrm{LS}})^t\mathbf{X}^t\mathbf{X}(\boldsymbol{\beta} - \widehat{\boldsymbol{\beta}}^{\mathrm{LS}}) = \phi_0$$

for some constant ϕ_0. The enforced constrain guarantees a relatively small residual sum of squares $Q(\boldsymbol{\beta})$ when compared to its minimum $Q_{\min}$. Fig. 7.2(a) depicts the contours of residual sum of squares together with the L_2 ridge constraint in the two-dimensional case. As a Lagrangian problem, it is equivalent to minimizing

$$Q^{\star}(\boldsymbol{\beta}) = \parallel \boldsymbol{\beta} \parallel^2 + (1/k)\left\{(\boldsymbol{\beta} - \widehat{\boldsymbol{\beta}}^{\mathrm{LS}})^t\mathbf{X}^t\mathbf{X}(\boldsymbol{\beta} - \widehat{\boldsymbol{\beta}}^{\mathrm{LS}}) - \phi_0\right\},$$

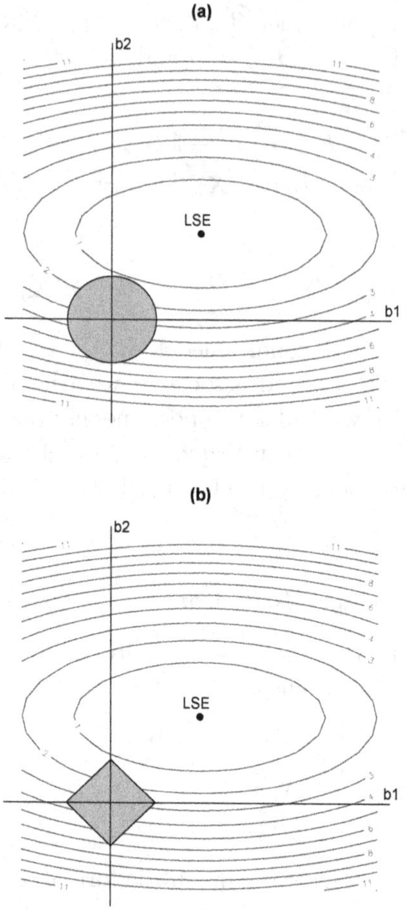

Fig. 7.2 Estimation in Constrained Least Squares in the Two-dimensional Case: Contours of the Residual Sum of Squares and the Constraint Functions in (a) Ridge Regression; (b) LASSO.

where $1/k$ is the multiplier chosen to satisfy the constraint. Thus

$$\frac{\partial Q^\star}{\partial \boldsymbol{\beta}} = 2\boldsymbol{\beta} + (1/k)\left\{2(\mathbf{X}^t\mathbf{X})\boldsymbol{\beta} - 2(\mathbf{X}^t\mathbf{X})\widehat{\boldsymbol{\beta}}^{\mathrm{LS}}\right\} = 0,$$

which yields the ridge estimator

$$\widehat{\boldsymbol{\beta}}^{\mathrm{R}} = \left\{\mathbf{X}^t\mathbf{X} + k\mathbf{I}\right\}^{-1}\mathbf{X}^t\mathbf{y}. \tag{7.45}$$

An equivalent way is to write the ridge problem in the penalized or constrained least squares form by

$$\text{minimizing} \;\; \| \mathbf{y} - \mathbf{X}\boldsymbol{\beta} \|^2, \text{ subject to } \;\; \| \boldsymbol{\beta} \|^2 \leq s, \tag{7.46}$$

for some constant s. The Lagrangian problem becomes minimizing

$$\| \mathbf{y} - \mathbf{X}\boldsymbol{\beta} \|^2 + \lambda \cdot \| \boldsymbol{\beta} \|^2,$$

which yields the same estimator given in (7.45). The penalty parameter $\lambda \geq 0$ controls the amount of shrinkage in $\| \boldsymbol{\beta} \|^2$. The larger value of λ, the greater amount of shrinkage. For this reason, the ridge estimator is also called the shrinkage estimator. There is a one-to-one correspondence among λ, s, k, and ϕ_0.

It is important to note that the ridge solution is not invariant under scaling of the inputs. Thus one should standardize both the inputs and the response

$$x'_{ij} = \frac{x_{ij} - \bar{x}_j}{s_{x_j}} \quad \text{and} \quad y'_i = \frac{y_i - \bar{y}}{s_y}$$

before computing the shrinkage estimator in (7.45). With the standardized variables, the matrices $\mathbf{X}^t\mathbf{X}$ and $\mathbf{X}^t\mathbf{y}$ become

$$\mathbf{X}^t\mathbf{X} = \mathbf{R}_{XX} \quad \text{and} \quad \mathbf{X}^t\mathbf{y} = \mathbf{r}_{XY},$$

where $\mathbf{R}_{XX}$ denotes the correlation matrix among X_j's and $\mathbf{r}_{YX}$ denotes the correlation vector between Y and all X_j's. Hence the ridge estimator becomes

$$\widehat{\boldsymbol{\beta}}^{\mathrm{R}} = \{\mathbf{R}_{XX} + k\mathbf{I}\}^{-1} \mathbf{r}_{YX}. \tag{7.47}$$

In the case of orthogonal predictors where $\mathbf{X}^t\mathbf{X} = \mathbf{I}$, it can be easily seen that the ridge estimates are just a scaled version of LSE, i.e., $\widehat{\boldsymbol{\beta}}^{\mathrm{R}} = 1/(1 + k) \cdot \widehat{\boldsymbol{\beta}}^{\mathrm{LS}}$ for some constant shrinkage $0 \leq 1/(1+k) \leq 1$. Besides, the intercept β_0 is automatically suppressed as 0 when working with standardized data.

Given a ridge estimator $\widehat{\boldsymbol{\beta}}^{\mathrm{R}}$, one needs transform its components back in order to get the fitted linear equation between the original Y and X_j values. It is convenient to express in matrix form the normalization and its inverse transformation involved. Let X_0 be the original design matrix. Its centered version is

$$\mathbf{X}_c = (\mathbf{I} - \mathbf{j}_n\mathbf{j}_n^t/n)\mathbf{X}_0$$

and its normalized version is

$$\mathbf{X} = \mathbf{X}_c\mathbf{L}^{-1/2}$$

where $\mathbf{j}_n$ is the n-dimensional vector with elements ones and $\mathbf{L}$ is a diagonal matrix with diagonal elements from the matrix $\mathbf{X}_c^t\mathbf{X}_c$, i.e.,

$$\mathbf{L} = \mathbf{diag}\left(\mathbf{X}_c^t\mathbf{X}_c\right).$$

Similarly, the original response vector $\mathbf{y}_0$ can be normalized as

$$\mathbf{y} = \frac{(\mathbf{I} - \mathbf{j}_n\mathbf{j}_n'/n)\,\mathbf{y}_0}{s_y},$$

where s_y denotes the sample standard deviation of $\mathbf{y}_0$. Given the ridge estimator $\widehat{\boldsymbol{\beta}}^{\mathrm{R}}$ in (7.45), suppose that we want to predict with a new data matrix $\mathbf{X}_{\text{new}}$, which is $m \times p$ on the original data scale. Note that we do not need to add $\mathbf{j}_m$ as the first column of $\mathbf{X}_{\text{new}}$ as the intercept has been suppressed. The predicted vector $\hat{\mathbf{y}}_{\text{new}}$ is then given as

$$\hat{\mathbf{y}}_{\text{new}} = s_y \cdot \left\{\left(\mathbf{X}_{\text{new}} - \mathbf{j}_m\mathbf{j}_n^t\mathbf{X}/n\right)\mathbf{L}^{-1/2}\widehat{\boldsymbol{\beta}}^{\mathrm{R}} + \mathbf{j}_m\mathbf{j}_n^t\mathbf{y}/n\right\}. \tag{7.48}$$

Next, we shall compute the expectation and variance of $\widehat{\boldsymbol{\beta}}^{\mathrm{R}}$. Eventually, we want to compare $\widehat{\boldsymbol{\beta}}^{\mathrm{R}}$ with $\widehat{\boldsymbol{\beta}}^{\mathrm{LS}}$ to see whether a smaller MSE can be achieved by $\widehat{\boldsymbol{\beta}}^{\mathrm{R}}$ for certain values of k. Denote

$$\mathbf{Z} = \left\{\mathbf{I} + k(\mathbf{X}^t\mathbf{X})^{-1}\right\}^{-1}. \tag{7.49}$$

Then

$$\widehat{\boldsymbol{\beta}}^{\mathrm{R}} = \mathbf{Z}\widehat{\boldsymbol{\beta}}^{\mathrm{LS}}. \tag{7.50}$$

It follows that

$$E(\widehat{\boldsymbol{\beta}}^{\mathrm{R}}) = \mathbf{Z}\boldsymbol{\beta} \tag{7.51}$$

$$\mathrm{Cov}(\widehat{\boldsymbol{\beta}}^{\mathrm{R}}) = \sigma^2 \cdot \mathbf{Z}(\mathbf{X}^{\mathbf{t}}\mathbf{X})^{-1}\mathbf{Z}^{\mathbf{t}}. \tag{7.52}$$

Let

$$\lambda_{\max} = \lambda_1 \geq \lambda_2 \geq \cdots \geq \lambda_m = \lambda_{\min} > 0 \tag{7.53}$$

denote the eigenvalues of $\mathbf{X}^{\mathbf{t}}\mathbf{X}$, then the corresponding eigenvalues of $\mathbf{Z}$ are $\lambda_j/(\lambda_j + k)$. From (7.42),

$$MSE(\widehat{\boldsymbol{\beta}}^{\mathrm{LS}}) = \sigma^2 \cdot \sum_j 1/\lambda_j \tag{7.54}$$

For the components in ridge estimator, it can be found from (7.51) and (7.52) that the sum of their squared biases is

$$\begin{aligned}
\sum_j \left\{E(\hat{\beta}_j^{\mathrm{R}}) - \beta_j\right\}^2 &= \left\{E(\widehat{\boldsymbol{\beta}}^{\mathrm{R}}) - \boldsymbol{\beta}\right\}^t \left\{E(\widehat{\boldsymbol{\beta}}^{\mathrm{R}}) - \boldsymbol{\beta}\right\} \\
&= \boldsymbol{\beta}^t(\mathbf{I} - \mathbf{Z})^t(\mathbf{I} - \mathbf{Z})\boldsymbol{\beta} \\
&= k^2\boldsymbol{\beta}^t(\mathbf{X}^{\mathbf{t}}\mathbf{X} + \mathbf{kI})^{-2}, \boldsymbol{\beta} \quad \text{using Problem 7.7.4.2,}
\end{aligned}$$

and the sum of their variances is

$$\begin{aligned}\mathrm{tr}\{\mathrm{Cov}(\widehat{\boldsymbol{\beta}}^{\mathrm{R}})\} &= \sigma^2 \cdot \mathrm{tr}\left\{(\mathbf{X}^{\mathrm{t}}\mathbf{X})^{-1}\mathbf{Z}^{\mathrm{t}}\mathbf{Z}\right\} \\ &= \sigma^2 \sum_j \left\{\frac{1}{\lambda_j} \cdot \frac{\lambda_j^2}{(\lambda_j + k)^2}\right\} \\ &= \sigma^2 \cdot \sum_j \frac{\lambda_j}{(\lambda_j + k)^2}.\end{aligned}$$

Therefore, the MSE for the ridge estimator is

$$\begin{aligned}MSE(\widehat{\boldsymbol{\beta}}^{\mathrm{R}}, k) &= \sigma^2 \cdot \sum_j \frac{\lambda_j}{(\lambda_j + k)^2} + k^2 \boldsymbol{\beta}^t(\mathbf{X}^{\mathrm{t}}\mathbf{X} + \mathbf{kI})^{-2}\boldsymbol{\beta} \\ &= \gamma_1(k) + \gamma_2(k). \end{aligned} \tag{7.55}$$

The function $\gamma_1(k)$ is a monotonic decreasing function of k while $\gamma_2(k)$ is monotonically increasing. The constant k reflects the amount of bias increased and the variance reduced. When $k = 0$, it becomes the LSE. Hoerl and Kennard (1970) showed that there always exists a $k > 0$ such that

$$MSE(\widehat{\boldsymbol{\beta}}^{\mathrm{R}}, k) < MSE(\widehat{\boldsymbol{\beta}}^{\mathrm{R}}, 0) = MSE(\widehat{\boldsymbol{\beta}}^{\mathrm{LS}}).$$

In other words, the ridge estimator can outperform the LSE in terms of providing a smaller MSE. Nevertheless, in practice the choice of k is yet to be determined and hence there is no guarantee that a smaller MSE always be attained by ridge regression.

7.3.2 *Connection with PCA*

The singular value decomposition (SVD) of the design matrix $\mathbf{X}$ can provide further insights into the nature of ridge regression. We shall establish, following Hastie, Tibshirani, and Friedman (2002), a connection between ridge regression and principal component analysis (PCA).

Assume the variables are standardized or centered so that the matrix

$$\mathbf{S} = \mathbf{X}^{\mathrm{t}}\mathbf{X}/n \tag{7.56}$$

gives either the sample variance-covariance matrix or the sample correlation matrix among predictors X_j's. The SVD of the $n \times p$ design matrix $\mathbf{X}$ has the form

$$\mathbf{X} = \mathbf{U}\mathbf{D}\mathbf{V}^{\mathrm{t}}, \tag{7.57}$$

where $\mathbf{U}$ and $\mathbf{V}$ are $n \times p$ and $p \times p$ orthogonal matrices such that the columns of $\mathbf{U}$ form an orthonormal basis of the column space of $\mathbf{X}$ and the columns of $\mathbf{V}$ form an orthonormal basis of the row space of $\mathbf{X}$; the $p \times p$ diagonal matrix $\mathbf{D} = \text{diag}(d_j)$ with diagonal entries d_j, $|d_1| \geq |d_2| \geq \cdots \geq |d_p|$ being the singular values of $\mathbf{X}$. It follows that

$$\mathbf{X^t X} = \mathbf{V D^2 V^t}, \tag{7.58}$$

which provides the spectral decomposition of $\mathbf{X^t X}$. Referring to (7.53), we have the correspondence $\lambda_j = d_j^2$. In addition, the spectral decomposition for $\mathbf{S}$ in (7.56) would be

$$\mathbf{S} = (1/n) \cdot \mathbf{V D^2 V^t}, \tag{7.59}$$

with eigenvalues d_j^2/n and eigenvectors $\mathbf{v}_j$ (i.e., the j-th column of $\mathbf{V}$).

We first briefly outline the main results of principal component analysis (PCA). A PCA is concerned with explaining the variance-covariance structure through uncorrelated linear combinations of original variables. We consider the sample version of PCA. Let $\mathbf{x}_i$ denote the i-th row in the centered or standardized design matrix $\mathbf{X}$ for $i = 1, \ldots, n$. It can be easily seen that the n values $\{\mathbf{q}^t\mathbf{x}_1, \mathbf{q}^t\mathbf{x}_2, \ldots, \mathbf{q}^t\mathbf{x}_n\}$ of any linear combination $\mathbf{q}^t\mathbf{x}_i$ has sample variance $\mathbf{q}^t\mathbf{S}\mathbf{q}$. These n values are elements in vector $\mathbf{Xq}$ induced by the direction $\mathbf{q}$.

The problem of PCA can be formulated as follows. We seek the direction $\mathbf{q}_1$ subject to $\mathbf{q}_1^t\mathbf{q}_1 = 1$ such that values in $\mathbf{Xq}_1$, referred to the first sample principal component, has the largest sample variance. Next, we seek the second sample principal component $\mathbf{Xq}_2$ with $\mathbf{q}_2^t\mathbf{q}_2 = 1$ such that its values have the largest sample variance and zero sample covariance with values of the first principal component. In general, at the j-th step for $j = 1, \ldots, p$, we seek the j-th principal component $\mathbf{Xq}_j$ subject to $\mathbf{q}_j^t\mathbf{q}_j = 1$ such that its values have the largest sample variance and zero sample covariance with values in each of the preceding principal components. It turns out that the principal component directions are exactly provided by the eigenvectors of $\mathbf{S}$, i.e., $\mathbf{q}_j = \mathbf{v}_j$. The j-th principal component is $\mathbf{Xv}_j$, which has a sample variance of d_j^2/n. A small singular value d_j^2 is associated with PC directions having small variance. Since $\mathbf{Xv}_j = d_j \cdot \mathbf{u_j}$ according to (7.57), $\mathbf{u_j}$, the j-th column of $\mathbf{U}$, is the j-th normalized principal component. The proof and detailed elaboration of PCA can be found in, e.g., Johnson and Wichern (2001).

Using the SVD of $\mathbf{X}$ in (7.57), the least squares fitted vector $\widehat{\boldsymbol{\mu}}^{\text{LS}}$ can be rewritten as

$$\begin{aligned}\widehat{\boldsymbol{\mu}}^{\text{LS}} = \mathbf{X}\widehat{\boldsymbol{\beta}}^{\text{LS}} &= \mathbf{X}(\mathbf{X}^t\mathbf{X})^{-1}\mathbf{X}^t\mathbf{y} \\ &= \mathbf{UDV}^t(\mathbf{VD}^t\mathbf{U}^t\mathbf{UDV}^t)^{-1}\mathbf{VD}^t\mathbf{U}^t\mathbf{y} \\ &= \mathbf{UDV}^t(\mathbf{VD}^{-2}\mathbf{V}^t)^{-1}\mathbf{VD}^t\mathbf{U}^t\mathbf{y} \\ &= \mathbf{UDV}^t\mathbf{VD}^2\mathbf{V}^t\mathbf{VD}^t\mathbf{U}^t\mathbf{y} \\ &= \mathbf{UU}^t\mathbf{y} \\ &= \sum_{j=1}^{p}(\mathbf{u}_j^t\mathbf{y})\cdot\mathbf{u}_j \end{aligned} \tag{7.60}$$

Note that $\mathbf{U}^t\mathbf{y}$ or $\mathbf{u}_j^t\mathbf{y}$'s are the coordinates of $\mathbf{y}$ with respect to the columns of $\mathbf{U}$. Recall that $\mathbf{u_j}$'s are the orthonormal basis spanning the column space of $\mathbf{X}$ and also the normalized sample principal components.

Another approach to deal with multicollinearity is called *principal components regression* (PCR). In this approach, y is regressed on the first m principal components by rejecting the last $(p-m)$ components that explain a relatively small portion of variation in $\mathbf{X}$. Thus the fitted response vector would be

$$\widehat{\boldsymbol{\mu}}^{\text{PCR}} = \sum_{j=1}^{m}(\mathbf{u}_j^t\mathbf{y})\cdot\mathbf{u}_j. \tag{7.61}$$

The implicit assumption is that the response tends to vary most in the directions where the predictors have large variations.

The fitted vector based on ridge regression, after similar simplification, is

$$\begin{aligned}\widehat{\boldsymbol{\mu}}^{\text{R}} = \mathbf{X}\widehat{\boldsymbol{\beta}}^{\text{R}} &= \mathbf{X}(\mathbf{X}^t\mathbf{X}+k\mathbf{I})^{-1}\mathbf{X}^t\mathbf{y} \\ &= \mathbf{UD}(\mathbf{D}^2+k\mathbf{I})^{-1}\mathbf{DU}^t\mathbf{y} \\ &= \sum_{j=1}^{p}\frac{d_j^2}{d_j^2+k}(\mathbf{u}_j^t\mathbf{y})\cdot\mathbf{u}_j. \end{aligned} \tag{7.62}$$

Thus, similar to least squares regression, the ridge solution computes the coordinates of $\mathbf{y}$ with respect to the orthonormal basis $\mathbf{U}$ and then shrinks them by the factor $d_j^2/(d_j^2+k)$. Note that $d_j^2/(d_j^2+k) \leq 1$ as $k \geq 0$. With this strategy, a greater amount of shrinkage is applied to basis vectors or principal component vectors corresponding to smaller $d_j^2 = \lambda_j$. Instead of rejecting low-variance directions as in PCR, ridge regression keeps all principal component directions but weighs the coefficients by shrinking low-variance directions more.

7.3.3 LASSO and Other Extensions

The LASSO or lasso (least absolute shrinkage and selection operator) is another shrinkage method like ridge regression, yet with an important and attractive feature in variable selection.

To motivate, we continue with the comparison between ridge regression and PCR from the variable selection perspective. PCR, same as subset selection procedures, is a discrete selecting process - regressors or the principal components of them are either fully retained or completely dropped from the model. Comparatively, the ridge regression makes the selection process continuous by varying shrinkage parameter k and hence is more stable. On the other hand, since ridge regression does not set any coefficients to 0, it does not give an easily interpretable model as in subset selection. The lasso technique is intended to balance off in between and retains the favorable features of both subset selection and ridge regression by shrinking some coefficients and setting others to 0.

The lasso estimator of $\boldsymbol{\beta}$ is obtained by

$$\text{minimizing } \parallel \mathbf{y} - \mathbf{X}\boldsymbol{\beta} \parallel^2, \text{ subject to } \sum_{j=1}^{p} |\beta_j| \leq s. \tag{7.63}$$

Namely, the L_2 penalty $\sum_j \beta_j^2$ in ridge regression is replaced by the L_1 penalty $\sum_j |\beta_j|$ in lasso. If s is chosen greater than or equal to $\sum_j |\beta_j^{\text{LS}}|$, then the lasso estimates are the same as the LSE; if s is chosen to be smaller, then it will cause shrinkage of the solutions towards 0.

Fig. 7.2(b) portrays the lasso estimation problem in the two dimensional case. The constraint region in ridge regression has a disk shape while the constraint region in lasso is a diamond. Both methods find the first point at which the elliptical contours hit the constraint region. However, unlike the disk, the diamond has corners. If the solution occurs at a corner, then it has one coefficient $\hat{\beta}_j$ equal to zero.

The lasso solution is generally competitive with ridge solution yet with many zero coefficient estimates. Insight about the nature of the lasso can be further gleaned from orthonormal designs where $\mathbf{X}^t\mathbf{X} = \mathbf{I}$. In this case, the lasso estimator can be shown to be

$$\hat{\beta}_j^{\text{lasso}} = \text{sign}(\hat{\beta}_j^{\text{LS}}) \left\{ |\hat{\beta}_j^{\text{LS}}| - \gamma \right\}_+, \tag{7.64}$$

where γ is determined by the condition $\sum_j |\hat{\beta}_j^{\text{lasso}}| = s$. Thus, coefficients less than the threshold γ would be automatically suppressed to 0 while coefficients larger than γ would be shrunk by a unit of γ. Hence, the lasso technique performs as a variable selection operator. By increasing s in discrete steps, one obtains a sequence of regression coefficients where those nonzero coefficients at each step correspond to selected predictors.

In general, the nonsmooth nature of the lasso constraint makes the solutions nonlinear in $\mathbf{y}$. In the initial proposal of lasso by Tibshirani (1996), quadratic programming was employed to solve the optimization problem by using the fact that the condition $\sum_j |\beta_j| \leq s$ is equivalent to $\boldsymbol{\delta}_i^t \boldsymbol{\beta} \leq s$ for all $i = 1, 2, \ldots, 2^p$, where $\boldsymbol{\delta}_i$ is the p-tuples of form $(\pm 1, \pm 1, \ldots, \pm 1)$. Later, Osborne, Presnell, and Turlach (2000a, 2000b) developed a compact descent method for solving the constrained lasso problem for any fixed s and a "homotopy method" that completely describe the possible selection regimes in the lasso solution. In the same vein, Efron et al. (2004) derived a parallel variant, called the least angle regression (LARS). LARS facilitates a variable selection method in its own right. More importantly, the entire path of lasso solutions as s varies from 0 to $+\infty$ can be extracted with a slight modification on LARS.

The LARS method works with normalized data and iteratively builds up the predicted response $\widehat{\boldsymbol{\mu}}$ with updating steps, analogous to boosting (see, e.g., Freund and Schapire, 1997). The main steps of LARS are first briefly outlined, with some details following up. Initially all coefficients are set to zero. The predictor that has highest correlation with the current residual, which is the response itself in this stage, is identified. A step is then taken in the direction of this predictor. The length of this step, which corresponds to the coefficient for this predictor, is chosen such that some other predictor (i.e., the second predictor entering the model) and the current predicted response have the same correlation with the current residual. Next, the predicted response moves in the direction that is equiangular between or equally correlated with these two predictors. Moving in this joint direction ensures that these two predictors continue to have a common correlation with the current residual. The predicted response moves in this direction until a third predictor has the same correlation with the current residual as the two predictors already in the model. A new joint direction that is equiangular between these three predictors is determined and the predicted response moves in this direction until a fourth predictor having the same correlation with the current residual joins the set. This process continues till all predictors have entered the model.

More specifically, start with $\widehat{\boldsymbol{\mu}}_0 = \mathbf{0}$. Let $\widehat{\boldsymbol{\mu}}_{\mathcal{A}}$ denote the current LARS estimate for the predicted vector, where $\mathcal{A}$ is the active set of indices corresponding to predictors that have the great absolute correlations with the current residuals, i.e., predictors in the current model. The sample correlation between the residuals and values of each predictor indexed in $\mathcal{A}$ is given as

$$\mathbf{c} = \mathbf{X}^t(\mathbf{y} - \widehat{\mathbf{y}}_{\mathcal{A}})$$

owing to normalization. Thus,

$$\mathcal{A} = \{j : \ |c_j| = C\} \ \text{ with } C = \max_j(|c_j|).$$

Let $s_j = \text{sign}(c_j)$ for $j \in \mathcal{A}$ be the sign of the correlation between X_j in the active set and the current residuals and let $\mathbf{X}_{\mathcal{A}} = (s_j \mathbf{x}_j)$, for $j \in \mathcal{A}$, be the design matrix containing all signed active predictors. Compute matrices

$$\mathbf{G}_{\mathcal{A}} = \mathbf{X}^t_{\mathcal{A}}\mathbf{X}_{\mathcal{A}}, \tag{7.65}$$

$$A_{\mathcal{A}} = (\mathbf{j}^t_{\mathcal{A}}\mathbf{G}^{-1}_{\mathcal{A}}\mathbf{j}_{\mathcal{A}})^{-1/2}, \tag{7.66}$$

$$\mathbf{w}_{\mathcal{A}} = A_{\mathcal{A}}\mathbf{G}^{-1}_{\mathcal{A}}\mathbf{j}_{\mathcal{A}}, \tag{7.67}$$

where $\mathbf{j}_{\mathcal{A}}$ is the vector of 1's of dimension equal to $|\mathcal{A}|$, the cardinality of $\mathcal{A}$. Then the equiangular vector

$$\mathbf{u}_{\mathcal{A}} = \mathbf{X}_{\mathcal{A}}\mathbf{w}_{\mathcal{A}} \ \text{ with } \parallel \mathbf{u}_{\mathcal{A}} \parallel = 1 \tag{7.68}$$

makes equal angles, less than 90°, with each column of $\mathbf{X}_{\mathcal{A}}$, i.e.,

$$\mathbf{X}^t_{\mathcal{A}}\mathbf{u}_{\mathcal{A}} = A_{\mathcal{A}}\mathbf{j}_{\mathcal{A}}. \tag{7.69}$$

Also compute the correlations between each predictor with the equiangular vector $\mathbf{u}_{\mathcal{A}}$, given as

$$\mathbf{a} = \mathbf{X}^t\mathbf{u}_{\mathcal{A}}.$$

Then, the next updating step of the LARS algorithm is

$$\widehat{\boldsymbol{\mu}}_{\mathcal{A}+} = \widehat{\boldsymbol{\mu}}_{\mathcal{A}} + \hat{\gamma}\mathbf{u}_{\mathcal{A}}, \tag{7.70}$$

where

$$\hat{\gamma} = \min{}^+_{j \in \mathcal{A}^c} \left\{ \frac{C - c_j}{A_{\mathcal{A}} - a_j}, \ \frac{C + c_j}{A_{\mathcal{A}} + a_j} \right\} \tag{7.71}$$

and the $\min^+$ means that the minimum is taken over only these positive components. Let $\hat{j}$ be the minimizing index in (7.71). Then $X_{\hat{j}}$ is the variable added to the active set and the new maximum absolute correlation becomes $C - \hat{\gamma}A_{\mathcal{A}}$.

Two notable remarks are in order. First, LARS is rather thrifty in computation, simply requiring a total of p steps. Secondly, surprisingly, with a slight modification of LARS, one can obtain a sequence of lasso estimates, from which all other lasso solutions can be obtained by linear interpolation. The modification is that, if a non-zero coefficient turns into zero, it will be removed from the active set of predictors and the joint direction will be recomputed.

Extensions and further explorations of LASSO are currently under intensive research. For example, the BRIDGE regression, introduced by Frank and Friedman (1993), generalizes the penalty term by applying a L_q constraint with $q \geq 0$. The optimization problem becomes

$$\text{minimizing} \ \parallel \mathbf{y} - \mathbf{X}\boldsymbol{\beta} \parallel^2, \text{ subject to } \sum |\beta_j|^q \leq s, \tag{7.72}$$

or equivalently,

$$\text{minimizing} \ \parallel \mathbf{y} - \mathbf{X}\boldsymbol{\beta} \parallel^2 + \lambda \cdot \sum |\beta_j|^q.$$

The case when $q = 0$ corresponds to variable subset selection as the constraint simply counts the number of nonzero parameters and hence penalizes for model complexity; $q = 1$ corresponds to the lasso penalty and $q = 2$ corresponds to the ridge regression; when $q > 2$, the constraint region becomes a polyhedron with many corners, flat edges, and faces, and hence it is more likely to have zero-valued coefficients.

Table 7.3 The Macroeconomic Data Set in Longley (1967).

	GNP.deflator	GNP	Unemployed	Armed.Forces	Population	Year	Employed
	83.0	234.289	235.6	159.0	107.608	1947	60.323
	88.5	259.426	232.5	145.6	108.632	1948	61.122
	88.2	258.054	368.2	161.6	109.773	1949	60.171
	89.5	284.599	335.1	165.0	110.929	1950	61.187
	96.2	328.975	209.9	309.9	112.075	1951	63.221
	98.1	346.999	193.2	359.4	113.270	1952	63.639
	99.0	365.385	187.0	354.7	115.094	1953	64.989
	100.0	363.112	357.8	335.0	116.219	1954	63.761
	101.2	397.469	290.4	304.8	117.388	1955	66.019
	104.6	419.180	282.2	285.7	118.734	1956	67.857
	108.4	442.769	293.6	279.8	120.445	1957	68.169
	110.8	444.546	468.1	263.7	121.950	1958	66.513
	112.6	482.704	381.3	255.2	123.366	1959	68.655
	114.2	502.601	393.1	251.4	125.368	1960	69.564
	115.7	518.173	480.6	257.2	127.852	1961	69.331
	116.9	554.894	400.7	282.7	130.081	1962	70.551
mean	101.681	387.698	319.331	260.669	117.424	8.5*	65.317
sd	10.792	99.395	93.446	69.592	6.956	4.761	3.512

7.3.4 *Example*

Implementation of ridge regression is widely available, e.g., in PROC REG of SAS and the function `lm.ridge` in R (`http://www.r-project.org/`) package `MASS`. LASSO is relatively new, available in the experimental SAS procedure PROC GLMSELECT and in the R packages: `lasso2` and `lars`. We shall discuss many practical issues involved in both ridge regression and LASSO through a macroeconomic data set in Longley (1967), which is a well-known example for highly collinear regression. The data set, as provided in Table 7.3, contains $n = 16$ observations on seven economical variables, observed yearly from 1947 to 1962. The response variable is `Employed`. A brief description of the data set is provided in Table 7.4.

Table 7.4 Variable Description for Longley's (1967) Macroeconomic Data.

Variable	Name	Description
X_1	`GNP.deflator`	GNP implicit price deflator (1954=100)
X_1	`GNP`	Gross National Product.
X_1	`Unemployed`	number of unemployed.
X_1	`Armed.Forces`	number of people in the armed forces.
X_1	`Population`	"noninstitutionalized" population $\geq$ 14 years of age.
X_1	`Year`	the year (time).
Y_1	`Employed`	number of people employed.

To proceed, we first take the transformation `Year` = `Year` − 1946 so that the time points are integers from 1 to 16 and then standardize or normalize every variable in the data set. Thus the intercept term is suppressed in all the following models that we consider.

For exploratory data analysis (EDA) purposes, Fig. 7.3 gives a scatter plot for every pair of the variables. It can be seen that several predictors have strong linear relationships with each other, as also evidenced by high correlation coefficients from the correlation matrix given in panel (a) of Table 7.5. Indeed, the variance inflation factor (VIF) values, given in the last column of panel (b) in Table 7.5, show that severe multicollinearity exists among all predictors except `Armed.Forces` (X_4). Recall that the VIF is a commonly used indicator for multicollinearity and computed as the coefficient of determination or R^2 in the linear model that regresses X_j on other predictors $(X_1, \ldots, X_{j-1}, X_{j+1}, \ldots, X_p)$ in the model, for $j = 1, \ldots, p$. It measures how large the variance of $\hat{\beta}_j^{\text{LS}}$ is relative to its variance when predictors are uncorrelated. Usually a VIF value larger than 10, in

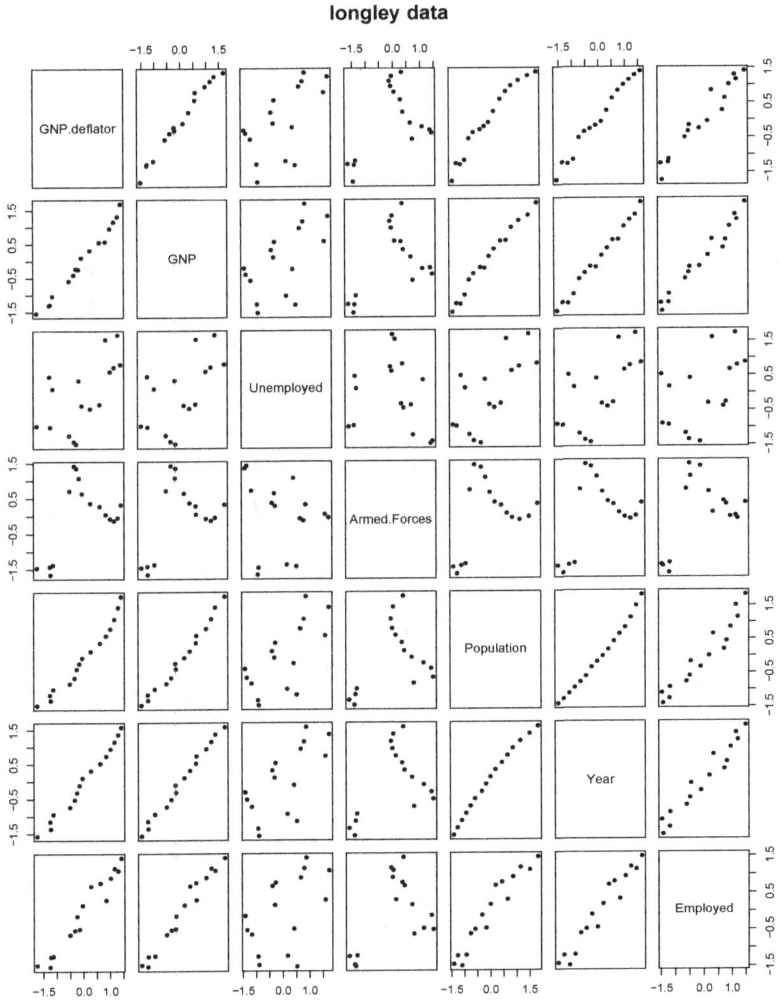

Fig. 7.3 Paired Scatter Plots for Longley's (1967) Economic Data.

which case $R^2 > .90$, causes concerns.

The least squares estimation shows insignificance of both `GNP` and `Population`. This contradicts the impression gained from the paired scatter plots. Applying the stepwise selection procedure leads to elimination

Table 7.5 Least Squares Results for Longley's (1967) Economic Data.

(a) Correlation Matrix

	X_1	X_2	X_3	X_4	X_5	X_6	Y
X_1	1.000	0.992	0.621	0.465	0.979	0.991	0.971
X_2	0.992	1.000	0.604	0.446	0.991	0.995	0.984
X_3	0.621	0.604	1.000	−0.177	0.687	0.668	0.502
X_4	0.465	0.446	−0.177	1.000	0.364	0.417	0.457
X_5	0.979	0.991	0.687	0.364	1.000	0.994	0.960
X_6	0.991	0.995	0.668	0.417	0.994	1.000	0.971
Y	0.971	0.984	0.502	0.457	0.960	0.971	1.000

(b) The Full Model with All Six Predictors

	Estimate	Standard Error	t Test Statistic	Two-Sided P-Value	VIF
`GNP.deflator`	0.0463	0.2475	0.19	0.85542	135.532
`GNP`	−1.0137	0.8992	−1.13	0.28591	1788.513
`Unemployed`	−0.5375	0.1233	−4.36	0.00142	33.619
`Armed.Forces`	−0.2047	0.0403	−5.08	0.00048	3.589
`Population`	−0.1012	0.4248	−0.24	0.81648	399.151
`Year`	2.4797	0.5858	4.23	0.00174	758.981

(c) The Best Subset Model Obtained by Stepwise Selection

	Estimate	Standard Error	t Test Statistic	Two-Sided P-Value	VIF
`GNP`	−1.1375	0.4464	−2.55	0.02555	515.124
`Unemployed`	−0.5557	0.0739	−7.52	0.0000	14.109
`Armed.Forces`	−0.2011	0.0349	−5.77	< 0.00001	3.142
`Year`	2.5586	0.4968	5.15	0.00024	638.128

of both `GNP.deflator` and `Population`. The best subset model with the remaining four predictors is presented in penal (c) of Table 7.5. Note that VIF values for `GNP` and `Year` remains to be strikingly high. In conclusion, the least squares estimates together with their associated standard errors are quite unstable and inference based on them is not very reliable.

The ridge regression may be preferred by providing a more stable estimator that has a small bias but is substantially more precise when predictors are highly correlated. By comparing ridge regression estimates with least squares estimates, we are able to assess how sensitive the fitted model is with small changes in the data.

To apply ridge regression, one needs to determine an optimal choice of the biasing constant k. A plot of the ridge trace is helpful. The ridge trace is a simultaneous plot of the standardized ridge estimates in (7.45) of the regression coefficient with different values of k. Panel (a) in Fig. 7.4 plots the ridge trace. Instead of k itself, the parameter s in (7.46), expressed as a fraction form

$$s = \frac{\sum \left(\hat{\beta}_j^{\mathrm{R}}\right)^2}{\sum \left(\hat{\beta}_j^{\mathrm{LS}}\right)^2}, \tag{7.73}$$

is used in the plot. Recall that, when $\sum \left(\hat{\beta}_j^{\mathrm{R}}\right)^2 \geq \sum \left(\hat{\beta}_j^{\mathrm{LS}}\right)^2$, the ridge regression estimates are the same as LSE. It can be seen from Fig. 7.4 that an estimated coefficient fluctuates widely, more or less, as s is changed slightly from 0, and may even change signs. Gradually, however, these wide fluctuations cease and the magnitude of the regression coefficient tends to vary slightly as s further increases.

The optimal choice of the biasing constant k is often a judgemental one. One convenient way is to apply an AIC-typed criterion, e.g., the generalized cross-validation (GCV). Recall that the GCV is intended to provide an approximation to PRESS, the sum of squared errors computed with the jackknife or leave-one-out cross-validation technique. Suppose that in a linear fitting method, the predicted vector can be expressed as

$$\hat{\mathbf{y}} = \mathbf{H}\mathbf{y}, \tag{7.74}$$

for some matrix $\mathbf{H}$. In ordinary linear regression, $\mathbf{H}$ is the projection matrix $\mathbf{H} = \mathbf{X}(\mathbf{X}^t\mathbf{X})^{-1}\mathbf{X}^t$. Let h_{ii} be the i-th diagonal element of $\mathbf{H}$. The jackknife-based SSE, which is often called PRESS in (5.8) and (5.9), is given by

$$\mathrm{PRESS} = (1/n)\sum_{i=1}^{n} \left(y_i - \hat{y}_{i(-i)}\right)^2 = (1/n)\sum_{i=1}^{n} \frac{(y_i - \hat{y}_i)^2}{1 - h_{ii}}, \tag{7.75}$$

where $\hat{y}_{i(-i)}$ is the predicted value for the i-th observation based on the estimated model that is fit without the i-th observation. Craven and Wahba (1979) proposed to replace h_i by its average $\mathrm{tr}(\mathbf{H})/n$ in (7.75), which leads to the GCV approximation given by

$$\mathrm{GCV} = \frac{1}{n}\frac{\sum_{i=1}^{n}(y_i - \hat{y}_i)^2}{1 - \mathrm{tr}(\mathbf{H})/n}. \tag{7.76}$$

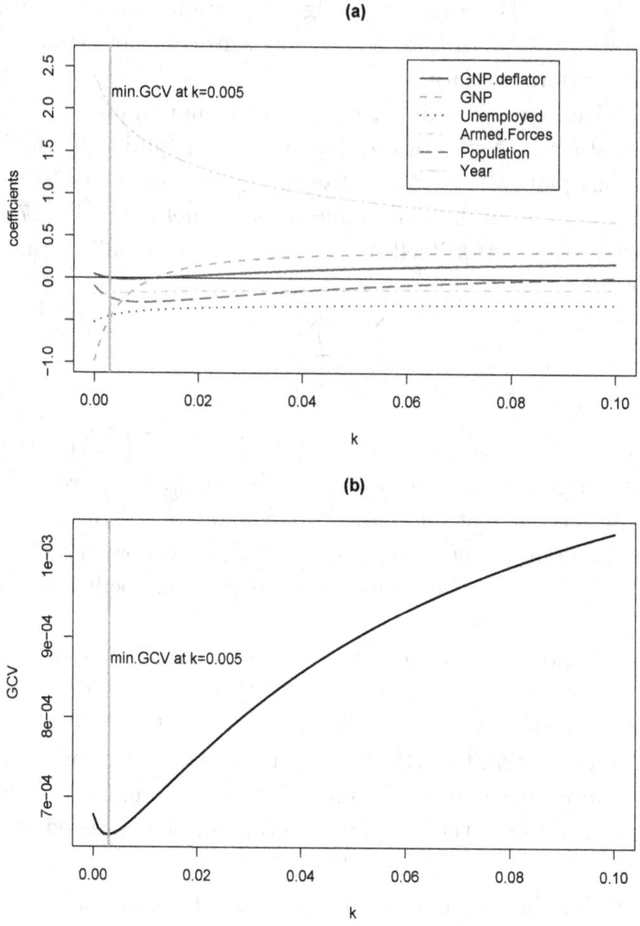

Fig. 7.4 Ridge Regression for Longley's (1967) Economic Data: (a) Plot of Parameter Estimates Versus s; (b) Plot of GCV Values Versus s.

The quantity $\text{tr}(\mathbf{H})$ is referred to the effective number of degrees of freedom. GCV, which is asymptotically equivalent to AIC and C_p, has been widely used as a model selection criterion in modern statistical applications. Again, it is important to note that equation (7.75) holds and hence the definition of GCV applies in a wide variety of problems that go far beyond ordinary linear regression. Similar results are also available for the leaving-

Table 7.6 Ridge Regression and LASSO for Longley's (1967) Data.

(a) The Ridge Regression Results When $k = 0.005$.

	Estimate	Standard Error	t Test Statistic	Two-Sided P-Value
GNP.deflator	−0.0031	0.0414	−0.0758	0.9396
GNP	−0.4768	0.3171	−1.5038	0.1326
Unemployed	−0.4505	0.0066	−68.0210	0.0000
Armed.Forces	−0.1836	0.0013	−145.6954	0.0000
Population	−0.2323	0.0968	−2.4009	0.0164
Year	2.0263	0.1732	11.6987	0.0000

(b) The Ridge Regression Results When $k = 0.03$.

	Estimate	Standard Error	t Test Statistic	Two-Sided P-Value
GNP.deflator	0.0627	0.0285	2.1996	0.0278
GNP	0.2191	0.0298	7.3520	0.0000
Unemployed	−0.3387	0.0022	−156.7838	0.0000
Armed.Forces	−0.1487	0.0014	−102.8985	0.0000
Population	−0.1954	0.0369	−5.2883	0.0000
Year	1.1408	0.0499	22.8567	0.0000

(c) LASSO Fit with $s = 1.904$ or Fraction of 43.43%. The Lagrangian for the Bound Is 0.02541.

	Estimate	Standard Error	t Test Statistic	Two-Sided P-Value
GNP.deflator	0.000	0.298	0.000	1.000
GNP	0.000	0.992	0.000	1.000
Unemployed	−0.378	0.132	−2.871	0.004
Armed.Forces	−0.146	0.041	−3.526	0.000
Population	−0.048	0.512	−0.095	0.925
Year	1.331	0.502	2.650	0.008

several-out cases or the general v-fold cross-validation. In ridge regression, for a given k, we have

$$\begin{aligned} \mathrm{df} = \mathrm{tr}(\mathbf{H}) &= \mathrm{tr}\{\mathbf{X}(\mathbf{X}^t\mathbf{X} + k\mathbf{I})^{-1}\mathbf{X}^t\} \\ &= \sum_{j=1}^{p} \frac{d_j^2}{d_j^2 + k}, \end{aligned} \tag{7.77}$$

in view of (7.62).

Figure 7.4(b) plots the GCV values for different choices of s in Longley's data example. The minimum GCV occurs at $s = 0.005$. Panel (a) of Table 7.6 gives the ridge estimates at $s = 0.005$. In contrast to the least squares estimates, there are two remarkable issues. First, the slope estimate for `GNP.deflator` has changed sign while remaining insignificant. Secondly, `Population` now becomes significant. Nevertheless, from the ridge trace plot in Fig. 7.4(a), the performance of ridge estimates is not stable yet at $s = 0.005$.

There are two other alternative ways of determining k in the literature. The HKB (Hoerl, Kennard, and Baldwin, 1975) estimator of k is given as

$$\hat{k} = \frac{n \cdot \hat{\sigma}^2}{\| \widehat{\boldsymbol{\beta}}^{\text{LS}} \|^2},$$

where $\hat{\sigma}^2$ is the unbiased estimate of the error variance from ordinary least squares (OLS). The rationale behind this choice of k is to provide a smaller MSE than that of OLS, which occurs when $k < \sigma^2 / \| \boldsymbol{\beta} \|^2$. Farebrother (1975) observed that if $\mathbf{X}^t\mathbf{X} = \mathbf{I}$, then MSE is minimized at this value of k. In various other examples, $\hat{k}$ generated approximately the minimal value of MSE even when $\mathbf{X}^t\mathbf{X} \neq \mathbf{I}$. Hoerl, Kennard, and Baldwin (1975) shows that significant improvement over OLS in terms of MSE is obtained with $\hat{k}$. Lawless and Wang (1976) also proposed a method, the L-M estimator of k, based on the connection between ridge regression and principal components analysis. All these three selection methods for k are available in the R function `lm.ridge`.

The HKB estimate of s in this example is 0.0038, which is close to the choice supplied by minimum GCV. The L-M estimate of s is 0.03. After all, the ridge trace in Fig. 7.4(a) provides an ultimate demonstration of how sensitive the coefficient estimates are with different choices of s. Quite a few interesting observations can be taken from the plot of the ridge trace. The slope for `Armed.Forces`, which is the least collinear variable, remains unaltered with varying s. `Year` has a constantly positive strong effect while `Unemployed` has a negative effect. The wrong negative sign of `GNP` has changed to positive as s increases. Accordingly, Hoerl and Kennard (1970) suggest to pick up a k value from the ridge trace plot by inspecting when the system stabilizes and has the general characteristics of an orthogonal system. They deem this to be the best method for selecting an admissible value of s. In this example, $s = 0.03$, also suggested by L-M, seems an appropriate and favorable choice. Panel (b) of Table 7.6 gives the ridge estimates at $s = 0.03$, the choice of the L-M method. Interestingly, all predictors become significant in the model. All slopes except the one for

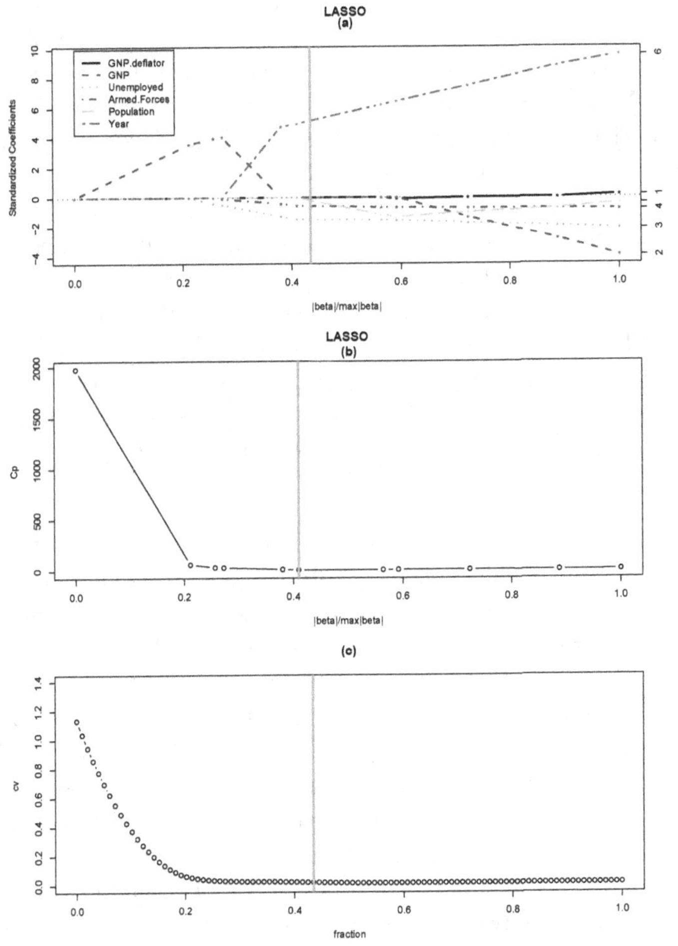

Fig. 7.5 LASSO Procedure for Longley's (1967) Economic Data: (a) Plot of Standardized Parameter Estimates Versus Fraction of the L_1 Shrinkage; (b) Plot of Cross-validated SSE Versus Fraction.

`Population` now have the correct signs. The correct sign of association can be roughly seen from the paired scatter plot in Fig. 7.3. Clearly `Population` has a very strong positive association with the response `Employed`. However, there is also a nearly perfect linear relationship between `Population` and `Year`. This is perhaps the reason that accounts for the weird behavior of the slope estimate for `Population`, which shrinks to 0 quickly.

Finally, we obtained the LASSO estimation via the LARS implementation. Figure 7.5(a) depicts the LASSO trace, which is the plot of the standardized coefficients versus s, expressed in the form of a fraction $\sum_j |\hat{\beta}_j^{\text{lasso}}| / \sum_j |\hat{\beta}_j^{\text{LS}}|$. Compared to the ridge trace, we see some similar performance of the slope estimates, e.g., the sign change of `GNP`, the suppression of `GNP.deflator`, the relative stability of Armed.Forces and `Unemployed`, the prominent positive effect of `Year`. The most striking difference, however, is that there are always some zero coefficients for a given value of $s \in [0, 1]$, which renders lasso an automatic variable selector.

Two methods are suggested by Efron et al. (2004) to determine an optimal s. The first is to apply a C_p estimate for prediction error. Efron et al. (2004) found that the effective number of degrees of freedom involved in lasso can be well approximated simply by k_0, the number of predictors with nonzero coefficients. The C_p criterion is then given by

$$C_p(s) = \frac{SSE(s)}{\hat{\sigma}^2} - n + 2k_0,$$

where, $SSE(s)$ is the resulting sum of squared errors from LASSO fit for a given value of s and $\hat{\sigma}^2$ is the unbiased estimate of σ^2 from OLS fitting of the full model. The second method is through v-fold cross validation. In this approach, the data set is randomly divided into v equally-sized subsets. For observations in each subset, their predicted values $\hat{y}^{\text{cv}}$ are computed based on the model fit using the other $v - 1$ subsets. Then the cross-validated sum of squared errors is $\sum_i (y_i - \hat{y}_i^{\text{cv}})^2$. Figure 7.5 (b) and (c) plot the C_p values and 10-fold cross-validated sum of squared errors, both versus s. The minimum C_p yields an optimal fraction of 0.41, which is very close to the choice, 0.43, selected by minimum cross-validated SSE.

With a fixed s, the standard errors of LASSO estimates can be computed either via bootstrap (Efron and Tibshirani, 1993) or by analytical approximation. In the bootstrap or resampling approach, one generates B replicates, called bootstrap samples, from the original data set by sampling with replacement. For the b-th bootstrap sample, the LASSO estimate $\widehat{\boldsymbol{\beta}}_b^{\text{lasso}}$ is computed. Then the standard errors of lasso estimates are computed as the sample standard deviation of $\widehat{\boldsymbol{\beta}}_b^{\text{lasso}}$'s. Tibshirani (1996) derived an approximate close form for the standard errors by rewriting the lasso penalty $\sum |\beta_j|$ as $\sum \beta_j^2 / |\beta_j|$. The lasso solution can then be approximated by a ridge regression of form

$$\widehat{\boldsymbol{\beta}}^{\text{lasso}} \approx (\mathbf{X}^t\mathbf{X} + k\mathbf{W}^-)^{-1}\mathbf{X}^t\mathbf{y}, \tag{7.78}$$

where $\mathbf{W}$ is a diagonal matrix with diagonal elements $|\hat{\beta}_j^{\text{lasso}}|$; $\mathbf{W}^-$ denotes a generalized inverse of $\mathbf{W}$; and k is chosen so that $\sum |\hat{\beta}_j^{\text{lasso}}| = s$. The covariance matrix of the lasso estimates can then be approximated by

$$\hat{\sigma}^2_{\text{lasso}} \cdot (\mathbf{X}^t\mathbf{X} + k\mathbf{W}^-)^{-1}\mathbf{X}^t\mathbf{X}(\mathbf{X}^t\mathbf{X} + k\mathbf{W}^-)^{-1}, \tag{7.79}$$

where $\hat{\sigma}^2_{\text{lasso}}$ is an estimate of the error variance. Clearly, this approximation suggests an iterative way to compute the lasso estimates by alternating the estimates of $\mathbf{W}$ and $\widehat{\boldsymbol{\beta}}^{\text{lasso}}$. Besides, equation (7.78) implies a linear form of the predicted vector; hence GCV in (7.76) can be applied. A difficulty with the formula (7.79), however, is that the standard errors for zero coefficient estimates are all zeroes. To get around this difficulty, Osborne, Presnell, and Turlach (2000b) proposed a smooth approximation method for estimating the lasso standard errors.

Panel (c) in Table 7.5 presents the LASSO fit at $s = 1.904$ with a fraction 43% out of the saturated $|\widehat{\boldsymbol{\beta}}^{\text{LS}}|$, a choice suggested by minimum cross-validated SSE. The standard errors are estimated via bootstrap. The model fit roughly corresponds to keeping just four of the predictors: `Unemployed`, `Armer.Forces`, `Population`, and `Year`, although `Population` appears to be rather insignificant. This selection is somewhat different from the best subset selection in Table 7.4 (c). The variable `GNP` is included in the best subset, significant with a wrong sign, but does not show up in the lasso fit. Notice that the coefficients in the lasso fit have been shrunk in absolute value, which results in reduced significance.

7.4 Parametric Nonlinear Regression

Given data $\{(y_i, \mathbf{x}_i) : i = 1, \ldots, n\}$, where y_i is a continuous response and $\mathbf{x}_i$ is the predictor vector containing mixed types of variables, one may consider model

$$y_i = h(\boldsymbol{\beta}, \mathbf{x}_i) + \varepsilon_i \text{ with } \varepsilon_i \overset{\text{i.i.d.}}{\sim} \mathcal{N}(0, \sigma^2), \tag{7.80}$$

where the function $h(\cdot)$ is assumed to have a fully specified nonlinear form up to the parameter vector $\boldsymbol{\beta} \in \mathcal{R}^p$. This model is generally termed as a parametric nonlinear model. If, on the other hand, the form of $h(\cdot)$ is left completely or partially unspecified, it would result in a *nonparametric* or *semiparametric* nonlinear model, which are often approached with kernel- or spline-based smoothing techniques. The additive model (AM) discussed in Section 6.2 is of the semiparametric type. In this section, we focus on parametric nonlinear models only. Detailed treatment of nonparametric nonlinear models is beyond the scope of this text.

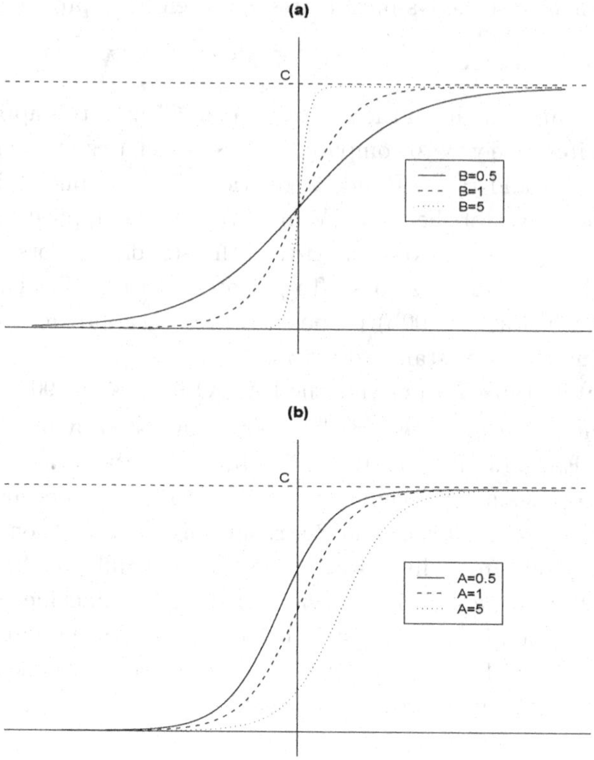

Fig. 7.6 Plot of the Logistic Function With $C = 1$: (a) When B Varies With $A = 1$; (b) When A Varies With $B = 1$.

Linear regression may fit poorly to data that show a strong nonlinear pattern, which calls for the need of fitting nonlinear models. There are a broad range of nonlinear functions available. The specific functional forms employed in a study are often derived on the basis of both scientific considerations and empirical evidences. The parameters involved usually have direct and meaningful interpretation in terms of the application background. For example, the general logistic function has three parameters (A, B, C):

$$f(x) = \frac{C}{1 + A \cdot \exp(-Bx)} = \frac{C \cdot \exp(Bx)}{\exp(Bx) + A}, \tag{7.81}$$

where C is the limiting value of the curve; the parameter B controls the monotonicity of the curve, a positive B corresponding to an increasing function and a negative B for a decreasing relationship; the parameter A controls the rate of change. Figure 7.6 plots the logistic function for some selected parameter values. The logistic function has a sigmoid shape. This type of functions are often seen in describing certain kinds of growth, which the growth rate of growth becomes progressively smaller over time.

Another advantage of using a parametric nonlinear model is that it has the convenience to enforce some desired constraints that are, otherwise, hard to do for linear regression. For example, the mean response may be desired to be positive or stay within certain range, which can be easily built in a nonlinear model. Many nonlinear functions can also achieve an asymptote or limiting value as X approaches a certain value.

7.4.1 *Least Squares Estimation in Nonlinear Regression*

The least squares (LS) method can be used to fit nonlinear models, in which case it is termed as *nonlinear least squares*. The nonlinear LSE $\widehat{\boldsymbol{\beta}}$ is obtained by minimizing the same objective function

$$Q(\boldsymbol{\beta}) = \sum_{i=1}^{n} \{y_i - h(\boldsymbol{\beta}, \mathbf{x}_i)\}^2$$

as in linear regression. However, fitting a nonlinear model is more involved. Due to the nonlinearity of $h(\cdot)$, there is no general close-form solution of $\widehat{\boldsymbol{\beta}}$. The optimization of $Q(\boldsymbol{\beta})$ is typically done with numerical iterative algorithms.

Assume that Q has continuous first and second derivatives in the sense that its gradient vector of Q

$$\mathbf{g} = Q'(\boldsymbol{\beta}) = \frac{\partial Q}{\partial \boldsymbol{\beta}} = -2\sum_{i=1}^{n} \{y_i - h(\boldsymbol{\beta}, \mathbf{x}_i)\} \frac{\partial h}{\partial \boldsymbol{\beta}}$$

and its Hessian matrix

$$\mathbf{H} = Q''(\boldsymbol{\beta}) = \frac{\partial^2 Q}{\partial \boldsymbol{\beta} \partial \boldsymbol{\beta}'} = \left(\frac{\partial^2 Q}{\partial \beta_j \partial \beta_{j'}} \right) \tag{7.82}$$

both have continuous components. The nonlinear LSE $\widehat{\boldsymbol{\beta}}$ is basically a *stationary point* of Q, satisfying $\mathbf{g}(\widehat{\boldsymbol{\beta}}) = \mathbf{0}$. In order for $\widehat{\boldsymbol{\beta}}$ to be a local minimizer of Q, we further assume that the Hessian matrix $\mathbf{H}$ is positive definite at $\widehat{\boldsymbol{\beta}}$.

A numerical algorithm updates an initial guess $\boldsymbol{\beta}^{(0)}$ iteratively with the general form

$$\boldsymbol{\beta}^{(k+1)} = \boldsymbol{\beta}^{(k)} + a^{(k)}\mathbf{d}^{(k)},$$

where $\mathbf{d}^{(k)}$ satisfying $\| \mathbf{d} \| = 1$ is the *step direction* and the scalar $a^{(k)}$ is called the *step size* or *step length*. If there exists a δ such that $Q(\boldsymbol{\beta}^{(k+1)}) = \boldsymbol{\beta}^{(k)} + a^{(k)}\mathbf{d}^{(k)} < Q(\boldsymbol{\beta}^{(k)})$ for any $a \in (0, \delta)$, we say that the step $\mathbf{d}^{(k)}$ is acceptable. The following theorem (Bard, 1974) characterizes acceptable directions by establishing a necessary and sufficient condition.

Theorem 7.5. *A direction* $\mathbf{d}^{(k)}$ *is acceptable if and only if there exists a positive definite matrix* $\mathbf{R}$ *such that*

$$\mathbf{d}^{(k)} = -\mathbf{R}\,\mathbf{g}(\boldsymbol{\beta}^{(k)}).$$

Various optimization methods usually differ in their criteria used to find the step directions and step sizes. In particular, methods that have step directions of the form given in Theorem 7.5 are called the *gradient* methods.

The Newton-Raphson method is the most commonly used gradient algorithm for optimization. To motivate, consider the Taylor's expansion of $\mathbf{g}(\widehat{\boldsymbol{\beta}})$ at an initial guess $\boldsymbol{\beta}$:

$$\mathbf{0} = \mathbf{g}(\widehat{\boldsymbol{\beta}}) = \mathbf{g}(\boldsymbol{\beta}) + \mathbf{H}(\boldsymbol{\beta}) \left(\widehat{\boldsymbol{\beta}} - \boldsymbol{\beta}\right).$$

Solving it for $\widehat{\boldsymbol{\beta}}$ leads to the updating formula:

$$\boldsymbol{\beta}^{(k+1)} = \boldsymbol{\beta}^{(k)} - \left\{\mathbf{H}(\boldsymbol{\beta}^{(k)})\right\}^{-1} \mathbf{g}(\boldsymbol{\beta}^{(k)}). \tag{7.83}$$

However, it can be difficult to supply explicit derivative formula for a complicated function $h(\cdot)$. One solution, known as a DUD (Does not Use Derivatives) algorithm, approximates the derivatives by a finite-difference method. Take a one-dimensional function $g(x)$ for instance. The finite difference method numerically approximate its first derivative

$$g'(x) = \lim_{t \to 0} \frac{g(x+t) - g(x)}{t}$$

by

$$g'(x) \approx \frac{g(x+t) - g(x)}{t}$$

for a small value of t. Nevertheless, the approximation from the DUD algorithm is known to perform poorly when the model function $h(\cdot)$ is not well behaved.

Under some mild regularity conditions as given in Amemiya (1985), the nonlinear LSE $\widehat{\boldsymbol{\beta}}$ is asymptotically normally distributed as

$$\widehat{\boldsymbol{\beta}} \sim \mathcal{N}\left\{\boldsymbol{\beta}, \mathrm{Cov}(\widehat{\boldsymbol{\beta}})\right\}, \tag{7.84}$$

where $\text{Cov}(\widehat{\boldsymbol{\beta}})$ denotes the asymptotic variance-covariance matrix of $\widehat{\boldsymbol{\beta}}$. Denote

$$\mathbf{F} = (F_{ij})_{n\times p} = \left(\frac{\partial h(\boldsymbol{\beta}, \mathbf{x}_i)}{\partial \beta_j}\right).$$

It can be shown to be

$$\text{Cov}(\widehat{\boldsymbol{\beta}}) = \sigma^2 \left(\mathbf{F}^t\mathbf{F}\right)^{-1} \quad (7.85)$$

The error variance σ^2 can be estimated in the similar manner as

$$\hat{\sigma}^2 = \frac{SSE}{n-p} = \frac{\sum_{i=1}^n \left\{y_i - h(\widehat{\boldsymbol{\beta}}, \mathbf{x}_i)\right\}^2}{n-p}.$$

Substituting $(\widehat{\boldsymbol{\beta}}, \hat{\sigma}^2)$ for $(\boldsymbol{\beta}, \sigma^2)$ in (7.85) yields estimates for the variance-covariance matrix for $\widehat{\boldsymbol{\beta}}$. Statistical inference on $\boldsymbol{\beta}$, as well as $\mathbf{a}^t\boldsymbol{\beta}$, can be made accordingly. At a given $\mathbf{x}$, confidence intervals for prediction purposes, i.e., estimating $h(\boldsymbol{\beta}, \mathbf{x})$, are also available via the *delta method*, as we shall illustrate in the following example.

7.4.2 *Example*

To illustrate, we consider an example taken from (Fox, 2002), pertaining to the US population growth. The data set contains the decennial census population data in the United States of America from 1970 through 1990. It is interesting to study the growth pattern of population (in millions) over years. Figure 7.7 (a) plots the population size (Y) versus year (X), exhibiting a strong curvature. The superimposed LS straight line $y = -35.65409 + 12.13828x$ clearly does not seem to provide a satisfactory fit, as also evidenced by the systematic pattern shown in the residual plot given by Fig. 7.7(b).

We consider a logistic function of special form

$$y_i = \frac{\beta_1}{1 + \exp(\beta_2 + \beta_3 x_i)} + \varepsilon_i,$$

where y_i is the population size at time x_i; β_1 is the asymptote towards which the population grows; β_2 reflects the size of the population at time $x = 0$ relative to its asymptote; and β_3 controls the growth rate of the population.

Nonlinear regression can be fit with PROC NLIN in SAS or the R function `nls()` in the `nls` library. In this illustrate, we shall only discuss PROC

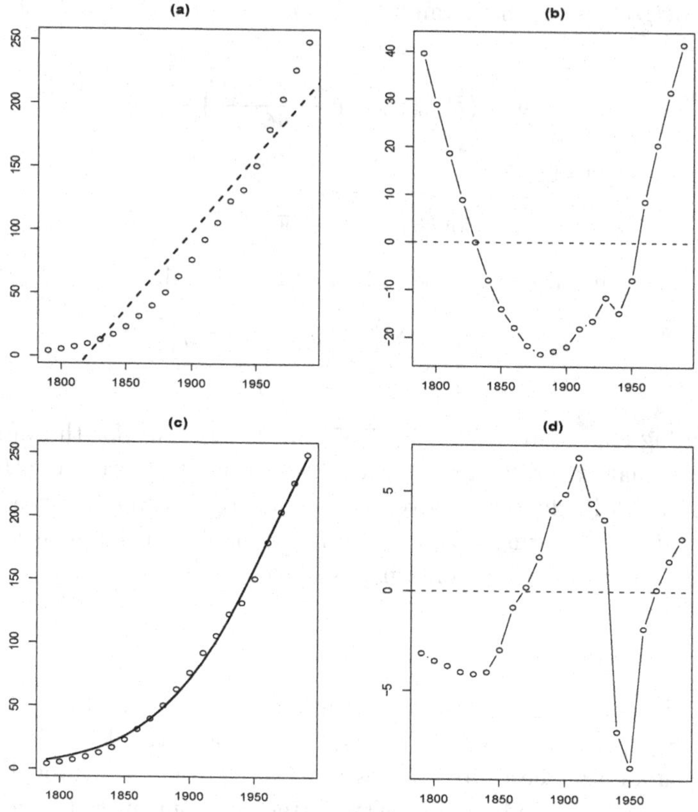

Fig. 7.7 Parametric Nonlinear Regression with the US Population Growth Data Reproduced from Fox (2002): (a) Scatter Plot of the Data Superimposed by the LS Fitted Straight Line; (b) Plot of the Residuals From the Linear Model Versus Year; (c) Scatter Plot of the Data Superimposed by the Fitted Nonlinear Curve; (d) Plot of the Residuals From the Nonlinear Fit Versus Year.

NLIN in SAS. One is referred to Fox (1997) and Fox (2002) for an R example. The PROC NLIN procedure has the METHOD= option for the user to supply derivatives or let SAS calculate them. By default, the derivatives are computed with numerical differentiation. The common suggestion in SAS is to avoid DUD if possible and, instead, choose METHOD=NEWTON, or METHOD=MARQUARDT when the parameter estimates are highly correlated. The latter option METHOD=MARQUARDT applies the Levenberg-Marquardt method to handle the singularity problem of the Hessian matrix $\mathbf{H}(\boldsymbol{\beta}^{(k)})$ during iteration.

Another important issue in numerical optimization is the starting points for the parameters. PROC NLIN allows the user to specify a grid of starting values. SAS calculates the initial sum of squared errors for each combination and then starts the iterations with the best set.

Table 7.7 Nonlinear Regression Results of the US Population Data from PROC NLIN and PROC AUTOREG.

(a) Parameter Estimates from the NLIN Procedure

Parameter	Estimate	Approx Std Error	Approximate 95% Confidence Limits	
β_1	389.2	30.8121	324.4	453.9
β_2	3.9903	0.0703	3.8426	4.1381
β_3	−0.2266	0.0109	−0.2494	−0.2038

(b) Approximate Correlation Matrix

	$\hat{\beta}_1$	$\hat{\beta}_2$	$\hat{\beta}_3$
$\hat{\beta}_1$	1.0000000	−0.1662384	0.9145423
$\hat{\beta}_2$	−0.1662384	1.0000000	−0.5406492
$\hat{\beta}_3$	0.9145423	−0.5406492	1.0000000

(c) Durbin-Watson Statistics for First-order Autocorrelation Among Residuals: PROC AUTOREG.

Order	DW	Pr < DW	Pr > DW
1	0.6254	< .0001	1.0000

Table 7.7 presents the fitted results. The fitted logistic growth model for US population size is found to be

$$\hat{h}(x) = \frac{389.2}{1 + \exp(3.99 - 0.23 \cdot x)},$$

which has been added to the scatter plot in Fig. 7.4(c). It gives a much improved fit to the data. The above equation can be used for prediction purpose. In particular, the confidence intervals for prediction can be obtained via the so-called *delta method*:

$$\hat{h}(\widehat{\boldsymbol{\beta}}, x) \pm z_{1-\alpha/2} \cdot \sqrt{\{\mathbf{g}(\widehat{\boldsymbol{\beta}}, x)\}^t \, \widehat{\text{Cov}}(\widehat{\boldsymbol{\beta}}) \, \mathbf{g}(\widehat{\boldsymbol{\beta}}, x)},$$

in view of the Taylor expansion

$$h(\widehat{\boldsymbol{\beta}}, x) = h(\boldsymbol{\beta}, x) + \{\mathbf{g}(\boldsymbol{\beta}, x)\}^t \, (\widehat{\boldsymbol{\beta}} - \boldsymbol{\beta}) + o\left(\| \widehat{\boldsymbol{\beta}} - \boldsymbol{\beta} \|\right).$$

The estimated correlation matrix is presented in Penal (b) of Table 7.7. Multiplying it by $\hat{\sigma}^2 = MSE = 19.8025$ would give the estimated covariance matrix.

Figure 7.7 (d) gives the residual plot from the nonlinear model fit. It can be seen that the systematic pattern due to lack-of-fit in the mean function has mostly been eliminated. Nevertheless, the residual seems to fluctuate cyclically with time. The Durbin-Watson statistic is used to test for autocorrelation and lack of independence of residuals, which are commonly seen in time series data. The test statistic provided by PROC AUTOREG in SAS ranges from 0 to 4. A value close to 2.0 indicates no strong evidence against the null hypothesis of no autocorrelation. The first-order Durbin-Watson statistic is printed by default. This statistic can be used to test for first-order autocorrelation. The DWPROB option prints the p-values for the Durbin-Watson test. Since these p-values are computationally expensive, they are not reported by default. One can also use the DW= option to request higher-order Durbin-Watson statistics, called *generalized Durbin-Watson* statistics, for testing higher-order autocorrelations.

Panel (c) Table 7.7 presents the results from PROC AUTOREG. Note that $Pr < DW$ is the p-value for testing positive autocorrelation, and $Pr > DW$ is the p-value for testing negative autocorrelation. Thus the residuals from the nonlinear fit shows a statistically significant (with p-value $< .0001$) first-order positive autocorrelation.

Problems

1. Consider the non-full-rank model $\mathbf{y} = \mathbf{X}\boldsymbol{\beta} + \boldsymbol{\varepsilon}$, where $\mathbf{X}$ is $n \times (k+1)$ of rank p and $\varepsilon \sim (\mathbf{0}, \sigma^2\mathbf{I})$. Show that a linear function of $\boldsymbol{\beta}$, $\boldsymbol{\lambda}^t\boldsymbol{\beta}$, is estimable if and only if any of the following equivalent conditions is met:

   (a) $\boldsymbol{\lambda} \in C(\mathbf{X}^t\mathbf{X})$, the column space of $\mathbf{X}^t\mathbf{X}$; or, equivalently, $\boldsymbol{\lambda}^t \in R(\mathbf{X}^t\mathbf{X})$, the row space of $\mathbf{X}^t\mathbf{X}$.

   (b) $\boldsymbol{\lambda}$ such that
   $$\mathbf{X}^t\mathbf{X}(\mathbf{X}^t\mathbf{X})^-\boldsymbol{\lambda} = \boldsymbol{\lambda} \text{ or } \boldsymbol{\lambda}^t(\mathbf{X}^t\mathbf{X})^-\mathbf{X}^t\mathbf{X} = \boldsymbol{\lambda}^t$$

2. Using Guass-Markov Thoerem, prove Corollary 7.2.
3. Consider estimable functions $\boldsymbol{\Lambda}\boldsymbol{\beta}$, where $\boldsymbol{\Lambda}$ be an $m \times (k+1)$ matrix with rows λ_i^t for $i = 1, \ldots, m$.

(a) Show that $\mathbf{\Lambda}\boldsymbol{\beta}$ if and only if $\mathbf{\Lambda} = \mathbf{AX}$ for some $m \times n$ matrix $\mathbf{A}$. In this way, $\mathbf{\Lambda}\boldsymbol{\beta} = \mathbf{AX}\boldsymbol{\beta} = \mathbf{A}\boldsymbol{\mu}$.

(b) Let $\mathbf{By}$ be an LUE of $\mathbf{A}\boldsymbol{\mu}$. Show that $\text{Cov}(\mathbf{By}) - \text{Cov}(\mathbf{A}\widehat{\boldsymbol{\mu}}) \geq 0$, i.e., is semi-positive definite. In other words, $\mathbf{A}\widehat{\boldsymbol{\mu}}$ is the BLUE of $\mathbf{A}\boldsymbol{\mu}$. (Hint: consider the problem of estimating $\mathbf{a}^{\mathbf{t}}\mathbf{A}\boldsymbol{\mu}$, for any nonzero vector $\mathbf{a}$. According to Theorem 7.4, $\mathbf{a}^{\mathbf{t}}\mathbf{A}\widehat{\boldsymbol{\mu}}$ is the BLUE of $\mathbf{a}^{\mathbf{t}}\mathbf{A}\boldsymbol{\mu}$. Next, show that $\mathbf{a}^{\mathbf{t}}\mathbf{By}$ is an LUE of $\mathbf{a}^{\mathbf{t}}\mathbf{A}\boldsymbol{\mu}$. Thus $\text{Var}(\mathbf{a}^{\mathbf{t}}\mathbf{A}\widehat{\boldsymbol{\mu}}) \leq \text{Var}(\mathbf{a}^{\mathbf{t}}\mathbf{By})$, which leads to the desired conclusion by the definition of semi-positive definite matrices.)

(c) If $\mathbf{\Lambda}$ is of rank $m < (k+1)$ and $\Lambda\boldsymbol{\beta}$ is estimable, show that matrix $\Lambda(\mathbf{X}^{\mathbf{t}}\mathbf{X})^{-}\mathbf{\Lambda}^{t}$ is nonsingular.

4. Show that the difference of $\text{Cov}(\widehat{\boldsymbol{\beta}}^{\star})$ in (7.35) and $\text{Cov}(\widehat{\boldsymbol{\beta}})$ in (7.26), i.e., $\text{Cov}(\widehat{\boldsymbol{\beta}}^{\star}) - \text{Cov}(\widehat{\boldsymbol{\beta}})$ is a positive definite matrix. (Hint: Use the fact that $\boldsymbol{\lambda}^{t}\widehat{\boldsymbol{\beta}}$ is the BLUE and $\boldsymbol{\lambda}^{t}\widehat{\boldsymbol{\beta}}^{\star}$ is an LUE of $\boldsymbol{\lambda}^{t}\boldsymbol{\beta}$ for any $\boldsymbol{\lambda}$.)
5. Referring to (7.49) for the definition of matrix $\mathbf{Z}$ in ridge regression, show that the eigenvalues of $\mathbf{Z}$ are $\lambda_j/(\lambda_j + k)$.
6. Define $\mathbf{W} = \{\mathbf{X}^{t}\mathbf{X} + k\mathbf{I}\}^{-1}$. Show that the eigenvalues of $\mathbf{W}$ are $1/(\lambda_i + k)$ and $\mathbf{Z} = \mathbf{I} - k\mathbf{W}$. (Hint: To establish $\mathbf{Z} = \mathbf{I} - k\mathbf{W}$, it suffices to show $\mathbf{Z}\mathbf{W}^{-1} = \mathbf{W}^{-1} - k\mathbf{I} = (\mathbf{X}^{\mathbf{t}}\mathbf{X} + k\mathbf{I}) - k\mathbf{I} = \mathbf{X}^{\mathbf{t}}\mathbf{X}$. First $\mathbf{Z}^{-1} = \mathbf{I} + k(\mathbf{X}^{\mathbf{t}}\mathbf{X})^{-1}$. Thus $(\mathbf{X}^{\mathbf{t}}\mathbf{X})\mathbf{Z}^{-1} = \mathbf{X}^{\mathbf{t}}\mathbf{X} + k\mathbf{I} = \mathbf{W}^{-1}$. Q.E.D.)
7. Using (7.50), establish that
$$\| \widehat{\boldsymbol{\beta}}^{\text{R}} \| \leq \frac{\lambda_1}{\lambda_1 + k} \| \widehat{\boldsymbol{\beta}}^{\text{LS}} \|,$$
where λ_1 is the largest eigenvalue of $\mathbf{X}^{t}\mathbf{X}$. Now since $k > 0$, it follows that
$$\| \widehat{\boldsymbol{\beta}}^{\text{R}} \|^2 < \| \widehat{\boldsymbol{\beta}}^{\text{LS}} \|^2 .$$
8. Referring to Section 7.3.3, check that the normalized equianglular vector $\mathbf{u}_{\mathcal{A}}$ that makes equal angles with each predictor in $\mathbf{X}_{\mathcal{A}}$ is given by (7.68). (Hint: Start with
$$\mathbf{X}_{\mathcal{A}}^{t}\mathbf{X}_{\mathcal{A}}(\mathbf{X}_{\mathcal{A}}^{t}\mathbf{X}_{\mathcal{A}})^{-1}\mathbf{j}_{\mathcal{A}} = \mathbf{X}_{\mathcal{A}}^{t}\left\{\mathbf{X}_{\mathcal{A}}(\mathbf{X}_{\mathcal{A}}^{t}\mathbf{X}_{\mathcal{A}})^{-1}\mathbf{j}_{\mathcal{A}}\right\} = \mathbf{I}\mathbf{j}_{\mathcal{A}} = \mathbf{j}_{\mathcal{A}}.$$
Let $\mathbf{u} = \left\{\mathbf{X}_{\mathcal{A}}(\mathbf{X}_{\mathcal{A}}^{t}\mathbf{X}_{\mathcal{A}})^{-1}\mathbf{j}_{\mathcal{A}}\right\}$, which can then be normalized to have length 1 by the scalar $A_{\mathcal{A}}$ given in (7.66). Also, use the fact that
$$\cos(\theta) = \frac{\langle \mathbf{a}, \mathbf{b} \rangle}{\| \mathbf{a} \| \cdot \| \mathbf{b} \|}$$
if θ is the angle (less than 90°) formed by $\mathbf{a}$ and $\mathbf{b}$ for any two vectors $\mathbf{a}$ and $\mathbf{b}$.)

Chapter 8

Generalized Linear Models

Linear regression has wide and fundamental applications in various fields. Its popularity can be attributed to its simple form, sound theoretical support, efficient computation in estimation, great flexibility to incorporate interactions, dummy variables, and other transformations, and easy interpretations. For all the linear models discussed so far, the response Y is a continuous variable. Many studies or experiments are often involve responses of other types. Consider, for example, evaluation of the academic performance of college students. Instead of the continuous grade point average (GPA) score, the categorical grade A–F may have been used. Generalized linear models (GLM; McCullagh and Nelder, 1983) extends linear regression to encompass other types of response while, at the same time, enjoying nearly all the merits of linear modeling.

In this chapter, we study how linear regression is generalized to handle data with different types of responses. We first motivate the problem using an example on simple logistic regression in Section 8.1, followed by general discussion on the basic components (Section 8.2), estimation (Section 8.3), statistical inference, and other issues (Section 8.4) in GLM. We then introduce two important and very commonly used GLMs, logistic regression models for binary responses in Section 8.5 and log-linear models for count data in Section 8.6.

8.1 Introduction: A Motivating Example

We first reproduce an example from Hosmer and Lemeshow (2000) that gives an excellent motivation to the problem with a real application. The data set, which can be downloaded at

`ftp://ftp.wiley.com/public/sci_tech_med/logistic/alr.zip`,

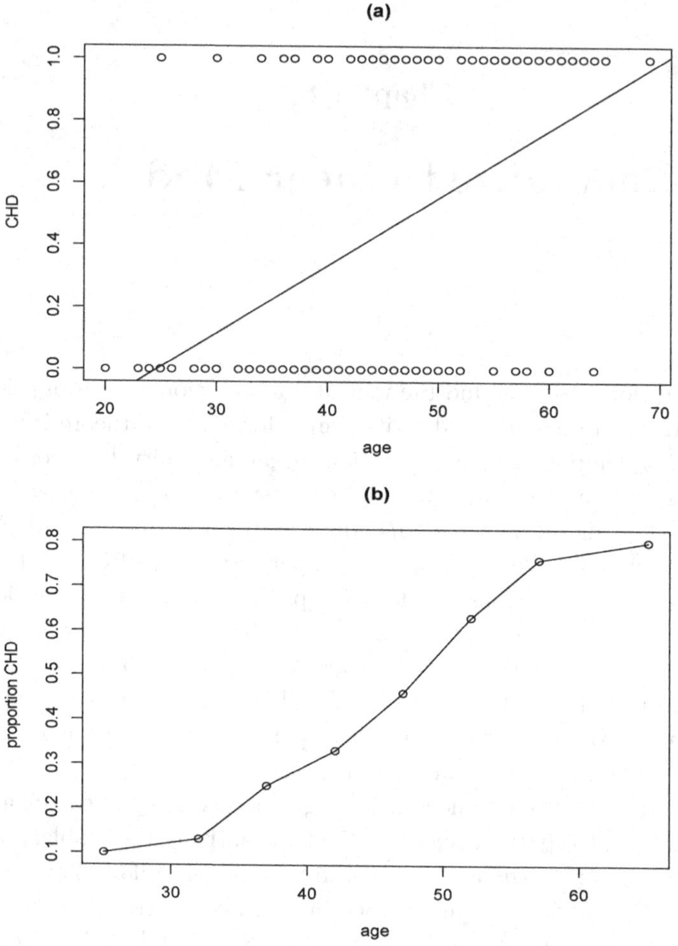

Fig. 8.1 (a) Scatterplot of the CHD Data, Superimposed With the Straight Line from Least Squares Fit; (b) Plot of the Percentage of Subjects With CHD Within Each Age Group, Superimposed by LOWESS Smoothed Curve.

was collected from a retrospective coronary heart disease (CHD) study. It contains three variables: ID for each subject, AGE (X), and a binary indicator CHD (Y) indicating whether CHD occurs to the subject. The objective is to explore the relationship between age and prevalence of CHD.

A scatterplot of the data is given in Fig. 8.1(a). Although one can see larger values of AGE tend to be more associated with "1"'s of CHD, the plot is not very informative due to discreteness of the response Y.

Recall that linear regression relates the conditional mean of the response, $E(Y|X)$, to a linear combination of predictors. Can we do the same with binary data? Let the binary response $y_i = 1$ if CHD is found in the ith individual and 0 otherwise for $i = 1, \ldots, n$. Denote $\pi_i = \Pr(y_i = 1)$. Thus, the conditional distribution of y_i given x_i is a Bernoulli trial with parameter π_i. It can be found that $E(y_i) = \pi_i$. Hence, the linear model would be

$$E(y_i) = \pi_i = \beta_0 + \beta_1 x_i. \tag{8.1}$$

Figure 8.1(a) plots the straight line fitted by least squares. Clearly, it is not a good fit. There is another inherent problem with Model (8.1). The left-hand side π_i ranges from 0 to 1, which does not mathematically match well with the range $(-\infty, \infty)$ of the linear equation on the right-hand side. A transformation on π_i, $g(\cdot)$, which maps $[0, 1]$ onto $(-\infty, \infty)$, would help. This transformation function is referred to the link function.

In order to explore the functional form between π_i and x_i, we must have available estimates of the proportions π_i. One approach is group the data by categorizing AGE into several intervals and record the relatively frequency of CHD within each interval. Table 8.1 shows the worksheet for this calculation.

Table 8.1 Frequency Table of AGE Group by CHD.

		CHD		
Age Group	n	Absent	Present	Proportion
20-29	10	9	1	0.10
30-34	15	23	2	0.13
35-39	12	9	3	0.25
40-44	15	10	5	0.33
45-49	13	7	6	0.46
50-54	8	3	5	0.63
55-59	17	4	13	0.76
60-69	10	2	8	0.80
Total	100	57	43	0.43

Figure 8.1(b) plots the proportions of subjects with CHD in each age interval versus the middle value of the interval. It can be seen that the conditional mean of y_i or proportion gradually approaches zero and one to each end. The plot shows an 'S'-shaped or sigmoid nonlinear relationship, in which the change in $\pi(x)$ with per unit increase in x grows quickly at first, but gradually slows down and then eventually levels off. Such a pattern is

rather representative and can be generally seen in many other applications. It is often expected to see that a fixed change in x has less impact when $\pi(x)$ is near 0 or 1 than when $\pi(x)$ is near 0.5. Suppose, for example, that $\pi(x)$ denotes the probability to pass away for a person of age x. An increase of five years in age would have less effect on $\pi(x)$ when $x = 70$, in which case $\pi(x)$ is perhaps close to 1, than when $x = 40$.

In sum, a suitable link function $g(\pi_i)$ is desired to satisfy two conditions: it maps $[0, 1]$ onto the whole real line and has the sigmoid shape. A natural choice for $g(\cdot)$ would be a cumulative distribution function of a random variable. In particular, the logistic distribution, whose CDF is the simplified logistic function $g(x) = \exp(x)/\{1 + \exp(x)\}$ in (7.81), yields the most popular link. Under the logistic link, the relationship between the CHD prevalence rate and AGE can be formulated by the following simple model

$$\text{logit}(\pi_i) = \log\left(\frac{\pi_i}{1-\pi_i}\right) = \beta_0 + \beta_1 x_i.$$

When several predictors $\{X_1, \ldots, X_p\}$ are involved, the multiple logistic regression can be generally expressed as

$$\log\left(\frac{\pi_i}{1-\pi_i}\right) = \mathbf{x}_i'\boldsymbol{\beta}.$$

We shall explore more on logistic regression in Section 8.5.

8.2 Components of GLM

The logistic regression model is one of the generalized linear models (GLM). Many models in the class had been well studied by the time when Nelder and Wedderburn (1972) introduced the unified GLM family. The specification of a GLM generally consists of three components: a *random component* specifies the probability distribution of the response; a *systematic component* forms the linear combination of predictors; and a link function relates the mean response to the systematic component.

8.2.1 *Exponential Family*

The random component assumes a probability distribution for the response y_i. This distribution is taken from the natural exponential distribution family of form

$$f(y_i; \theta_i, \phi) = \exp\left\{\frac{y_i\theta_i - b(\theta_i)}{a(\phi)} + c(y_i; \phi)\right\}, \tag{8.2}$$

where θ_i is the *natural parameter* and ϕ is termed as the dispersion parameter. It can be shown (see Exercise 8.6.2) that

$$E(y_i) = \mu_i = b'(\theta_i) \quad \text{and} \quad \text{Var}(y_i) = b''(\theta_i)a(\phi), \tag{8.3}$$

both moments determined by the function $b(\cdot)$.

Take the Gaussian distribution for example. The probability density function (pdf) of $N(\mu, \sigma^2)$ can be rewritten as

$$\begin{aligned} f_Y(y) &= \frac{1}{\sqrt{2\pi\sigma^2}} \exp\left\{-\frac{(y-\mu)^2}{2\sigma^2}\right\} \\ &= \exp\left\{\frac{y\mu - \mu^2/2}{\sigma^2} - \frac{y^2/\sigma^2 + \log(2\pi\sigma^2)}{2}\right\}. \end{aligned}$$

Therefore, $\theta = \mu$, $\phi = \sigma^2$, $a(\phi) = \phi$, $b(\theta) = \theta^2/2$, and $c(y, \phi) = -\left\{y^2/\phi + \log(2\pi\phi)\right\}/2$.

8.2.2 *Linear Predictor and Link Functions*

The systematic component is the *linear predictor*, denoted as

$$\eta_i = \sum_j \beta_j x_{ij} = \mathbf{x}_i \boldsymbol{\beta},$$

for $i = 1, \ldots, n$ and $j = 0, 1, \ldots, p$ with $x_{i0} = 1$ to account for the intercept. Similar to ordinary linear regression, this specification allows for incorporation of interaction, polynomial terms, and dummy variables.

The *link function* in GLM relates the linear predictor η_i to the mean response μ_i. Thus

$$g(\mu_i) = \eta_i$$

or inversely

$$\mu_i = g^{-1}(\eta_i).$$

In classical Gaussian linear models, the identity link $g(\mu_i) = \mu_i$ is applied. A preferable link function usually not only maps the range of μ_i onto the whole real line, but also provides good empirical approximation and carries meaningful interpretation when it comes to real applications.

As an important special case, the link function g such that

$$g(\mu_i) = \theta_i$$

is called the *canonical link.* Under this link, the direct relationship $\theta_i = \eta_i$ occurs. Since $\mu_i = b'(\theta_i)$, we have $\theta_i = (b')^{-1}(\mu_i)$. Namely, the canonical link is the inverse of $b'(\cdot)$:

$$(b')^{-1}(\mu_i) = \eta_i = \sum_j \beta_j x_{ij}. \tag{8.4}$$

In Gaussian linear models, the canonical link is the identity function. With the canonical link, the sufficient statistic is $\mathbf{X}^t\mathbf{y}$ in vector notation with components $\sum_i x_{ij}y_i$ for $j = 0, 1, \ldots, p$. The canonical link provides mathematical convenience in deriving statistical properties of the model; at the same time, they are also often found eminently sensible on scientific grounds.

8.3 Maximum Likelihood Estimation of GLM

The least squares (LS) method is no longer directly appropriate when the response variable Y is not continuous. Estimation of GLM is processed within the maximum likelihood (ML) framework. However, as we will see, the ML estimation in GLM has a close connection with an iteratively weighted least squares method.

Given data $\{(y_i, \mathbf{x}_i) : i = 1, \ldots, n\}$, the log likelihood function is

$$L(\boldsymbol{\beta}) = \sum_i L_i = \sum_i \log f_Y(y_i; \theta_i, \phi) = \sum_i \frac{y_i\theta_i - b(\theta_i)}{a(\phi)} + \sum_i c(y_i, \phi). \quad (8.5)$$

8.3.1 *Likelihood Equations*

The likelihood equations are

$$\frac{\partial L(\boldsymbol{\beta})}{\partial \beta_j} = \sum_i \frac{\partial L_i}{\partial \beta_j} = 0$$

for $j = 0, 1, \ldots, p$. Using the chain rule, we have

$$\frac{\partial L_i}{\partial \beta_j} = \frac{\partial L_i}{\partial \theta_i}\frac{\partial \theta_i}{\partial \mu_i}\frac{\partial \mu_i}{\partial \eta_i}\frac{\partial \eta_i}{\partial \beta_j},$$

where

$$\frac{\partial L_i}{\partial \theta_i} = \frac{y_i - b'(\theta_i)}{a(\phi)} = \frac{y_i - \mu_i}{a(\phi)} \quad \text{using} \quad \mu_i = b'(\theta_i);$$

$$\frac{\partial \theta_i}{\partial \mu_i} = b''(\theta_i) = \mathrm{Var}(y_i)/a(\phi) \quad \text{using} \quad \mathrm{Var}(y_i) = b''(\theta_i)a(\phi);$$

$$\frac{\partial \mu_i}{\partial \eta_i} = (g^{-1})'(\eta_i);$$

$$\text{and} \quad \frac{\partial \eta_i}{\partial \beta_j} = x_{ij}.$$

Therefore, the likelihood equations for $\boldsymbol{\beta}$ become

$$\sum_{i=1}^{n} \frac{(y_i - \mu_i)\, x_{ij}}{\mathrm{Var}(y_i)} \cdot \frac{\partial \mu_i}{\partial \eta_i} = 0, \quad \text{for } j = 0, 1, \ldots, p. \quad (8.6)$$

In the case of a canonical link $\eta_i = \theta_i$, we have

$$\frac{\partial \mu_i}{\partial \eta_i} = \frac{\partial \mu_i}{\partial \theta_i} = \frac{\partial b'(\theta_i)}{\partial \theta_i} = b''(\theta_i).$$

Thus

$$\frac{\partial L_i}{\partial \beta_j} = \frac{y_i - \mu_i}{\text{Var}(y_i)} b''(\theta_i) x_{ij} = \frac{(y_i - \mu_i) x_{ij}}{a(\phi)} \tag{8.7}$$

using (8.3) and the likelihood equations simplify to

$$\sum_i x_{ij} y_i = \sum_i x_{ij} \mu_i, \text{ for } j = 0, 1, \ldots, p.$$

Or in matrix notations, $\mathbf{X}^t(\mathbf{y} - \boldsymbol{\mu}) = \mathbf{0}$, which is in the same form as seen in ordinary linear regression

$$\mathbf{X}^t\mathbf{X}\boldsymbol{\beta} = \mathbf{X}^t\mathbf{y} \implies \mathbf{X}^t(\mathbf{y} - \mathbf{X}\boldsymbol{\beta}) = \mathbf{X}^t(\mathbf{y} - \boldsymbol{\mu}) = \mathbf{0}.$$

8.3.2 *Fisher's Information Matrix*

Fisher's information matrix $\mathcal{I}$ is defined as the negative expectation of the second derivatives of the log-likelihood with respect to $\boldsymbol{\beta}$, i.e., $\mathcal{I} = E(-L'')$ with elements $E\{-\partial^2 L(\boldsymbol{\beta})/\partial\beta_j \partial\beta_{j'}\}$.

Using the general likelihood results

$$E\left(\frac{\partial^2 L_i}{\partial \beta_j \partial \beta_{j'}}\right) = -E\left(\frac{\partial L_i}{\partial \beta_j}\frac{\partial L_i}{\partial \beta_{j'}}\right),$$

which holds for distributions in the exponential family (Cox and Hinkley, 1974, Sec. 4.8), we have

$$\begin{aligned} E\left(\frac{\partial L_i}{\partial \beta_j}\frac{\partial L_i}{\partial \beta_{j'}}\right) &= -E\left\{\frac{(y_i - \mu_i)x_{ij}}{\text{Var}(y_i)}\frac{\partial \mu_i}{\partial \eta_i}\frac{(y_i - \mu_i)x_{ij'}}{\text{Var}(y_i)}\frac{\partial \mu_i}{\partial \eta_i}\right\} \quad \text{from (8.6)} \\ &= -\frac{x_{ij}x_{ij'}}{\text{Var}(y_i)}\left(\frac{\partial \mu_i}{\partial \eta_i}\right)^2 \end{aligned}$$

and hence

$$E\left(-\frac{\partial^2 L(\boldsymbol{\beta})}{\partial \beta_j \beta_{j'}}\right) = \sum_{i=1}^n \frac{x_{ij}x_{ij'}}{\text{Var}(y_i)}\left(\frac{\partial \mu_i}{\partial \eta_i}\right)^2. \tag{8.8}$$

Let $\mathbf{W} = \text{diag}(w_i)$ be the diagonal matrix with diagonal elements

$$w_i = \frac{1}{\text{Var}(y_i)}\left(\frac{\partial \mu_i}{\partial \eta_i}\right)^2. \tag{8.9}$$

Then the information matrix is given by

$$\mathcal{I} = \mathbf{X}^t\mathbf{W}\mathbf{X} \tag{8.10}$$

If the canonical link is used, then from equation (8.7)

$$\left(\frac{\partial^2 L(\boldsymbol{\beta})}{\partial \beta_j \beta_{j'}}\right) = -\frac{x_{ij}}{a(\phi)} \cdot \frac{\partial \mu_i}{\partial \beta_{j'}},$$

which does not involve the random variable y_i. This implies that

$$\left(\frac{\partial^2 L(\boldsymbol{\beta})}{\partial \beta_j \beta_{j'}}\right) = E\left(\frac{\partial^2 L(\boldsymbol{\beta})}{\partial \beta_j \beta_{j'}}\right). \tag{8.11}$$

In other words, the *observed information* matrix is equal to the *expected information* matrix with the canonical link.

8.3.3 *Optimization of the Likelihood*

The log-likelihood function of a GLM is typically nonlinear in $\boldsymbol{\beta}$. Optimization of the likelihood is usually done via iterative numerical algorithms such as the *Newton-Raphson method* or Fisher scoring method.

Denote the maximum likelihood estimate (MLE) of $\boldsymbol{\beta}$ as $\widehat{\boldsymbol{\beta}}$, which satisfies $L'(\widehat{\boldsymbol{\beta}}) = \mathbf{0}$. Applying the first-order Taylor series expansion on $L'(\widehat{\boldsymbol{\beta}})$ at the current estimate $\boldsymbol{\beta}^{(k)}$ gives

$$0 = L'(\widehat{\boldsymbol{\beta}}) = L'(\boldsymbol{\beta}^{(k)}) + L''(\boldsymbol{\beta}^{(k)})(\widehat{\boldsymbol{\beta}} - \boldsymbol{\beta}^{(k)}).$$

Solving the equation for $\widehat{\boldsymbol{\beta}}$ leads to the Newton-Raphson updating formula as in (7.83)

$$\boldsymbol{\beta}^{(k+1)} = \boldsymbol{\beta}^{(k)} - \left(\mathbf{H}^{(k)}\right)^{-1} \mathbf{u}^{(k)}, \tag{8.12}$$

where $\mathbf{u}^{(k)} = L'(\boldsymbol{\beta}^{(k)})$ is the first derivative or *gradient* of the log-likelihood evaluated at $\boldsymbol{\beta}^{(k)}$ and $\mathbf{H}^{(k)} = L''(\boldsymbol{\beta}^{(k)})$ is its second derivative or *Hessian matrix* evaluated at $\boldsymbol{\beta}^{(k)}$.

Fisher scoring resembles the Newton-Raphson method, except for that Fisher scoring uses the expected value of $-\mathbf{H}^{(k)}$, called the *expected information*, whereas Newton-Raphson applies the matrix directly, called the *observed information*. Note that $\mathcal{I}^{(k)} = E\left(-\mathbf{H}^{(k)}\right)$. Plugging it into expression (8.12) yields the updating formula for Fisher scoring

$$\boldsymbol{\beta}^{(k+1)} = \boldsymbol{\beta}^{(k)} + \left(\mathcal{I}^{(k)}\right)^{-1} \mathbf{u}^{(k)}. \tag{8.13}$$

It is worth noting that, from (8.11), the Newton-Raphson method is identical to Fisher scoring under the canonical link, in which case the observed information is non-random and hence equal to the expected information.

Next, we shall show that implementation of Fisher scoring in GLM takes the form of an *iteratively reweighted least squares* algorithm. The weighted least squares (WLS; see Section 7.2) estimator of $\boldsymbol{\beta}$ is referred to

$$\widehat{\boldsymbol{\beta}} = (\mathbf{X}^t\mathbf{V}^{-1}\mathbf{X})^{-1}\mathbf{X}^t\mathbf{V}^{-1}\mathbf{z}, \tag{8.14}$$

when the linear model is

$$\mathbf{z} = \mathbf{X}\boldsymbol{\beta} + \boldsymbol{\varepsilon}$$

with $\boldsymbol{\varepsilon} \sim (\mathbf{0}, \mathbf{V})$.

From (8.6), the component of the gradient $\mathbf{u}$ is

$$u_j = \sum_{i=1}^{n} \frac{(y_i - \mu_i)\,x_{ij}}{\mathrm{Var}(y_i)} \cdot \left(\frac{\partial \mu_i}{\partial \eta_i}\right)^2 \cdot \frac{\partial \eta_i}{\partial \mu_i}.$$

Hence, $\mathbf{u}$ can be rewritten in matrix form as

$$\mathbf{u} = \mathbf{X}^t\mathbf{W}\boldsymbol{\Delta}(\mathbf{y} - \boldsymbol{\mu}), \tag{8.15}$$

where $\boldsymbol{\Delta} = \mathrm{diag}\,(\partial\eta_i/\partial\mu_i)$ and $\mathbf{W}$ is given in equation (8.9).

Also, from equation (8.10), $\mathcal{I} = \mathbf{X}^t\mathbf{W}\mathbf{X}$. Therefore,

$$\begin{aligned}
\boldsymbol{\beta}^{(k+1)} &= \boldsymbol{\beta}^{(k)} + \left(\mathcal{I}^{(k)}\right)^{-1}\mathbf{u}^{(k)}, \text{ where } \mathcal{I}^{(k)} = \mathbf{X}^t\mathbf{W}^{(k)}\mathbf{X} \text{ from (8.10)}, \\
&= (\mathbf{X}^t\mathbf{W}^{(k)}\mathbf{X})^{-1}\mathbf{X}^t\mathbf{W}^{(k)}\mathbf{X}\boldsymbol{\beta}^{(k)} + \\
&\quad (\mathbf{X}^t\mathbf{W}^{(k)}\mathbf{X})^{-1}\mathbf{X}^t\mathbf{W}^{(k)}\boldsymbol{\Delta}^{(k)}(\mathbf{y} - \boldsymbol{\mu}^{(k)}) \\
&= \left\{\mathbf{X}^t\mathbf{W}^{(k)}\mathbf{X}\right\}^{-1}\mathbf{X}^t\mathbf{W}^{(k)}\left\{\mathbf{X}\boldsymbol{\beta}^{(k)} + \boldsymbol{\Delta}^{(k)}(\mathbf{y} - \boldsymbol{\mu}^{(k)})\right\} \\
&= \left\{\mathbf{X}^t\mathbf{W}^{(k)}\mathbf{X}\right\}^{-1}\mathbf{X}^t\mathbf{W}^{(k)}\left\{\boldsymbol{\eta}^{(k)} + \boldsymbol{\Delta}^{(k)}(\mathbf{y} - \boldsymbol{\mu}^{(k)})\right\} \\
&= \left\{\mathbf{X}^t\mathbf{W}^{(k)}\mathbf{X}\right\}^{-1}\mathbf{X}^t\mathbf{W}^{(k)}\mathbf{z}^{(k)}
\end{aligned} \tag{8.16}$$

if one defines an adjusted response

$$\mathbf{z}^{(k)} = \boldsymbol{\eta}^{(k)} + \boldsymbol{\Delta}^{(k)}(\mathbf{y} - \boldsymbol{\mu}^{(k)})$$

with components

$$z_i^{(k)} = \eta_i^{(k)} + \left(y_i - \mu_i^{(k)}\right) \cdot \left.\frac{\partial \eta_i}{\partial \mu_i}\right|_{\boldsymbol{\beta}=\boldsymbol{\beta}^{(k)}} \tag{8.17}$$

for $i = 1, \ldots, n$. Note that $z_i^{(k)}$'s are continuously scaled. Comparing (8.16) to (8.14), it is clear that $\boldsymbol{\beta}^{(k+1)}$ is the WLS solution for fitting ordinary linear model

$$\mathbf{z}^{(k)} = \mathbf{X}\boldsymbol{\beta}^{(k)} + \boldsymbol{\varepsilon}, \quad \text{with } \boldsymbol{\varepsilon} \sim \left\{\mathbf{0}, \left(\mathbf{W}^{(\mathbf{k})}\right)^{-1}\right\}.$$

8.4 Statistical Inference and Other Issues in GLM

8.4.1 *Wald, Likelihood Ratio, and Score Test*

The maximum likelihood framework is a well-established system for statistical inference. The maximum likelihood estimators have many attractive properties. For example, they are asymptotically consistent and efficient. Large-sample normality is also readily available under weak regularity conditions. Standard ML results supplies that

$$\widehat{\boldsymbol{\beta}} \xrightarrow{d} \mathcal{N}\left(\boldsymbol{\beta},\, \mathcal{I}^{-1}\right) \;\text{ or }\; \mathcal{N}\left\{\boldsymbol{\beta},\; \left(\mathbf{X}^t\mathbf{W}\mathbf{X}\right)^{-1}\right\} \;\text{ as } n \to \infty \tag{8.18}$$

where the notation $\xrightarrow{d}$ means "converges in distribution to." The asymptotic variance-covariance matrix of $\widehat{\boldsymbol{\beta}}$ is

$$\mathrm{Cov}(\widehat{\boldsymbol{\beta}}) = \mathcal{I}^{-1} = \left(\mathbf{X}^t\mathbf{W}\mathbf{X}\right)^{-1},$$

which can be estimated by replacing $\mathbf{W}$ with its estimate $\hat{\mathbf{W}}$. Another immediate ML result is the so-called delta method, which can be used to obtain asymptotic distributions for any smooth function of $\widehat{\boldsymbol{\beta}}$, $g(\widehat{\boldsymbol{\beta}})$. Let $\mathbf{g}' = \partial g(\boldsymbol{\beta})/\partial\boldsymbol{\beta}$. Then

$$g(\widehat{\boldsymbol{\beta}}) \xrightarrow{d} \mathcal{N}\left\{g(\boldsymbol{\beta}),\; (\mathbf{g}')^t\mathcal{I}^{-1}\mathbf{g}'\right\}, \tag{8.19}$$

where the asymptotic variance $\mathrm{Var}\{g(\widehat{\boldsymbol{\beta}})\} = \mathbf{g}'^t\mathcal{I}^{-1}\mathbf{g}'$ can be estimated by substituting $\boldsymbol{\beta}$ with its MLE $\widehat{\boldsymbol{\beta}}$. This result can be heuristically justified by the Taylor expansion

$$g(\widehat{\boldsymbol{\beta}}) - g(\boldsymbol{\beta}) \approx (\mathbf{g}')^t(\widehat{\boldsymbol{\beta}} - \boldsymbol{\beta}).$$

Within the ML framework, there are three commonly-used methods for making statistical inference on $\boldsymbol{\beta}$: the Wald test, the likelihood ratio test, and the score test, all exploiting the asymptotic normality of maximum likelihood estimation. We will briefly discuss each of them from the hypothesis testing prospective. Confidence intervals can be generally derived by inverting the testing procedures.

The Wald statistic for testing the null H_0: $\boldsymbol{\Lambda}\boldsymbol{\beta} = \mathbf{b}$ where $\boldsymbol{\Lambda}$ is $q \times (p+1)$ of rank q takes a similar form seen in ordinary linear regression:

$$G = \left(\boldsymbol{\Lambda}\widehat{\boldsymbol{\beta}} - \mathbf{b}\right)^t \left\{\boldsymbol{\Lambda}\left(\mathbf{X}^t\hat{\mathbf{W}}\mathbf{X}\right)^{-1}\boldsymbol{\Lambda}^t\right\}^{-1} \left(\boldsymbol{\Lambda}\widehat{\boldsymbol{\beta}} - \mathbf{b}\right). \tag{8.20}$$

The null distribution of G is referred to $\chi^2(q)$. As a special case, for H_0: $\beta_j = b$, the z test

$$z = \sqrt{G} = \frac{\hat{\beta}_j - b}{s.e.(\hat{\beta}_j)} \overset{H_0}{\sim} N(0,1)$$

can be used. It is most convenient to derive confidence intervals from the Wald test. For example, $(1-\alpha)\times 100\%$ CI for β_j is given by

$$\hat{\beta}_j \pm z_{1-\frac{\alpha}{2}}\; s.e.(\hat{\beta}_j).$$

Clearly, Wald-typed inference is also readily available for functions of $\boldsymbol{\beta}$, $g(\boldsymbol{\beta})$, using results in (8.19).

The likelihood ratio test (LRT) compares the maximized log-likelihood functions between two nested models. Suppose that we have a model, called the *full model*, and a null hypothesis H_0. Bringing the conditions in H_0 into the full model gives a *reduced* model. Let $\widehat{L}_{\text{full}}$ and $\widehat{L}_{\text{reduced}}$ denote their respective maximized log-likelihoods. The likelihood ratio test statistic is

$$LRT = -2\cdot\left(\widehat{L}_{\text{reduced}} - \widehat{L}_{\text{full}}\right). \tag{8.21}$$

The null distribution of LRT is again $\chi^2(\nu)$, where ν is the difference in number of degrees of freedom between two models.

The third alternative for testing $H_0 : \boldsymbol{\beta} = \mathbf{b}$ is the score test, which is also the Lagrange multiplier test. It is based on the slope and expected curvature of $L(\boldsymbol{\beta})$ at the null value $\mathbf{b}$. Let

$$\mathbf{u}_0 = \mathbf{u}(\mathbf{b}) = \left.\frac{\partial L}{\partial \boldsymbol{\beta}}\right|_{\boldsymbol{\beta}\,=\,\mathbf{b}}$$

be the score function of the full model evaluated at the null value $\mathbf{b}$ (recall that $\mathbf{u}(\hat{\boldsymbol{\beta}}) = 0$) and

$$\mathcal{I}_0 = \mathcal{I}(\mathbf{b}) = -\left. E\left(\frac{\partial^2 L}{\partial \boldsymbol{\beta}\,\partial \boldsymbol{\beta}^t}\right)\right|_{\boldsymbol{\beta}\,=\,\mathbf{b}}$$

be the Fisher's information matrix of the full model evaluated at $\mathbf{b}$. The score test statistic is a quadratic form given as

$$S = \mathbf{u}_0^t\, \mathcal{I}_0^{-1}\, \mathbf{u}_0. \tag{8.22}$$

Under H_0, the score test statistic follows the same chi-squared null distribution. Consider another illustration, which is more practically useful. Suppose that $\boldsymbol{\beta}$ can be partitioned into $(\boldsymbol{\beta}_1, \boldsymbol{\beta}_2)^t$ and we want to test $H_0 : \boldsymbol{\beta}_1 = \mathbf{b}_1$. Let $\hat{\boldsymbol{\beta}}_2^{(0)}$ denote the MLE of $\boldsymbol{\beta}_2$ obtained from fitting the reduced model or the null model. The score test statistic is then given by

$$S = \mathbf{u}_1^t \left(\mathbf{I}_{11} - \mathbf{I}_{12}\mathbf{I}_{22}^{-1}\mathbf{I}_{21}\right)^{-1} \left.\mathbf{u}_1\right|_{\boldsymbol{\beta}=\,(\mathbf{b}_1,\,\hat{\boldsymbol{\beta}}_2^{(0)})^t}, \tag{8.23}$$

where

$$\mathbf{u}_1 = \frac{\partial L}{\partial \boldsymbol{\beta}_1} \text{ and } \mathbf{I}_{jj'} = \frac{\partial^2 L}{\partial \boldsymbol{\beta}_j\, \partial \boldsymbol{\beta}_{j'}^t}$$

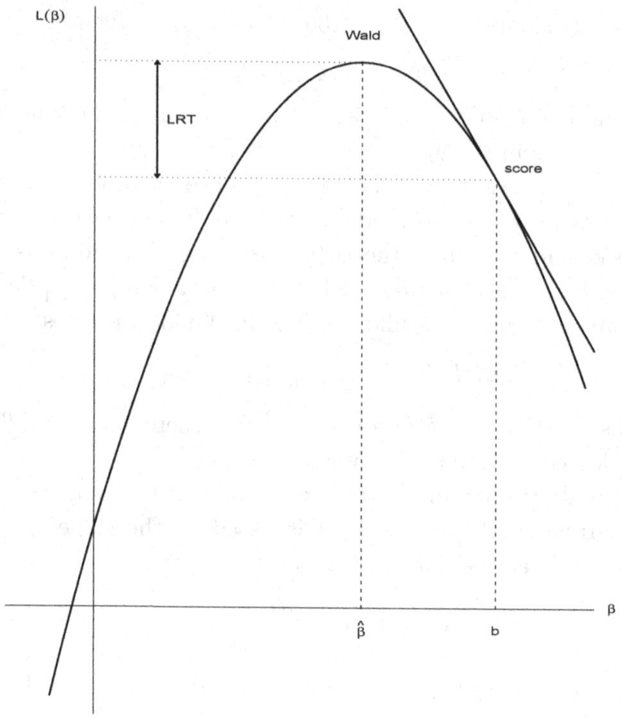

Fig. 8.2 Plot of the Log-likelihood Function: Comparison of Wald, LRT, and Score Tests on $H_0 : \beta = b$.

for $j, j' = 1, 2$. Note that all the quantities involved in the above expression are derived from the full model but evaluated at $\boldsymbol{\beta} = (\mathbf{b}_1, \hat{\boldsymbol{\beta}}_2^{(0)})^t$, which are obtained from fitting the null model. The null distribution of S is χ^2 with df equal to the dimension of $\boldsymbol{\beta}_1$.

The Wald, LRT, and score tests all refer to the same null chi-squared distribution and they can be shown to be asymptotically equivalent for large sample sizes. On the other hand, they show different empirical performances. Fig. 8.2 illustrated the comparison of these three tests for testing $H_0 : \beta = b$ in the one-dimensional setting $L(\beta)$. The Wald test utilizes estimation of the full model. Its form is very similar to what we have in ordinary linear regression and hence easy to comprehend. For this reason, confidence intervals are often derived from Wald tests. The score test is solely based on the null or reduced model estimation. A heuristic derivation of the score test can be carried out as follows. At the null point b, the tangent line of the loglikelihood function $L(\cdot)$ is given by

$$y - L(b) = u(b)(x - b) \tag{8.24}$$

with slope

$$u(b) = \left.\frac{\partial L}{\partial \beta}\right|_{\beta=b}$$

is the score function or the first derivative of $L(\cdot)$, evaluated at b. Note that $u(\hat{\beta}) = 0$ as the tangent line at $\hat{\beta}$ is flat. Thus how different $u(b)$ is from $u(\hat{\beta}) = 0$ naturally signalizes how different $\hat{\beta}$ is from b. Furthermore, applying Taylor expansion

$$0 = u(\hat{\beta}) \approx u(b) + \mathcal{I}(b)(\hat{\beta} - b)$$

yields $u(b) = -\mathcal{I}(b)(\hat{\beta} - b)$. Under $H_0 : \beta = b$, $\hat{\beta} \xrightarrow{d} \mathcal{N}\{b, \mathcal{I}^{-1}(b)\}$. It follows that, under H_0, $u(b) \xrightarrow{d} \mathcal{N}\{0, \mathcal{I}(b)\}$. Therefore, the score test is $S = u(b)^2/\mathcal{I}(b) \overset{H_0}{\sim} \chi^2(1)$. The LRT combines information from both models and is the more resourceful. In fact, the LRT can be shown to be the most powerful test asymptotically. Often more complex in its specific form, the score test is advantageous computationally as its calculation only requires estimation of the null model, i.e., the reduced model under the null hypothesis. This property renders the score test a very attractive technique in many scenarios where computational efficiency is a major concern. Examples include evaluation of the added variables in stepwise selection and evaluation of allowable splits in recursive partitioning.

8.4.2 *Other Model Fitting Issues*

Many other methods and procedures of linear regression are readily extended to generalized linear models. In the following, we shall briefly discuss several important aspects. First, the sum of squares error (SSE), also named as residual sum of squares (RSS), is an important lack-of-fit measure in linear regression. In GLM, the *deviance* plays the same role as SSE. Let $\hat{L}$ denote the maximized log likelihood score for the current model and $\hat{L}_{\max}$ denote the maximum possible log likelihood score for the given data, achieved by the so-called *saturated model.* The deviance is defined as

$$D = 2 \times \left(\hat{L}_{\max} - \hat{L}\right). \tag{8.25}$$

For nested models, a larger deviance is always associated with a reduced or simpler model. The *analysis of deviance* compares two nested models with an LRT chi-squared test, analogous to the use of analysis of variance (ANOVA) in linear regression.

Similar to linear regression, categorical predictors are handled by defining dummy variables. Interaction among variables are often quantified by cross-product terms. The model selection technique in Chapter 5 can be incorporated in its entirety into GLM as well. The AIC criterion, for example, is

$$AIC = -2\,\hat{L} + 2 \times \text{number of parameters}.$$

For model diagnostic purposes, two types of residuals are commonly used in GLM. The first type uses components of the deviance contributed by individual observations. Let $D = \sum_{i=1}^{n} d_i$. The *deviance residual* for ith observation is $\sqrt{d_i} \cdot \text{sign}(y_i - \hat{\mu}_i)$. An alternative is the *Pearson residuals*, defined as

$$e_i = \frac{y_i - \hat{\mu}_i}{\sqrt{\widehat{\text{Var}}(y_i)}}. \tag{8.26}$$

Besides, the *hat matrix*, whose diagonal elements supply the *leverage*, is given by

$$\mathbf{H} = \mathbf{W}^{1/2}\mathbf{X}\left(\mathbf{X}^t\mathbf{W}\mathbf{X}\right)^{-1}\mathbf{X}^t\mathbf{W}^{1/2}$$

following (7.29). Various jackknife-based diagnostic measures, such as Cook's distance, also find their natural extensions in GLM.

A detailed description of all GLM fitting aspects is beyond the scope of this book. we refer interested readers to McCullagh and Nelder (1989) for a full account.

8.5 Logistic Regression for Binary Data

8.5.1 *Interpreting the Logistic Model*

A logistic regression model is a GLM for modeling data with binary responses. Consider data that contain n observations $\{(y_i, \mathbf{x}_i) : i = 1, \ldots, n\}$, where y_i is the binary 0-1 response for the ith individual and $\mathbf{x}_i$ is its associated predictor vector.

A natural model for y_i is the Bernoulli trial with parameter $\pi_i = E(y_i) = \mu_i = \text{P}\{y_i = 1\}$. The probability distribution function of y_i can be written in the exponential family form

$$\begin{aligned} f_Y(y_i) &= \pi_i^{y_i}(1-\pi_i)^{1-y_i} = (1-\pi_i)\left\{\frac{\pi_i}{1-\pi_i}\right\}^{y_i} \\ &= \exp\left\{y_i \log\frac{\pi_i}{1-\pi_i} + \log(1-\pi_i)\right\} \\ &= \exp\left\{\frac{y_i\log\{\pi_i/(1-\pi_i)\} - \{-\log(1-\pi_i)\}}{1} + 0\right\} \end{aligned}$$

Correspondingly, we have $\theta_i = \log\{\pi_i/(1-\pi_i)\}$, $a(\phi) = 1$, $c(y_i, \phi) = 0$, and

$$b(\theta_i) = -\log(1-\pi_i) = \log\left(\frac{1}{1-\pi_i}\right) = \log\left(1+\frac{\pi_i}{1-\pi_i}\right) = \log\left(1+e^{\theta_i}\right).$$

It follows that $E(y_i|\mathbf{x}_i) = \pi_i$ and $\mathrm{Var}(y_i|\mathbf{x}_i) = \pi_i(1-\pi_i)$.

The logistic regression model applies the canonical link $\theta_i = \eta_i$, which leads to the following formulation:

$$\mathrm{logit}(\pi_i) = \log\left(\frac{\pi_i}{1-\pi_i}\right) = \mathbf{x}_i^t\boldsymbol{\beta}, \tag{8.27}$$

or, equivalently,

$$\pi_i = \mathrm{logistic}\left(\mathbf{x}_i^t\boldsymbol{\beta}\right) = \frac{\exp(\mathbf{x}_i^t\boldsymbol{\beta})}{1+\exp(\mathbf{x}_i^t\boldsymbol{\beta})}. \tag{8.28}$$

Interpretation of the regression coefficients in $\boldsymbol{\beta}$ in logistic regression, extracted analogously as in linear regression, has to do with the so-called *odds ratio*. The quantity

$$\frac{\pi}{1-\pi} = \frac{\mathrm{P}(Y=1|\mathbf{x}_i)}{\mathrm{P}(Y=0|\mathbf{x}_i)}$$

is often referred to as the *odds* of having $Y = 1$ conditioning on $\mathbf{x}_i$, which is a critical risk measure in many applications. The logistic model can be expressed in terms of odds

$$\log(\mathrm{odds}) = \mathbf{x}_i^t\boldsymbol{\beta}. \tag{8.29}$$

The ratio of two odds, each from a different scenario, is termed as the *odds ratio* (OR). The odds ratio is an appealing measure for comparing risks. For example,

$$OR = \frac{\mathrm{P}(Y=1|X=1)/\mathrm{P}(Y=0|X=1)}{\mathrm{P}(Y=1|X=0)/\mathrm{P}(Y=0|X=0)}$$

is the ratio of odds for having $Y = 1$ between two different states: $X = 1$ vs. $X = 0$. If $Y = 1\{\text{lung cancer is present}\}$ indicates the status of lung cancer for an individual and $X = 1\{\text{smoker}\}$ indicates whether he or she is a smoker, then $OR = 3$ implies that the odds of developing lung cancer for smokers is three times as much as that for non-smokers.

With similar arguments in linear regression, model (8.29) implies that every one-unit increase in X_j, while holding other predictors fixed, would lead to an amount of β_j change in the logarithm of the odds. That is

$$\beta_j = \log\left(\text{Odds}_{X_j=x+1}\right) - \log\left(\text{Odds}_{X_j=x}\right) = \log\left(OR_{(x+1):x}\right).$$

In other words, the odds ratio comparing $X_j = x+1$ vs. $X_j = x$, with other predictor fixed, is $OR_{(x+1):x} = \exp(\beta_j)$.

8.5.2 *Estimation of the Logistic Model*

The log likelihood for the logistic model (8.27) is

$$L(\boldsymbol{\beta}) = \sum_i^n \left\{ y_i \log \frac{\pi_i}{1-\pi_i} + \log(1-\pi_i) \right\}.$$

The regression coefficients in $\boldsymbol{\beta}$ enter the log likelihood through its relationship with π_i in (8.27). But it is often a tactical manoeuvre not to make the direct substitution for π_i. Instead, differentiation of the log-likelihood with respect to $\boldsymbol{\beta}$ is done via the chain rule.

It can be found from equations (8.6) and (8.7) that the gradient

$$\mathbf{u} = \frac{\partial L}{\partial \boldsymbol{\beta}} = \mathbf{X}^t(\mathbf{y} - \boldsymbol{\pi}),$$

where $\boldsymbol{\pi} = (\pi_i)$ is the vector of expected values. From equations (8.9) and (8.10), the Fisher's information matrix is

$$\mathcal{I} = \mathbf{X}^t \text{diag}\{\pi_i(1-\pi_i)\}\mathbf{X}.$$

Thus the updating formula in Fisher scoring becomes, according to (8.13),

$$\boldsymbol{\beta}^{(k+1)} = \boldsymbol{\beta}^{(k)} + \left\{ \mathbf{X}^t \text{diag}\{\pi_i^{(k)}(1-\pi_i^{(k)})\}\mathbf{X} \right\}^{-1} (\mathbf{y} - \boldsymbol{\pi}^{(k)}).$$

When implemented with the iterative reweighted least squares, the redefined response in (8.17) at each intermediate step would be

$$\mathbf{z} = \mathbf{X}\widehat{\boldsymbol{\beta}}^{(k)} + \text{diag}\left\{ \left(\pi_i^{(k)}(1-\pi_i^{(k)}) \right)^{-1} \right\} (\mathbf{y} - \boldsymbol{\pi}^{(k)}),$$

with components

$$z_i^{(k)} = \log \frac{\pi_i^{(k)}}{1-\pi_i^{(k)}} + \frac{y_i - \pi_i^{(k)}}{\pi_i^{(k)}(1-\pi_i^{(k)})}.$$

The asymptotic variance-covariance matrix of the MLE $\widehat{\boldsymbol{\beta}}$ is given by

$$\text{Cov}\left(\widehat{\boldsymbol{\beta}}\right) = \left[\mathbf{X}^t \text{diag}\{\pi_i(1-\pi_i)\}\mathbf{X}\right]^{-1}, \tag{8.30}$$

which can be estimated by substituting π_i with $\hat{\pi}_i$.

8.5.3 *Example*

To illustrate, we consider the kyphosis data (Chambers and Hastie, 1992) from a study of children who have had corrective spinal surgery. The data set contains 81 observations and four variables. A brief variable description is given below. The binary response, `kyphosis`, indicates whether kyphosis, a type of deformation was found on the child after the operation.

Table 8.2 Variable Description for the Kyphosis Data: Logistic Regression Example.

`kyphosis`	indicating if kyphosis is absent or present;
`age`	age of the child (in months);
`number`	number of vertebrae involved;
`start`	number of the first (topmost) vertebra operated on.

Logistic regression models can be fit using PROC LOGISTIC, PROC GLM, PROC CATMOD, and PROC GENMOD in SAS. In R, the function `glm` in the `base` library can be used. Another R implementation is also available in the package `Design`. In particular, function `lrm` provides penalized maximum likelihood estimation, i.e., the ridge estimator, for logistic regression.

Table 8.1 presents some selected fitting results from PROC LOGISTIC for model

$$\log\left\{\frac{\text{P}(\texttt{kyphosis}=1)}{\text{P}(\texttt{kyphosis}=0)}\right\} = \beta_0 + \beta_1 \cdot \texttt{age} + \beta_2 \cdot \texttt{number} + \beta_3 \cdot \texttt{start}.$$

Panel (a) gives the table of parameter estimates. The fitted logistic model is

$$\begin{aligned}\text{logit}\{\text{P}(\texttt{kyphosis}=1)\} &= -2.0369 + 0.0109 \times \texttt{age} + 0.4106 \times \texttt{number} \\ &\quad -0.2065 \times \texttt{start}.\end{aligned}$$

Accordingly, prediction of $\text{P}(\texttt{kyphosis}=1)$ can be obtained using (8.28). Panel (b) provides the estimates for the odds ratios (OR), $\exp(\hat{\beta}_j)$, and the associated 95% confidence intervals. The confidence interval for OR is constructed by taking the exponential of the lower and upper bounds of the confidence interval for β. Based on the results, we can conclude with 95% confidence that the odds of having kyphosis would be within $[0.712, 0.929]$ times if the number of the first (topmost) vertebra operated on, `start`, increases by one, for children with same fixed `age` and `number` values. Since 1 is not included in this confidence interval, the effect of

Table 8.3 Analysis Results for the Kyphosis Data from PROC LOGISTIC.

(a) Analysis of Maximum Likelihood Estimates

Parameter	DF	Estimate	Standard Error	Wald Chi-Square	Pr > ChiSq
Intercept	1	−2.0369	1.4496	1.9744	0.1600
Age	1	0.0109	0.00645	2.8748	0.0900
Number	1	0.4106	0.2249	3.3340	0.0679
Start	1	−0.2065	0.0677	9.3045	0.0023

(b) Odds Ratio Estimates

Effect	Point Estimate	95% Wald Confidence Limits	
Age	1.011	0.998	1.024
Number	1.508	0.970	2.343
Start	0.813	0.712	0.929

(d) Wald Test for $H_0 : \beta_1 = \beta_2 = 0$.

Wald Chi-Square	DF	Pr > ChiSq
5.0422	2	0.0804

(c) Estimated Covariance Matrix

	Intercept	Age	Number	Start
Intercept	2.101364	−0.00433	−0.27646	−0.0371
Age	−0.00433	0.000042	0.000337	−0.00012
Number	−0.27646	0.000337	0.050565	0.001681
Start	−0.0371	−0.00012	0.001681	0.004583

`start` is significant at $\alpha = 0.05$, which is consistent with the Wald test in Panel (a). Panel (c) gives an example of the Wald test in its more general form for testing $H_0 : \beta_1 = \beta_2 = 0$. In this case,

$$\mathbf{\Lambda} = \begin{bmatrix} 0\,1\,0\,0 \\ 0\,0\,1\,0 \end{bmatrix}$$

of rank 2 when applying equation (8.20). The calculation also involves the estimated variance-covariance matrix for $\widehat{\boldsymbol{\beta}}$, as shown in Panel (d). Referred to $\chi^2(2)$, the resultant p-value, 0.0804, is rather marginal. At the significance level $\alpha = 0.05$, one might consider dropping both `age` and `number` from the model.

8.6 Poisson Regression for Count Data

We next study the loglinear or Poisson models for count data as another example of GLM. Counts are frequencies of some event. Examples include the number of car accidents in different cities over a given period of time, the number of daily phone calls received in a call center, the number of students graduated from a high school, and so on. Count data are also commonly encountered in contingency tables.

8.6.1 *The Loglinear Model*

The Poisson or log-linear model is a popular GLM for count data, ideally when successive events occur independently and at the same rate. When the response Y_i follows Poisson(μ_i) for $i = 1, \ldots, n$, its probability distribution function is

$$
\begin{aligned}
f_{Y_i}(y_i) &= e^{-\mu_i}\mu_i^{y_i}/y_i! = \exp\{y_i \log \mu_i - \mu_i - \log y_i!\} \\
&= \exp\left\{\frac{y_i\theta_i - \exp(\theta_i)}{1} + (-\log y_i!)\right\}
\end{aligned}
$$

with the natural parameter $\theta_i = \log(\mu_i)$. Thus, in exponential family form, $b(\theta_i) = \exp(\theta_i)$, $a(\phi) = 1$, and $c(y_i, \phi) = -\log(y_i!)$. It follows from (8.3) that

$$
\begin{aligned}
E(Y_i) &= b'(\theta_i) = \exp(\theta_i) = \mu_i \\
\text{Var}(Y_i) &= b''(\theta_i) = \exp(\theta_i) = \mu_i.
\end{aligned}
$$

The log-linear model is specified as

$$\log(\mu_i) = \eta_i = \mathbf{x}_i^t\boldsymbol{\beta} \text{ for } i = 1, \ldots, n. \tag{8.31}$$

The canonical link, i.e., the logarithm function, is applied. In this model, one-unit increase in X_j has a multiplicative impact $\exp(\beta_j)$ on the mean response, holding other predictors fixed. It is worth mentioning that the log-linear or Poisson model in (8.31) is different from the ordinary Gaussian linear model with logarithm transformation on the response

$$\log y_i = \mathbf{x}_i^t\boldsymbol{\beta} + \varepsilon_i \text{ with } \varepsilon_i \sim N(0, \sigma^2).$$

The log likelihood for the Poisson model (8.31) is given as

$$L(\boldsymbol{\beta}) = \sum_i y_i \log \mu_i - \sum_i \mu_i - \sum_i \log y_i!.$$

The last term $\sum_i \log y_i!$ can be ignored as it does not involve any parameter. It follows from (8.6), (8.7), and (8.10) that the gradient and the Fisher's information matrix are

$$\mathbf{u} = \frac{\partial L}{\partial \boldsymbol{\beta}} = \mathbf{X}^t(\mathbf{y} - \boldsymbol{\mu})$$

$$\mathcal{I} = \mathbf{X}^t \text{diag}(\boldsymbol{\mu})\mathbf{X} = \left(\sum_{i=1}^{n} \mu_i x_{ij} x_{ij'} \right),$$

where $\boldsymbol{\mu} = (\mu_i)$ denote the mean response vector and $\mathbf{W} = \text{diag}(\boldsymbol{\mu})$ is a diagonal matrix with diagonal elements μ_i. Therefore, in the iterative reweighted least squares algorithm for Fisher scoring, the intermediate response $\mathbf{z}^{(k)}$ in (8.17) has components

$$z_i^{(k)} = \log \mu_i^{(k)} + \frac{y_i - \mu_i^{(k)}}{\mu_i^{(k)}}.$$

The resulting MLE $\widehat{\boldsymbol{\beta}}$ has asymptotic variance-covariance matrix

$$\text{Cov}\left(\widehat{\boldsymbol{\beta}}\right) = \left\{\mathbf{X}^t \text{diag}(\boldsymbol{\mu})\mathbf{X}\right\}^{-1}.$$

8.6.2 *Example*

We consider a school attendance data from Aitkin (1978), in which 146 children from Walgett, New South Wales, Australia, were classified by Culture, Age, Sex and Learner status. The response variable (Y) is the number of days absent from school in a particular school year. A brief description of the variables is provided in Table 8.4.

Table 8.4 Variable Description for the Log-Linear Regression Example.

Variable	Description	Levels
`Eth`	Ethnic background	Aboriginal ("A") or Not ("N")
`Sex`	Sex	"F" or "M"
`Age`	Age group	"F0", "F1", "F2", or "F3"
`Lrn`	Learner status: Average or Slow	"AL" or "SL"
`Days`	Days absent from school in the year	Integer

The objective is to explore the relationship between the four categorical predictors and school absence. To account for the four levels of `Age`, three dummy variables $\{Z_1^{\texttt{Age}}, Z_2^{\texttt{Age}}, Z_3^{\texttt{Age}}\}$ are introduced using the reference cell coding scheme such that

$$Z_1^{\texttt{Age}} = \begin{cases} 1 & \text{if the child is in the "F3" age group;} \\ 0 & \text{otherwise.} \end{cases}$$

$$Z_2^{\texttt{Age}} = \begin{cases} 1 & \text{if the child is in the "F2" age group;} \\ 0 & \text{otherwise.} \end{cases}$$

$$Z_3^{\texttt{Age}} = \begin{cases} 1 & \text{if the child is in the "F1" age group;} \\ 0 & \text{otherwise.} \end{cases}$$

The other three predictors $\{\texttt{Sex}, \texttt{Lrn}, \texttt{Days}\}$ are all binary and 0-1 coded. The specific coding information for them is given in Panel (a) of Table 8.5. We consider the following log-linear model

$$\log(\texttt{days}) = \beta_0 + \beta_1 \cdot \texttt{Eth}\beta_2 \cdot \texttt{Sex} + \beta_{31} \cdot Z_1^{\texttt{Age}} + \beta_{32} \cdot Z_2^{\texttt{Age}} + \beta_{33} \cdot Z_3^{\texttt{Age}} + \beta_4 \cdot \texttt{Lrn}.$$

Panel (b) of Table 8.5 presents the fitting results from PROC GENMOD. The ML estimates $\widehat{\beta}$s are shown in the first column, followed by their standard errors, the Wald 95% confidence intervals, and the χ^2 test of H_0: $\beta_j = 0$ for each of the individual parameters.

Therefore, given a set of predictor values, the predicted response can be obtained by equation

$$\widehat{\mu} = \exp\Big\{2.7154 - 0.5336 \cdot \texttt{Eth} + 0.1616 \cdot \texttt{Sex} + 0.4277 \cdot Z_1^{\texttt{Age}} + 0.2578 \cdot Z_2^{\texttt{Age}} - 0.3339 \cdot Z_3^{\texttt{Age}} + 0.3489 \cdot \texttt{Lrn}.\Big\}$$

The model can be interpreted in the following way. Take the slope estimate for Sex for example. One may make the following statement: given boys and girls who are of the same ethnicity, same age, same learning status, the average number of absence days for boys is estimated to be $\exp(0.1616) = 1.1754$ times of that for girls, associated with a 95% confidence interval $\{\exp(0.0782), \exp(0.2450)\} = (1.0813, 1.2776)$.

Problems

1. Given that Y follows a distribution in the exponential family of form in (8.2), show that $E(Y) = b'(\theta)$ and $\text{Var}(Y) = b''(\theta)a(\phi)$ by using the following two general likelihood results

$$E\left(\frac{\partial L}{\partial \theta}\right) = 0 \quad \text{and} \quad -E\left(\frac{\partial^2 L}{\partial \theta^2}\right) = E\left(\frac{\partial L}{\partial \theta}\right)^2, \qquad (8.32)$$

Table 8.5 Analysis Results for the School Absence Data from PROC GENMOD.

(a) Class Level Information

Class	Value	Design Variables		
Eth	N	1		
	A	0		
Sex	M	1		
	F	0		
Age	F3	1	0	0
	F2	0	1	0
	F1	0	0	1
	F0	0	0	0
Lrn	SL	1		
	AL	0		

(b) Analysis of Parameter Estimates

Parameter		DF	Estimate	Standard Error	Wald 95% Confidence Limits		Chi-Square
Intercept		1	2.7154	0.0647	2.5886	2.8422	1762.30
Eth	N	1	−0.5336	0.0419	−0.6157	−0.4515	162.32
Sex	M	1	0.1616	0.0425	0.0782	0.2450	14.43
Age	F3	1	0.4277	0.0677	0.2950	0.5604	39.93
	F2	1	0.2578	0.0624	0.1355	0.3802	17.06
	F1	1	−0.3339	0.0701	−0.4713	−0.1965	22.69
Lrn	SL	1	0.3489	0.0520	0.2469	0.4509	44.96

where $L(\theta, \phi) = \log f_Y(y) = \{y\theta - b(\theta)\}/a(\phi) + c(y, \phi)$ is the log likelihood. The two equations in (8.32) hold under some regularity conditions given in Sec. 4.8 of Cox and Hinkley (1974).

2. Let $y \sim \text{binomial}(n, \pi)$. We consider the maximum likelihood based inference on the parameter π.

   (a) Write down the log-likelihood function $L(\pi)$.

   (b) Find the score function
   $$u(\pi) = \frac{y}{\pi} - \frac{n-y}{1-\pi}$$
   and the Fisher's information
   $$I(\pi) = \frac{n}{\pi \cdot (1-\pi)}.$$

   (c) Find the MLE of π, $\hat{\pi}$. Show that $E(\hat{\pi}) = \pi$ and
   $$\text{Var}(\hat{\pi}) = \frac{\pi(1-\pi)}{n} = \frac{1}{I(\pi)}.$$

(d) Consider the test of $H_0 : \pi = \pi_0$. Show that the Wald test statistic has the form of

$$G = \frac{(\hat{\pi} - \pi_0)^2}{\hat{\pi}(1 - \hat{\pi})/n};$$

the score test is given by

$$S = \frac{(\hat{\pi} - \pi_0)^2}{\pi_0(1 - \pi_0)/n};$$

and the LRT can be simplified as

$$LRT = 2\left(y \log \frac{y}{n\pi_0} + (n - y) \log \frac{n - y}{n - n\pi_0}\right).$$

3. Log-linear model is often applied in the analysis of contingency tables, especially when more than one of the classification factors can be regarded as response variables. Consider a $2 \times 2 \times K$ contingency table induced by classification factors (X, Y, V). Namely, both X and Y have two levels and V has K levels. The cell frequencies can be modeled by Poisson models. Consider the following model

$$\log \mu = \beta_0 + \beta_1 Z^X + \beta_2 Z^Y + \beta_{31} Z_1^V + \beta_{32} Z_2^V + \cdots + \beta_{3,K-1} Z_{K-1}^V + \beta_4 Z^X \cdot Z^Y. \quad (8.34)$$

Let μ_{ijk} denote the expected frequency in the ijk-th cell for $i = 1, 2$, $j = 1, 2$, and $k = 1, \ldots, K$. Express the conditional odds ratio $\theta_{XY(k)}$ between X and Y, conditioning on $V = k$,

$$\log \theta_{XY(k)} = \log \frac{\mu_{00k}\, \mu_{11k}}{\mu_{01k}\, \mu_{10k}}$$

in terms of the regression parameters β's. Then argue that, under model (8.34), X and Y are independent when conditioning on V.

4. Consider data involves a binary response Y and a categorical predictor X that has K levels. We define $(K - 1)$ dummy variables $\{Z_1, Z_2, \ldots, Z_{K-1}\}$ using the reference cell coding scheme such that

$$Z_{ik} = \begin{cases} 1 & \text{if the } i\text{th subject falls into the } k\text{th category,} \\ 0 & \text{otherwise,} \end{cases}$$

for $k = 1, 2, \ldots, K - 1$. In this manner, the last category, level K, is left as the baseline. Denote $\pi_i = \Pr\{Y_i = 1 | Z_i)$ and consider model

$$\log\left(\frac{\pi_i}{1 - \pi_i}\right) = \beta_0 + \beta_1 Z_{i1} + \cdots + \beta_{K-1} Z_{i(K-1)}.$$

Let $OR_{k:k'}$ denote the odds ratio of having $Y = 1$ that compares Level k and Level k'.

(a) Express $OR_{k:K}$ in terms of β_j, for $k = 1, \ldots, (K-1)$ and describe how to obtain a confidence interval for $OR_{k:K}$.
(b) Express $OR_{k:k'}$ in terms of β_j, for $k \neq k' = 1, \ldots, (K-1)$ and describe how to obtain a confidence interval for $OR_{k:k'}$.

5. In studying the association between smoking (X) and the incidence of some disease (Y), let π_1 denote the probability of getting this disease for smokers and π_2 denote the probability of getting this disease for non-smokers. Note that both π_1 and π_2 are conditional probabilities. The collected data from this study are presented in the following 2×2 table.

		Disease	
		Yes (1)	No (0)
Smoke	Yes (1)	13	137
	No (0)	6	286

(a) Obtain the sample estimates of π_1 and π_2 based on the above data.
(b) Let $\pi = \Pr\{Y = 1|X\}$. Suppose that the logistic regression model

$$\text{logit}(\pi) = \beta_0 + \beta_1 X$$

is used to quantify their association. Construct a 95% confidence interval for the slope β_1.
(c) Obtain a 95% confidence interval for the odds ratio θ for comparing π_1 with π_2 and interpret.

6. We consider a data set collected from a study on prostate cancer, available from

`http://www.biostat.au.dk/teaching/postreg/AllData.htm.`

One of study objectives is to see if information collected at a baseline exam of a patient with prostate cancer, could predict whether the tumor has penetrated the prostatic capsule. The data set contains 380 observations and 5 variables. A brief description is given in Table 8.6. Denote $\pi = \Pr\{Y = 1|X_1, X_2, X_3, X_4\}$. Logistic regression is applied to model the relationship. To account for DPROS, three dummy variables $\{Z_1, Z_2, Z_3\}$ are introduced with the reference cell coding scheme.

(a) We first fit an additive logistic model

$$\text{Model I:} \quad \text{logit}(\pi) = \beta_0 + \beta_1 Z_1 + \beta_2 Z_2 + \beta_3 Z_3 + \beta_4 X_3 + \beta_5 X_4.$$

Table 8.6 Variable Description for the Prostate Cancer Data.

Var	Name	Description	Type	Values
Y	CAPSULE	Tumor penetration of prostatic capsule	binary	0 - no penetration 1 - penetration
X_1	AGE	Age in years	continuous	32 distinct values
X_2	DPROS	Results of the digital rectal exam	categorical	1 No Nodule 2 Unilobar Nodule (Left) 3 Unilobar Nodule (Right) 4 Bilobar Nodule
X_3	DCAPS	Detection of capsular involvement in rectal exam	binary	1 - No 0 - Yes
X_4	PSA	Prostatic Specific Antigen Value mg/ml	continuous	211 distinct values

Provided in Table 8.7 are some of ML fitting results from SAS PROC LOGISTIC. Answer the following question(s) based on Model I.

i. Suppose that there are two individuals, A and B, both having detection of capsular involvement (DCAPS, X_3) but no nodule (DPROS, X_2) found in the rectal exam. Patient A has a PSA value of 10 mg/ml higher than Patient B. Obtain a point estimate of the odds ratio for having tumor penetration that compares A versus B. Construct a Wald 95% confidence interval for this odds ratio.

ii. Given an individual who has detection of capsular involvement (DCAPS, X_3) but no nodule (DPROS, X_2) found in the rectal exam and a PSA value of 20 mg/ml, apply Model I to predict the probability for this individual to have tumor penetration of prostatic capsule. Construct a Wald 95% confidence interval for this probability.

(b) Next, we consider an interaction model

Model II: $\text{logit}(\pi) = \beta_0+\beta_1 Z_1+\beta_2 Z_2+\beta_3 Z_3+\beta_4 X_3+\beta_5 X_4+\beta_6 X_3 \cdot X_4.$

Provided in Table 8.8 are some of ML fitting results from SAS PROC LOGISTIC. Answer the following question(s) based on Model II.

i. Suppose that there are two individuals, A and B, both having detection of capsular involvement (DCAPS, X_3) but no nodule

Table 8.7 Table of Parameter Estimates and the Estimated Covariance Matrix for Model I: the Prostate Cancer Data.

(a) Analysis of Maximum Likelihood Estimates

Parameter		DF	Estimate	Standard Error	Wald Chi-Square	Pr > ChiSq
Intercept		1	0.4069	0.5144	0.6258	0.4289
DPROS	3	1	0.0977	0.3970	0.0605	0.8056
DPROS	2	1	−0.5910	0.3855	2.3503	0.1253
DPROS	1	1	−1.5724	0.4293	13.4169	0.0002
DCAPS		1	−1.0309	0.4284	5.7899	0.0161
PSA		1	0.0468	0.00967	23.4280	< .0001

(b) Estimated Covariance Matrix for $\widehat{\boldsymbol{\beta}}$.

	Intercept	DPROS 3	DPROS 2	DPROS 1	DCAPS	PSA
Intercept	0.2646	−0.10357	−0.09627	−0.06	−0.1561	−0.00182
DPROS 3	−0.1036	0.15762	0.11179	0.10964	−0.0150	0.00051
DPROS 2	−0.0963	0.11179	0.14864	0.11032	−0.0212	0.00036
DPROS 1	−0.0899	0.10964	0.11032	0.18429	−0.0216	−0.00001
DCAPS	−0.1561	−0.01502	−0.02117	−0.02156	0.1836	0.00038
PSA	−0.0018	0.00051	0.00036	−0.00001	0.0004	0.00009

(DPROS, X_2) found in the rectal exam. A has a PSA score of 10 mg/ml higher than B. Obtain a point estimate of the odds ratio for having tumor penetration that compares A versus B. Construct a Wald 95% confidence interval for this odds ratio.

7. Table 8.9 presents minus two times of the maximized log-likelihood score for several candidate models for the prostate cancer data.

    i. Compute the AIC and BIC for each of the above models. Based on these two criteria, which one is the best?
    ii. Concerning Model I, use the likelihood ratio test to see at significance level $\alpha = 0.05$ if the categorical predictor DPROS (X_2) can be dropped.

Table 8.8 Table of Parameter Estimates and the Estimated Covariance Matrix for Model II: the Prostate Cancer Data.

(a) Analysis of Maximum Likelihood Estimates

Parameter		DF	Estimate	Standard Error	Wald Chi-Square	Pr > ChiSq
Intercept		1	1.0626	0.6222	2.9163	0.0877
DPROS	3	1	0.0667	0.3982	0.0280	0.8670
DPROS	2	1	−0.6133	0.3854	2.5321	0.1116
DPROS	1	1	−1.6477	0.4353	14.3295	0.0002
DCAPS		1	−1.7478	0.5787	9.1221	0.0025
PSA		1	0.0149	0.0164	0.8187	0.3656
DCAPS*PSA		1	0.0397	0.0198	4.0215	0.0449

(b) Estimated Covariance Matrix for $\widehat{\boldsymbol{\beta}}$

	Intercept	DPROS 3	DPROS 2	DPROS 1	DCAPS	PSA	DCAPSPSA
Intercept	0.387192	−0.10879	−0.1008	−0.10779	−0.29216	−0.00696	0.006791
DPROS 3	−0.10879	0.158585	0.111772	0.110618	−0.00932	0.000783	−0.00037
DPROS 2	−0.1008	0.111772	0.14857	0.110755	−0.01573	0.000581	−0.0003
DPROS 1	−0.10779	0.110618	0.110755	0.189455	−0.00201	0.000847	−0.0011
DCAPS	−0.29216	−0.00932	−0.01573	−0.00201	0.334893	0.006327	−0.00782
PSA	−0.00696	0.000783	0.000581	0.000847	0.006327	0.00027	−0.00027
DCAPSPSA	0.006791	−0.00037	−0.0003	−0.0011	−0.00782	−0.00027	0.000392

Table 8.9 The Maximized Loglikelihood Scores for Several Fitted Models with the Prostate Cancer Data.

Model	Form	$-2\log$ L	AIC	BIC
I	$\text{logit}(\pi) = \beta_0 + \beta_1 Z_1 + \beta_2 Z_2 + \beta_3 Z_3 + \beta_4 X_3 + \beta_5 X_4$	424.018	(1)	(2)
II	$\text{logit}(\pi) = \beta_0 + \beta_1 Z_1 + \beta_2 Z_2 + \beta_3 Z_3 + \beta_4 X_3 + \beta_5 X_4 + \beta_6 X_3 \cdot X_4.$	420.897	(3)	(4)
III	$\text{logit}(\pi) = \beta_0 + \beta_1 X_1 + \beta_2 Z_1 + \beta_3 Z_2 + \beta_4 Z_3 + \beta_5 X_3 + \beta_6 X_4$	423.964	(5)	(6)
IV	$\text{logit}(\pi) = \beta_0 + \beta_1 X_3 + \beta_2 X_4$	452.461	(7)	(8)

Chapter 9

Bayesian Linear Regression

The Bayesian approach, offering different viewpoints towards statistical modeling, takes additional or historical evidences into account of the inference and analysis of current data. The advances in Bayesian statistics have been innovative and fruitful in both theories and computation over decades. In this chapter, we shall discuss the Bayesian approaches commonly used in linear regression. In order to miss small, we shall aim small in our coverage of this topic. Specifically we first discuss Bayesian linear models with conjugate normal-gamma priors. Then we discuss a relatively new development, Bayesian model averaging (BMA), for model selection and evaluation.

9.1 Bayesian Linear Models

9.1.1 *Bayesian Inference in General*

There are two main streams of viewpoints and approaches in statistics, underlaid respectively by the Fisherian and the Bayesian philosophies. So far we have mainly followed the Fisherian approach, named after Sir Ronald Fisher, to present the linear regression techniques. One key feature in this paradigm is that the regression coefficients in $\boldsymbol{\beta}$ and the error variance σ^2 are considered as fixed parameters or unknown constants. Therefore concepts such as unbiased estimators, hypothesis testing, confidence intervals, etc. follow. The Fisherian inference is often combined with likelihood methods. On the other hand, the Bayesian philosophy, named after the Reverend Thomas Bayes, allows for parameters to be treated as random variables as well, under the premise that all uncertainties should be probabilistically modelled. An important advantage for doing so is to facilitate incorporation of prior knowledge and beliefs from scientific field experts and information from historic studies into the inference and analysis. The Bayesian approach involves concepts such as prior, posterior distributions, and Bayes estimation or decision rules. One is referred to Berger (1993) for a full account of Bayesian inference.

The main idea of the Bayesian analysis is outlined in the following. In Bayesian approach, all information or uncertainties are integrated with various types of distribution functions, from which inference and conclusion are then made, obeying the laws of probability theories such as Bayes' rule. Let D be a generic notation for the observed data. To draw conclusions on parameter vector $\boldsymbol{\theta}$ of interest, the Bayesian inference is based on its its conditional distribution conditioning on D, i.e., the *posterior* distribution, given by

$$f(\boldsymbol{\theta}|D) = \frac{f(\boldsymbol{\theta}, D)}{f(D)}, \tag{9.1}$$

where $f(\boldsymbol{\theta}, D)$ is the joint density of D and $\boldsymbol{\theta}$ and $f(D)$ is the marginal density of D. With slight abuse of notations, we have used $f(\cdot)$ as a generic symbol for all density functions for the sake of convenience. Applying the Bayes' rule, we have

$$f(\boldsymbol{\theta}, D) = f(D|\boldsymbol{\theta})f(\theta), \tag{9.2}$$

where $f(D|\boldsymbol{\theta})$ is the conditional density of D given $\boldsymbol{\theta}$, i.e., the likelihood function, and $f(\theta)$ is the prior density of θ which integrates prior beliefs about $\boldsymbol{\theta}$. The marginal density of D, $f(D)$, can be obtained by integrating $\boldsymbol{\theta}$ out of $f(\boldsymbol{\theta}, D)$. Thus, the posterior distribution in (9.1) becomes

$$\begin{aligned} f(\theta|D) &= \frac{f(D|\boldsymbol{\theta})f(\theta)}{f(D)} \\ &= \frac{f(D|\boldsymbol{\theta})f(\boldsymbol{\theta})}{\int_\Omega f(D|\boldsymbol{\theta})f(\theta)d\boldsymbol{\theta}} \\ &= c \cdot f(D|\boldsymbol{\theta})f(\boldsymbol{\theta}) \\ &\propto f(D|\boldsymbol{\theta})f(\boldsymbol{\theta}), \end{aligned} \tag{9.3}$$

where Ω denotes the range of $\boldsymbol{\theta}$; the notation $\propto$ means "up to a constant"; and the normalizing constant $c = \int_\Omega f(D|\boldsymbol{\theta})f(\theta)d\boldsymbol{\theta}$, which ensures that $f(\theta|D)$ is a valid density function satisfying $\int_\Omega f(\theta|D)d\boldsymbol{\theta} = 1$, does not involve $\boldsymbol{\theta}$. Point estimates and confidence intervals of $\boldsymbol{\theta}$ can then be obtained from the joint posterior density $f(\boldsymbol{\theta}|D)$ and its associated marginal posteriors $f(\theta_j|D)$ for each component θ_j of $\boldsymbol{\theta}$. A marginal posterior $f(\theta_j|D)$ is derived by integrating other components out of the joint posterior $f(\boldsymbol{\theta}|D)$.

As opposed to likelihood optimization in the Fisherian approach, Bayesian computation essentially involves derivations of assorted probability distributions by solving integrals. The difficulty level of a Bayesian analysis depends on prior specification besides many other aspects. Sometimes, a prior is chosen for mathematical convenience to have easily solvable integrals. *Conjugate* priors, for instance, result in posteriors from the same family. Most often, the integrals involved are complicated or even intractable. In this case, approximations of multiple integration can be made either analytically by techniques such as Laplace transforms or computationally by numerical integration. One is referred to Bayesian texts, e.g., Tanner (1998) and Carlin and Louis (2000), for detailed implementation and computation issues in Bayesian analysis.

9.1.2 Conjugate Normal-Gamma Priors

We now discuss the Bayesian approach for fitting normal linear regression models given by $\mathbf{y} \sim \mathcal{N}\left\{\mathbf{X}\boldsymbol{\beta},\, \sigma^2\mathbf{I}\right\}$, with $\mathbf{X}$ being $n \times (k+1)$ of full column rank $(k+1)$. We first reparametrize the model by introducing the precision parameter $\nu = 1/\sigma^2$. Then the likelihood function can be written as, up to a constant

$$L(\boldsymbol{\beta}, \sigma^2) = f(\mathbf{y}|\boldsymbol{\beta}, \nu) \propto \nu^{n/2} \exp\left\{-\nu \cdot (\mathbf{y} - \mathbf{X}\boldsymbol{\beta})^t(\mathbf{y} - \mathbf{X}\boldsymbol{\beta})/2\right\}. \tag{9.4}$$

To seek conjugate priors that yields posteriors of similar functional form, we rewrite the likelihood so that terms involving $\boldsymbol{\beta}$ are in a multivariate normal density function form. Using

$$\begin{aligned}(\mathbf{y} - \mathbf{X}\boldsymbol{\beta})^t(\mathbf{y} - \mathbf{X}\boldsymbol{\beta}) &= (\mathbf{y} - \mathbf{X}\widehat{\boldsymbol{\beta}})^t(\mathbf{y} - \mathbf{X}\widehat{\boldsymbol{\beta}}) + (\boldsymbol{\beta} - \widehat{\boldsymbol{\beta}})^t(\mathbf{X}^t\mathbf{X})(\boldsymbol{\beta} - \widehat{\boldsymbol{\beta}}) \\ &= \{n - (k+1)\} \cdot \hat{\sigma}^2 + (\boldsymbol{\beta} - \widehat{\boldsymbol{\beta}})^t(\mathbf{X}^t\mathbf{X})(\boldsymbol{\beta} - \widehat{\boldsymbol{\beta}}),\end{aligned}$$

the likelihood in (9.4) can be expressed as

$$\begin{aligned}f(\mathbf{y}|\boldsymbol{\beta}, \nu) \propto &\left\{\nu^{\frac{n-(k+1)}{2}} \exp\left(-\frac{\{n - (k+1)\}\hat{\sigma}^2}{2}\nu\right)\right\} \\ &\times \left\{\nu^{\frac{k+1}{2}} \exp\left(-\frac{(\boldsymbol{\beta} - \widehat{\boldsymbol{\beta}})^t(\nu \cdot \mathbf{X}^t\mathbf{X})(\boldsymbol{\beta} - \widehat{\boldsymbol{\beta}})}{2}\right)\right\}.\end{aligned} \tag{9.5}$$

It has two parts: the first part contains a kernel of a

$$\Gamma\left\{\frac{n - (k+1)}{2} - 1,\ \frac{\{n - (k+1)\}\hat{\sigma}^2}{2}\right\}$$

density for ν and the second part contains a kernel of a multivariate normal $\mathcal{N}\left(\widehat{\boldsymbol{\beta}}, (\nu \cdot \mathbf{X}^t\mathbf{X})^{-1}\right)$ distribution for $\boldsymbol{\beta}$.

In general, a positive random variable U follows a $\Gamma(a,b)$ distribution with a shape parameter $a>0$ and a rate parameter $b>0$, i.e., $U\sim\Gamma(a,b)$, if its density is given by

$$f(u;a,b)=\frac{b^a}{\Gamma(a)}u^{a-1}e^{-bu}, \tag{9.6}$$

where $\Gamma(a)=\int_0^\infty u^{a-1}e^{-u}du$ is the Gamma function. Alternatively, the scale parameter $1/b$ can be used to express the gamma density (9.6). Given $U\sim\Gamma(a,b)$, we have

$$E(U)=\frac{a}{b} \quad \text{and} \quad \text{Var}(U)=\frac{a}{b^2}.$$

The inverse of U, $1/U$, is said to follow the inverse Gamma distribution with mean and variance given by

$$E(1/U)=\frac{b}{a-1} \quad \text{and} \quad \text{Var}(1/U)=\frac{b^2}{(a-1)^2(a-2)}. \tag{9.7}$$

Note that in (9.5), the multivariate normal distribution $\mathcal{N}\left(\widehat{\boldsymbol{\beta}},(\nu\cdot\mathbf{X}^t\mathbf{X})^{-1}\right)$ for $\boldsymbol{\beta}$ depends on ν. This suggests a natural choice for the priors of form

$$f(\boldsymbol{\beta},\nu)=f(\boldsymbol{\beta}|\nu)f(\nu),$$

where $\nu\sim\Gamma(a,b)$ with density $f(\nu)$ given as in (9.6); $\boldsymbol{\beta}|\nu\sim N(\boldsymbol{\beta}_0,\mathbf{V}/\nu)$ with density

$$f(\boldsymbol{\beta}|\nu)=\nu^{(k+1)/2}\exp\left\{-\nu\cdot(\boldsymbol{\beta}-\boldsymbol{\beta}_0)^t\mathbf{V}^{-1}(\boldsymbol{\beta}-\boldsymbol{\beta}_0)/2\right\}; \tag{9.8}$$

and the quantities $(a,b,\boldsymbol{\beta}_0,\mathbf{V})$ are the so-called *super-parameters.*

To summarize, the Bayesian linear regression model we shall consider has the following specification:

$$\begin{cases}\mathbf{y}|(\boldsymbol{\beta},\nu) \sim N(\mathbf{X}\boldsymbol{\beta},\ \mathbf{I}/\nu)\\ \boldsymbol{\beta}|\nu \sim N_{k+1}(\boldsymbol{\beta}_0,\ \mathbf{V}/\nu)\\ \nu \sim \Gamma(a,b).\end{cases} \tag{9.9}$$

Under this model, the posterior joint density $f(\boldsymbol{\beta},\nu|\mathbf{y})$ is, using (9.4), (9.6), and (9.8),

$$f(\boldsymbol{\beta},\nu|\mathbf{y})\propto f(\mathbf{y}|\boldsymbol{\beta},\nu)f(\boldsymbol{\beta}|\nu)f(\nu) \tag{9.10}$$

$$\propto \nu^{(n+2a+k+1)/2-1}\exp\left[-(\nu/2)\cdot\left\{2b+(\boldsymbol{\beta}-\boldsymbol{\beta}_0)^t\mathbf{V}^{-1}(\boldsymbol{\beta}-\boldsymbol{\beta}_0)\right.\right.$$
$$\left.\left.+\,(\mathbf{y}-\mathbf{X}\boldsymbol{\beta})^t(\mathbf{y}-\mathbf{X}\boldsymbol{\beta})\right\}\right] \tag{9.11}$$

Subsequently, expand the two quadratic forms and combine them into one quadratic form on $\boldsymbol{\beta}$. After a tedious yet straightforward simplification, the posterior joint density of $(\boldsymbol{\beta}, \nu)$ becomes

$$f(\boldsymbol{\beta}, \nu|\mathbf{y}) \propto \nu^{(\frac{n}{2}+a)-1} \exp(-\nu \cdot \tilde{b}) \times \nu^{\frac{k+1}{2}} \exp\left\{-\frac{\nu}{2} \cdot (\boldsymbol{\beta} - \widetilde{\boldsymbol{\beta}})^t \left(\mathbf{X}^t\mathbf{X} + \mathbf{V}^{-1}\right) (\boldsymbol{\beta} - \widetilde{\boldsymbol{\beta}})\right\} \quad (9.12)$$

where

$$\tilde{a} = \frac{n}{2} + a; \quad (9.13)$$

$$\tilde{b} = b + \frac{\mathbf{y}^t\mathbf{y} - (\mathbf{X}^t\mathbf{y} + \mathbf{V}^{-1}\boldsymbol{\beta}_0)^t(\mathbf{X}^t\mathbf{X} + \mathbf{V}^{-1})^{-1}(\mathbf{X}^t\mathbf{y} + \mathbf{V}^{-1}\boldsymbol{\beta}_0)}{2}; \quad (9.14)$$

$$\widetilde{\boldsymbol{\beta}} = (\mathbf{X}^t\mathbf{X} + \mathbf{V}^{-1})^{-1}(\mathbf{X}^t\mathbf{y} + \mathbf{V}^{-1}\boldsymbol{\beta}_0). \quad (9.15)$$

Note that the posterior in (9.12) is still in the form of $f(\nu)f(\boldsymbol{\beta}|\nu)$. Thus the marginal posterior density of ν is $\Gamma(\tilde{a}, \tilde{b})$.

The marginal posterior density of $\boldsymbol{\beta}$, $f(\boldsymbol{\beta}|\mathbf{y})$, which can be obtained by integrating ν out of (9.12), is given by

$$f(\boldsymbol{\beta}|\mathbf{y}) \propto \left\{2\tilde{b} + (\boldsymbol{\beta} - \widetilde{\boldsymbol{\beta}})^t(\mathbf{X}^t\mathbf{X} + \mathbf{V}^{-1})(\boldsymbol{\beta} - \widetilde{\boldsymbol{\beta}})\right\}^{-(2\tilde{a}+k+1)/2} \propto \left[1 + \frac{(\boldsymbol{\beta} - \widetilde{\boldsymbol{\beta}})^t \left\{(\mathbf{X}^t\mathbf{X} + \mathbf{V}^{-1}) \cdot \frac{\tilde{a}}{\tilde{b}}\right\} (\boldsymbol{\beta} - \widetilde{\boldsymbol{\beta}})}{2\tilde{a}}\right]^{-(2\tilde{a}+k+1)/2}. \quad (9.16)$$

The above form is recognized as a multivariate t distribution with $2\tilde{a}$ degrees of freedom, mean vector $\widetilde{\boldsymbol{\beta}}$, and covariance matrix

$$\widetilde{\mathbf{V}} = \frac{\tilde{a}}{\tilde{a}-1} \cdot \frac{\tilde{b}}{\tilde{a}} \left(\mathbf{X}^t\mathbf{X} + \mathbf{V}^{-1}\right)^{-1} = \frac{\tilde{b}}{\tilde{a}-1} \left(\mathbf{X}^t\mathbf{X} + \mathbf{V}^{-1}\right)^{-1} = (\tilde{v}_{ij}). \quad (9.17)$$

We denote it as $\boldsymbol{\beta}|\mathbf{y} \sim t(2\tilde{a}, \widetilde{\boldsymbol{\beta}}, \widetilde{V} \cdot (\tilde{a}-1)/\tilde{a})$.

In general, a p-dimensional vector $\mathbf{x}$ is said to follow a non-singular multivariate t distribution (Kotz and Nadarajah, 2004) with v degrees of freedom (df), mean vector $\boldsymbol{\mu}$, and covariance matrix $\nu(\nu-2)^{-1} \cdot \boldsymbol{\Sigma}$ if it has joint probability density function given by

$$f(\mathbf{x}) = \frac{\Gamma\{(v+p)/2\}}{(\pi v)^{v/2}\Gamma(v/2)|\boldsymbol{\Sigma}|^{1/2}} \left\{1 + \frac{(\mathbf{x} - \boldsymbol{\mu})^t\boldsymbol{\Sigma}^{-1}(\mathbf{x} - \boldsymbol{\mu})}{v}\right\}^{-(v+p)/2}. \quad (9.18)$$

We denote it as $t_p(v, \boldsymbol{\mu}, \boldsymbol{\Sigma})$. The df parameter v is also called the shape parameter as it determines how peaked the distribution is. The distribution is said to be central if $\boldsymbol{\mu} = 0$; and noncentral otherwise. A multivariate t random vector can be characterized by the following representation: given that $\mathbf{y} \sim N_p(\boldsymbol{\mu}, \boldsymbol{\Sigma})$ is independent of $v \cdot s^2 \sim \chi^2(v)$, if we define $\mathbf{x} = \mathbf{y}/s$, then $\mathbf{x} \sim t_p(v, \boldsymbol{\mu}, \boldsymbol{\Sigma})$. With this representation, it follows an important property (Lin, 1971) that,

$$\mathbf{x} \sim t_p(v, \boldsymbol{\mu}, \boldsymbol{\Sigma}) \Longleftrightarrow \frac{\mathbf{a}^t(\mathbf{x} - \boldsymbol{\mu})}{\sqrt{\mathbf{a}^t \boldsymbol{\Sigma} \mathbf{a}}} \sim t(v) \tag{9.19}$$

for any p-dimensional nonzero vector $\mathbf{a}$. This property is useful in statistical inference about $\boldsymbol{\beta}$.

9.1.3 *Inference in Bayesian Linear Model*

The posterior mean can be used as the Bayesian point estimator. Since $\boldsymbol{\beta}|\mathbf{y} \sim t(2\tilde{a}, \widetilde{\boldsymbol{\beta}}, \widetilde{V} \cdot (\tilde{a} - 1)/\tilde{a})$, the Bayesian estimate of $\boldsymbol{\beta}$ in model (9.9) is

$$\widetilde{\boldsymbol{\beta}} = (\mathbf{X}^t\mathbf{X} + \mathbf{V}^{-1})^{-1}(\mathbf{X}^t\mathbf{y} + \mathbf{V}^{-1}\boldsymbol{\beta}_0) = (\beta_j) \tag{9.20}$$

as also given by (9.15). It is interesting to note that $\widetilde{\boldsymbol{\beta}}$ can be obtained as the generalized least squares estimator (see Section 7.2) of $\boldsymbol{\beta}$ in the following augmented linear model:

$$\begin{pmatrix} \mathbf{y} \\ \boldsymbol{\beta}_0 \end{pmatrix} = \begin{pmatrix} \mathbf{X} \\ \mathbf{I}_{k+1} \end{pmatrix} \boldsymbol{\beta} + \begin{pmatrix} \boldsymbol{\varepsilon} \\ \boldsymbol{\varepsilon}_0 \end{pmatrix}, \quad \text{where } \begin{pmatrix} \boldsymbol{\varepsilon} \\ \boldsymbol{\varepsilon}_0 \end{pmatrix} \sim N \begin{pmatrix} \mathbf{I} & \mathbf{O} \\ \mathbf{O} & \mathbf{V} \end{pmatrix}. \tag{9.21}$$

The implication is that the prior information on $\boldsymbol{\beta}$ can be integrated into the least squares estimation by adding some pseudo-data.

Two special cases are worth noting. The first situation, corresponding to noninformative priors, is when $\boldsymbol{\beta}_0 = \mathbf{0}$, $a = b \approx 0$, and $\mathbf{V}^{-1} \approx \mathbf{O}$. In this case, it can be easily verified that $2\tilde{b} \approx SSE = \mathbf{y}^t\mathbf{y} - \mathbf{y}^t\mathbf{X}(\mathbf{X}^t\mathbf{X})^{-1}\mathbf{X}^t\mathbf{y}$ and $\tilde{a} = n/2$. Thus the posterior mean vector and covariance matrix of $\boldsymbol{\beta}$ are

$$\begin{aligned} \widetilde{\boldsymbol{\beta}} &\approx (\mathbf{X}^t\mathbf{X})^{-1}\mathbf{X}^t\mathbf{y} \\ \widetilde{\mathbf{V}} &\approx \frac{SSE}{n-2}(\mathbf{X}^t\mathbf{X})^{-1}, \end{aligned}$$

which are similar to their counterparts in ordinary LS estimation. The second case is when $\boldsymbol{\beta}_0 = \mathbf{0}$ and $\mathbf{V}^{-1} = \lambda\mathbf{I}$ for some constant $\lambda > 0$. It can be seen that the Bayesian point estimator of $\boldsymbol{\beta}$ becomes

$$\widetilde{\boldsymbol{\beta}} = (\mathbf{X}^t\mathbf{X} + \lambda\mathbf{I})^{-1}\mathbf{X}^t\mathbf{y}, \tag{9.22}$$

a form known as the ridge or shrinkage estimator. The ridge or shrinkage estimator is initially proposed as a remedial measure for handling the multicollinearity problem encountered in linear regression. We shall further explore it in Section 7.3.

Interval inference and hypothesis testing about $\boldsymbol{\beta}$ and its components can be derived from property (9.19). The posterior distribution of a linear combination on $\boldsymbol{\beta}$ is a t distribution with df $2\tilde{a} = n + 2a$, i.e.,

$$\left.\frac{\mathbf{a}^t\boldsymbol{\beta} - \mathbf{a}^t\widetilde{\boldsymbol{\beta}}}{\mathbf{a}^t\widetilde{\mathbf{V}}\mathbf{a}}\right| \mathbf{y} \sim t(n+2a), \tag{9.23}$$

where $\widetilde{\boldsymbol{\beta}}$ and $\widetilde{\mathbf{V}}$ are given in (9.20) and (9.17). In particular, the posterior distribution of β_j is

$$\left.\frac{\beta_j - \tilde{\beta}_j}{\tilde{v}_{jj}}\right| \mathbf{y} \sim t(n+2a). \tag{9.24}$$

A $100 \times (1-\alpha)\%$ Bayesian confidence interval for β_j can be given as

$$\tilde{\beta}_j \pm t(1-\alpha/2,\, n+2a) \cdot \tilde{v}_{jj}, \tag{9.25}$$

where $t(1-\alpha/2,\, n+2a)$ denotes the $(1-\alpha/2) \times 100$-th percentile of the $t(n+2a)$ distribution. Since β's are random variables in Bayesian inference, it is simply and correctly safe to state that, with $(1-\alpha)$ probability, β_j falls into the $(1-\alpha)\times 100\%$ Bayesian confidence interval. Recall that in classical inference one has to be careful not to interpret confidence intervals in such a manner. Clearly, equation (9.23) can also be useful in Bayesian interval estimation of the expected response $E(y_p) = \mathbf{x}_p^t\boldsymbol{\beta}$ at any given predictor vector $\mathbf{x} = \mathbf{x}_p$.

In addition, since $\nu|\mathbf{y} \sim \Gamma(\tilde{a}, \tilde{b})$ where $\tilde{a}$ and $\tilde{b}$ are given in (9.13) and (9.14), the posterior distribution of $\sigma^2 = 1/\nu$ is an inverse Gamma distribution. Using (9.7), a Bayesian point estimator of σ^2 is the posterior mean $\tilde{b}/(\tilde{a}-1)$, i.e.,

$$\tilde{\sigma}^2 = \frac{b + \left\{\mathbf{y}^t\mathbf{y} - (\mathbf{X}^t\mathbf{y} + \mathbf{V}^{-1}\boldsymbol{\beta}_0)^t(\mathbf{X}^t\mathbf{X} + \mathbf{V}^{-1})^{-1}(\mathbf{X}^t\mathbf{y} + \mathbf{V}^{-1}\boldsymbol{\beta}_0)\right\}/2}{n/2 + a - 1}.$$

A $(1-\alpha)\times 100\%$ Bayesian confidence interval for σ^2 can also be constructed using percentiles from the inverse Gamma distribution.

9.1.4 *Bayesian Inference via MCMC*

More often the marginal posterior density function for a parameter of interest is not explicitly available in Bayesian analysis. In this case, the solutions are provided by Markov Chain Monte Carlo (MCMC) simulation methods. The Monte Carlo methods, traced back to Metropolis and Ulam (1949) and then extensively studied and expanded by many others, are a very general class of computational methods that perform complicated calculations, most notably those involving integration and optimization, by using repetitive random number generation. The Markov chain is referred to a stochastic process or a sequence of random variables, say, $X_1, X_2, \ldots,$ that satisfies the Markov property

$$\Pr(X_{m+1} \in A | X_1, \ldots, X_m) = \Pr(X_{m+1} \in A | X_m). \tag{9.26}$$

Namely, the distribution of the random variable at any state of the process only depends on the random variable at the immediately preceding step. In other words, the description of the present state fully captures all the information that could influence the future evolution of the process. The Markov property in (9.26) can be viewed as a generalization of the concept of independence. Samples obtained from MCMC can be used to approximate distributions or to compute integrals such as expected values. The ergodic theory states that, under some mild regularity conditions, the Markov chain process satisfies

$$\lim_{M \to \infty} \frac{1}{M} \sum_{m=1}^{M} h(X_m) = E\{h(X)\}. \tag{9.27}$$

The MCMC methods have fundamental use in many, if not all, scientific fields and encompass a prohibitively long list of related references in the literature. In Bayesian analysis, MCMC methods provide empirical realization and exploration of posterior distributions especially when their closed forms are not easily available.

While a few different ways to perform MCMC in Bayesian linear regression are available, we describe a very popular algorithm, called the *Gibbs sampler* or Gibbs sampling. Gibbs sampler, as a special case of the MCMC methods, is particularly useful when the conditional posterior distribution of each parameter is explicitly known. The idea of Gibbs sampler is to alternate a sequence of random draws from conditional distributions to eventually characterize the joint distribution of interest.

Consider the Bayesian linear model with specification given by (9.9). From the joint posterior density $f(\boldsymbol{\beta}, \nu | \mathbf{y})$ in (9.12), it can be seen that the conditional posterior density of $(\boldsymbol{\beta} | \nu, \mathbf{y})$ is multivariate normal with mean $\widetilde{\boldsymbol{\beta}}$ and covariance matrix $\nu^{-1}(\mathbf{X}^t\mathbf{X} + \mathbf{V}^{-1})^{-1}$, i.e.,

$$\boldsymbol{\beta} | (\nu, \mathbf{y}) \sim N_{k+1}\left\{\widetilde{\boldsymbol{\beta}},\ \nu^{-1}(\mathbf{X}^t\mathbf{X} + \mathbf{V}^{-1})^{-1}\right\}. \tag{9.28}$$

The conditional posterior density of $(\nu | \boldsymbol{\beta}, \mathbf{y})$ is

$$f(\nu | \boldsymbol{\beta}, \mathbf{y}) \propto \nu^{a_1 - 1} \exp(-\nu \times b_1),$$

where

$$a_1 = \frac{n+k+1}{2} + a, \tag{9.29}$$

$$b_1 = \tilde{b} + \frac{(\boldsymbol{\beta} - \widetilde{\boldsymbol{\beta}})^t \left(\mathbf{X}^t\mathbf{X} + \mathbf{V}^{-1}\right) (\boldsymbol{\beta} - \widetilde{\boldsymbol{\beta}})}{2}. \tag{9.30}$$

That is, $\nu|(\boldsymbol{\beta}, \mathbf{y}) \sim \Gamma(a_1, b_1)$.

Algorithm 9.1: Gibbas Sampler for Bayesian Linear Regression.

- Set $\boldsymbol{\beta}_{(0)} = \widehat{\boldsymbol{\beta}}$, the ordinary LSE of $\boldsymbol{\beta}$;
- Do $m = 1$ to M,
  - Given $\boldsymbol{\beta} = \boldsymbol{\beta}_{(m-1)}$, generate $\nu_{(m)} \sim \Gamma(a_1, b_1)$;
  - Given $\nu = \nu_{(m)}$, generate $\boldsymbol{\beta}_{(m)} \sim N_{k+1}\left\{\widetilde{\boldsymbol{\beta}},\ \nu^{-1}(\mathbf{X}^t\mathbf{X} + \mathbf{V}^{-1})^{-1}\right\}$;
  - (Optional) Prediction at $\mathbf{x}_p$: Generate $y_{(m)} \sim N(\mathbf{x}_p^t\boldsymbol{\beta}_{(m)},\ \nu_{(m)}^{-1})$;
- End do;
- Burn-In: Throw away the first N_0 observations of $(\boldsymbol{\beta}_{(m)}, \nu_{(m)})$.

The Gibbas sampler is outlined in the above algorithm. In the computation, $\boldsymbol{\beta}_0$ and $\mathbf{V}$ are super-parameters pre-specified by the user and $\widetilde{\boldsymbol{\beta}}$ is given in (9.20). Since each draw is always conditional on the past draw, the resultant sample is a Markov chain sequence. The very last step of discarding a few initial observations, known as the "burn-in" strategy, ensures a purer MCMC sample from the posterior densities and leads to improved performance. Once an MCMC sample (with large M) is available for $(\boldsymbol{\beta}, \nu)$, all their inferential properties can be obtained empirically. For example, the bounds of confidence intervals for β_j's can be simply found as appropriate percentiles from the corresponding MCMC sample values.

Note that prior specification plays a critical role in Bayesian analysis. It is often emphasized that priors should reflect a reasonable approximation of the actual information and those developed by statisticians should always refer to experts for validation. The natural normal-gamma conjugate prior, which has been chosen mainly for mathematical convenience, also has considerable flexibilities to reflect expert prior knowledge on the parameters. For example, these prior distributions could be devised with small variances to have strong influence on the posteriors. They could also be formulated with large variances so as to have little effect on the posterior distributions. In the latter case, the priors are termed as *noninformative*, *diffuse*, or *improper* priors.

As a final note, variations for prior specification exist in Bayesian linear models. For example, it is also widely popular to consider *independent* normal-gamma priors for $\boldsymbol{\beta}$ and ν, a scenario referred to as the *semi-conjugate* priors. The Bayesian model with a semi-conjugate prior becomes

$$\begin{cases} \mathbf{y}|(\boldsymbol{\beta}, \nu) \sim N(\mathbf{X}\boldsymbol{\beta},\ \mathbf{I}/\nu) \\ \boldsymbol{\beta} \sim N_{k+1}(\boldsymbol{\beta}_0,\ \mathbf{V}) \\ \nu \sim \Gamma(a, b). \end{cases} \tag{9.31}$$

The resultant joint posterior density of $(\boldsymbol{\beta}, \nu)$ is

$$\begin{aligned} f(\boldsymbol{\beta}, \nu|\mathbf{y}) &\propto f(\mathbf{y}|\boldsymbol{\beta}, \nu) f(\boldsymbol{\beta}) f(\nu) \\ &\propto \nu^{(n+2a)/2-1} \exp\left[-(\nu/2) \cdot \{2b + (\mathbf{y} - \mathbf{X}\boldsymbol{\beta})^t(\mathbf{y} - \mathbf{X}\boldsymbol{\beta})\}\right. \\ &\qquad \left. + (\boldsymbol{\beta} - \boldsymbol{\beta}_0)^t \mathbf{V}^{-1} (\boldsymbol{\beta} - \boldsymbol{\beta}_0)\right]. \end{aligned} \tag{9.32}$$

It has a different analytic form as in (9.12). However, the conditional posteriors of $\boldsymbol{\beta}$ and ν can be easily derived as

$$\begin{aligned} \boldsymbol{\beta}|\nu, \mathbf{y} &\sim N_{k+1}\left\{(\nu\mathbf{X}^t\mathbf{X} + \mathbf{V}^{-1})^{-1}(\nu\mathbf{X}^t\mathbf{y} + \mathbf{V}^{-1}\boldsymbol{\beta}_0),\ (\nu\mathbf{X}^t\mathbf{X} + \mathbf{V}^{-1})^{-1}\right\}, \\ \nu|\boldsymbol{\beta}, \mathbf{y} &\sim \Gamma\left\{n + a/2,\ b + \frac{(\mathbf{y} - \mathbf{X}\boldsymbol{\beta})^t(\mathbf{y} - \mathbf{X}\boldsymbol{\beta})}{2}\right\}, \end{aligned}$$

which allows for direct application of the Gibbs sampler.

9.1.5 *Prediction*

Prediction in Bayesian models is also made in a way that is quite different from conventional approaches. Given a vector $\mathbf{x}_p$, the Bayesian inference on the associated future response y_p is again based on its posterior distribution $f(y_p|\mathbf{y})$. Applying the Bayes' rule, the posterior density of y_p can be analytically found as,

$$\begin{aligned} f(y_p|\mathbf{y}) &= \int_\nu \int_{\boldsymbol{\beta}} f(y_p, \boldsymbol{\beta}, \nu|\mathbf{y}) d\boldsymbol{\beta} d\nu \\ &= \int_\nu \int_{\boldsymbol{\beta}} f(y_p|\boldsymbol{\beta}, \nu, \mathbf{y}) f(\boldsymbol{\beta}, \nu|\mathbf{y}) d\boldsymbol{\beta} d\nu \\ &= \int_\nu \int_{\boldsymbol{\beta}} f(y_p|\boldsymbol{\beta}, \nu) f(\boldsymbol{\beta}, \nu|\mathbf{y}) d\boldsymbol{\beta} d\nu. \end{aligned} \tag{9.33}$$

It is important to note that y_p is not independent of $\mathbf{y}$ in the Bayesian setting, as y_p depends on parameters $(\boldsymbol{\beta}, \nu)$ whose distribution is dependent on $\mathbf{y}$. Nevertheless, y_p depends on $\mathbf{y}$ only through $(\boldsymbol{\beta}, \nu)$. Functions

$f(y_p|\boldsymbol{\beta}, \nu)$ and $f(\boldsymbol{\beta}, \nu|\mathbf{y})$ in equation (9.33) can be obtained using

$$y_p|\boldsymbol{\beta}, \nu \sim N(\mathbf{x}_p^t\boldsymbol{\beta}, \nu^{-1}), \tag{9.34}$$

and the Bayesian linear model specification in (9.9). However, the integrations involved in (9.33) are still difficult to solve. Comparatively, it is more convenient to base the inference of y_p on Gibbs sampler, as outlined in Algorithm 9.1.

Table 9.1 Bayesian Linear Regression Results With the Shingle Data.

	Least Squares Estimation			Bayesian Analysis		
	Estimate	95% CI		Estimate	95% CI	
β_0	5.656147	4.777982	6.534312	5.656913	4.762665	6.548519
β_1	−0.060878	−0.122139	0.000382	−0.060839	−0.122124	0.001082
β_2	0.024041	0.014840	0.033242	0.024087	0.014737	0.033405
β_3	−0.159635	−0.211159	−0.108110	−0.159749	−0.211082	−0.107679
β_4	−0.013595	−0.040586	0.013397	−0.013858	−0.039959	0.013536
σ^2	0.038055	0.018579	0.117201	0.047764	0.018721	0.118779

9.1.6 *Example*

Implementation of Bayesian linear models techniques is available in both SAS (PROC MCMC and the BAYES statement in PROC GENMOD) and R (several packages available).

To illustrate, we analyze the shingle sales data in Table 4.1. The R package `MCMCpack`, which implements Bayesian linear regression with semi-conjugate priors in model (9.31), is used for analysis. A logarithm transformation is first applied to the response Y. Here we employ a set of non-informative priors with the default specification

$$\boldsymbol{\beta}_0 = \mathbf{0}, \ \mathbf{V} \approx \mathbf{O}, \ a = 0.0005, \text{ and } b = 0.0005.$$

Gibbs sampler is used to generate 10,000 MC observations after a "burn-in" of the first $N_0 = 1,000$ random draws.

Figure 9.1 plots the resultant marginal densities for regression coefficients $(\beta_0, \beta_1, \ldots, \beta_4)$ and σ^2, obtained by smoothing the empirical density of the MC samples. Table 9.1 shows the Bayesian estimate together with their 95% confidence intervals, which are very similar to the results from LS. This comes as no surprise considering the noninformative priors employed.

We next consider prediction at $\mathbf{x}_p = (1, 5.20, 47.07, 8.87, 8.93)^t$, which corresponds to sample averages of $(x_0 = 1, x_1, x_2, x_3, x_4)$. To do so, a MC

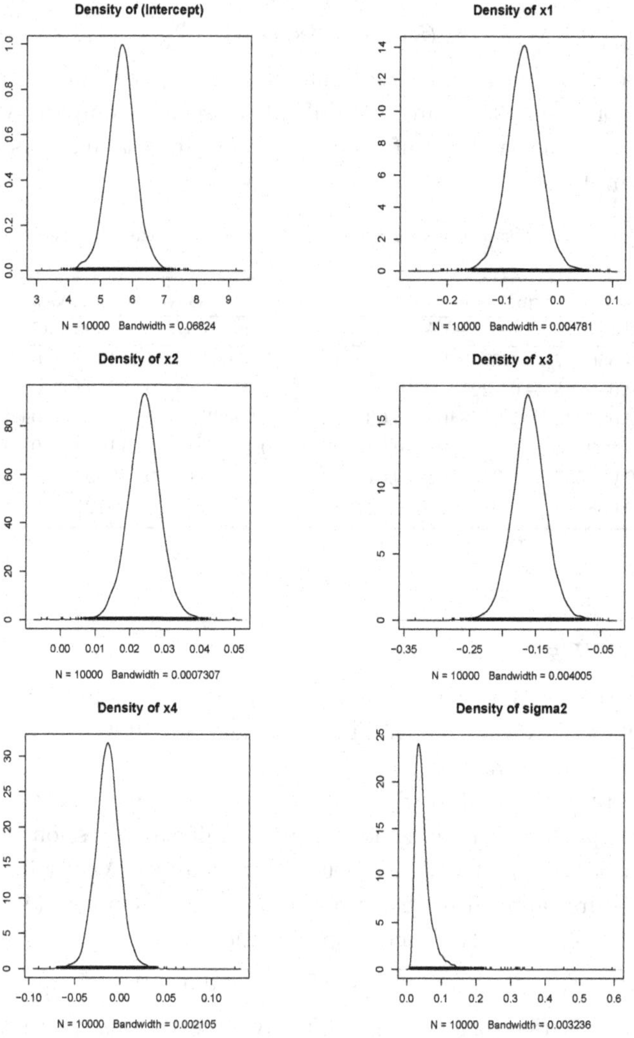

Fig. 9.1 Posterior Marginal Densities of $(\boldsymbol{\beta}, \sigma^2)$: Bayesian Linear Regression for Shingle Data.

sample for y_p can be formed by computing $\mathbf{x}_p^t\boldsymbol{\beta}_{(m)}$, where $\boldsymbol{\beta}_{(m)}$ denotes the m-th observation of $\boldsymbol{\beta}$ in the MC sample. Figure 9.2 plots the posterior density of y_p. Based on the MC sample for y_p, the point estimate for prediction is the sample mean, 4.934. The 95% prediction interval is $(4.823, 5.046)$,

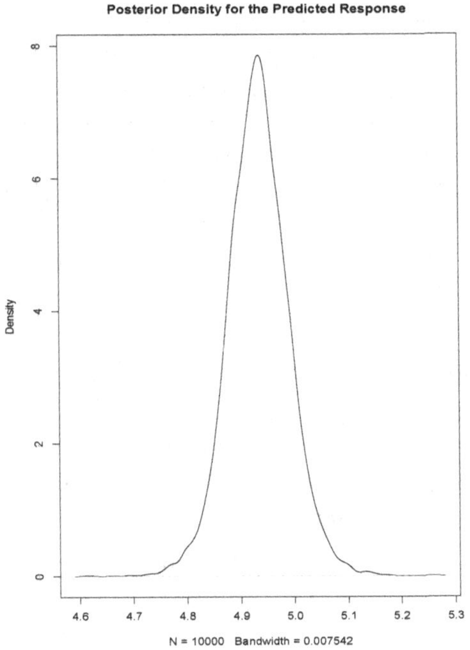

Fig. 9.2 Posterior Densities for Predicting y_p: the Shingle Data.

which are the 2.5% and 97.5% percentiles. Comparatively, the ordinary LSE yields a predicted value of 4.934, with 95% CI $(4.485, 5.383)$, at $\mathbf{x}_p$.

9.2 Bayesian Model Averaging

As we have demonstrated in Chapter 5, the term "best model" can only be vaguely defined. This is essentially because part of the information in the data has been spent to specify or estimate the model, which invokes considerable ambiguities revolving around different model choices. As argued by Hoeting et al. (1999), there may exist alternative models that provide a good fit to the data but leads to substantially different model forms or predictions. If one, however, bases all the inferences on a single selected model, it would lead to underestimated uncertainty and overly optimistic conclusions about the quantities of interest. Bayesian model averaging (BMA) provides a coherent mechanism to account for model uncertainties. The philosophy underlying BMA is to integrate the information from those reasonable models together in an effort to provide a more comprehensive assessment and a better understanding of the problems under study. Note that BMA is not meant to do model ensemble or have combined predictions, a common wrong impression carried by the term 'model averaging'.

The original idea of BMA can be traced back to the proposal of Leamer (1978). Let $\mathcal{M} = \{M_1, \ldots, M_K\}$ be the set of *all* models under consideration. Let Δ denote some quantity that is of central interest to the inference, such as a parameter or a future observation, then its posterior distribution given data D is

$$\text{pr}(\Delta|D) = \sum_{k=1}^{K} \text{pr}(\Delta|M_k, D)\text{pr}(M_k|D), \tag{9.35}$$

which is an average of the posterior distributions under each of the models, weighted by the posterior model probabilities. The posterior model probability of M_k can be computed as

$$\text{pr}(M_k|D) = \frac{\text{pr}(D|M_k)\text{pr}(M_k)}{\sum_{l=1}^{K} \text{pr}(D|M_l)\text{pr}(M_l)}, \tag{9.36}$$

where

$$\text{pr}(D|M_k) = \int \text{pr}(D|\boldsymbol{\beta}_k, M_k)\text{pr}(\boldsymbol{\beta}_k|M_k)d\boldsymbol{\beta}_k \tag{9.37}$$

is the integrated marginal likelihood of model M_k; $\boldsymbol{\beta}_k$ is the parameters involved in M_k; $\text{pr}(\boldsymbol{\beta}_k|M_k)$ is the prior distribution of $\boldsymbol{\beta}_k$; $\text{pr}(D|\boldsymbol{\beta}_k, M_k)$ is the likelihood; and $\text{pr}(M_k)$ is the prior model probability that M_k is the true model. Define the *Bayes factor*, B_{21}, for a model M_2 against another model M_1 given data D to be the ratio of the posterior to prior odds or the ratio of their marginal likelihoods. Namely,

$$B_{21} = \text{pr}(D|M_2)/\text{pr}(D|M_1). \tag{9.38}$$

Then clearly the posterior probability of M_k can be rewritten as

$$\text{pr}(M_k|D) = \frac{a_k B_{k1}}{\sum_{r=1}^{K} a_r B_{r1}}, \tag{9.39}$$

where $a_k = \text{pr}(M_k)/\text{pr}(M_1)$ is the prior odds for M_k against M_1 for $k = 1, \ldots, K$.

Once the posterior distribution of Δ is available, many of its properties can be derived. The posterior mean and variance of Δ, for instance, are given by

$$E(\Delta|D) = \sum_{k=1}^{K} \hat{\Delta}_k \cdot \text{pr}(M_k|D) \tag{9.40}$$

$$\text{Var}(\Delta|D) = \sum_{k=1}^{K} \left\{ \text{Var}(\Delta|D, M_k) + \hat{\Delta}_k^2 \right\} \cdot \text{pr}(M_k|D) - E(\Delta^2|D), \tag{9.41}$$

where $\hat{\Delta}_k = E(\Delta|D, M_k)$.

Nevertheless, implementation of BMA is practically easier said than done because of two major difficulties. First, the integrals in (9.37) can be hard to compute. In linear regression, explicit form for $\text{pr}(D|M_k)$ is available with appropriate prior specification. Consider model M_k of form

$$\mathbf{y} = \mathbf{X}\boldsymbol{\beta} + \boldsymbol{\varepsilon} \text{ with } \boldsymbol{\varepsilon} \sim \text{MVN}(\mathbf{0}, \sigma^2 \cdot \mathbf{I}),$$

where the design matrix $\mathbf{X}$ is of dimension $n \times (p+1)$. Raftery, Madigan, and Hoeting (1997) applied the standard normal-gamma conjugate class of priors for $(\boldsymbol{\beta}, \sigma^2)$:

$$\boldsymbol{\beta}|\sigma^2 \sim \text{MVN}(\mathbf{b}, \sigma^2\mathbf{V}) \text{ and } \frac{\nu \cdot \lambda}{\sigma^2} \sim \chi^2_\nu, \tag{9.42}$$

where $(\nu, \lambda, \mathbf{V}, \mathbf{b})$ are hyper-parameters to be chosen. In this case, the marginal likelihood for $\mathbf{y}$ under a model M_k can be explicitly given by

$$\text{pr}(\mathbf{y}|M_k) = \frac{\Gamma\{\frac{\nu+n}{2}\}(\nu\lambda)^{\frac{\nu}{2}}}{\pi^{\frac{n}{2}}\Gamma(\frac{\nu}{2})\, |\mathbf{I} + \mathbf{H}_k|^{1/2}} \cdot a_k^{-\frac{\nu+n}{2}} \tag{9.43}$$

where $\mathbf{H}_k = \mathbf{X}_k(\mathbf{X}_k^t\mathbf{X}_k)^{-1}\mathbf{X}_k^t$ is the projection matrix associated with $\mathbf{X}_k$, the design matrix for model M_k; $\mathbf{e}_k = \mathbf{y} - \mathbf{X}_k\mathbf{b}_k$ is the residual vector; and $a_k = \lambda\nu + \mathbf{e}_k^t(\mathbf{I}+\mathbf{H}_k)^{-1}\mathbf{e}_k$ for $k = 1, \ldots, K$. Thus, the Bayes factor, defined in (9.38), is

$$B_{12} = \left(\frac{|\mathbf{I} + \mathbf{H}_1|}{|\mathbf{I} + \mathbf{H}_2|}\right)^{1/2} \cdot \left\{\frac{a_1}{a_2}\right\}^{-\frac{\nu+n}{2}}. \tag{9.44}$$

In more general cases such as generalized linear models (see Chapter 8), an explicit form of the Bayes factor is seldom easily obtainable and its calculation must resort to various approximations.

The second difficulty stems from the large number K of possible models. Two solutions have been proposed to get around this problem. The first approach applies the Occam's window method, which averages over a set of parsimonious and data-supported models. Madigan and Raftery (1994) argued that if a model predicts the data far less well than the model which provides the best predictions, then it should be effectively discredited and hence not be considered. Under this principal, they exclude models in set

$$\mathcal{A}_1 = \left\{ M_k : \frac{\max_l\{\mathrm{pr}(M_l|D)\}}{\mathrm{pr}(M_k|D)} > C \right\}, \tag{9.45}$$

for some user-defined threshold C. Moreover, they suggested to exclude complex models which receive less support from the data than their simpler counterparts, a principle appealing to Occam's razor. Namely, models in set

$$\mathcal{A}_2 = \left\{ M_k : \exists\, M_l \notin \mathcal{A}_1, M_l \subset M_k \text{ and } \frac{\mathrm{pr}(M_l|D)}{\mathrm{pr}(M_k|D)} > 1 \right\} \tag{9.46}$$

could also be excluded from averaging. The notation $M_l \subset M_k$ means that M_l is nested into or a sub-model of M_k. This strategy greatly reduces the number of models and all the probabilities in (9.35) – (9.37) are then implicitly conditional on the reduced set. The second approach, called Markov chain Monte Carlo model composition (MC^3) due to Madigan and York (1995), approximates the posterior distribution in (9.35) numerically with a Markov chain Monte Carlo procedure.

One additional issue in implementation of BMA is how to assign the prior model probabilities. When little prior information about the relative plausibility of the models consider, a noninformative prior that treats all models equally likely is advised. When prior information about the importance of a variable is available for model structures that have a coefficient associated with each predictor (e.g., linear regression), a prior probability on model M_k can be specified as

$$\mathrm{pr}(M_k) = \prod_{j=1}^{p} \pi_j^{\delta_{kj}} (1 - \pi_j)^{1-\delta_{kj}}, \tag{9.47}$$

where $\pi_j \in [0, 1]$ is the prior probability that $\beta_j \neq 0$ in a regression model, and δ_{ki} is an indicator of whether or not X_j is included in model M_k. See Raftery, Madigan, and Hoeting (1997) and Hoeting et al. (1999) for more details and other issues in BMA implementation. BMA is available in the R package `BMA`.

Example We revisit the quasar example with an analysis of BMA in Chapter 5. The data set, given in Table 5.1, involves 5 predictors. Thus there are a total of $2^5 = 32$ linear models under consideration. With a threshold $C = 40$ in (9.46), 13 models are selected according to Occam's window. Figure 9.3 shows the variable selection information for all the

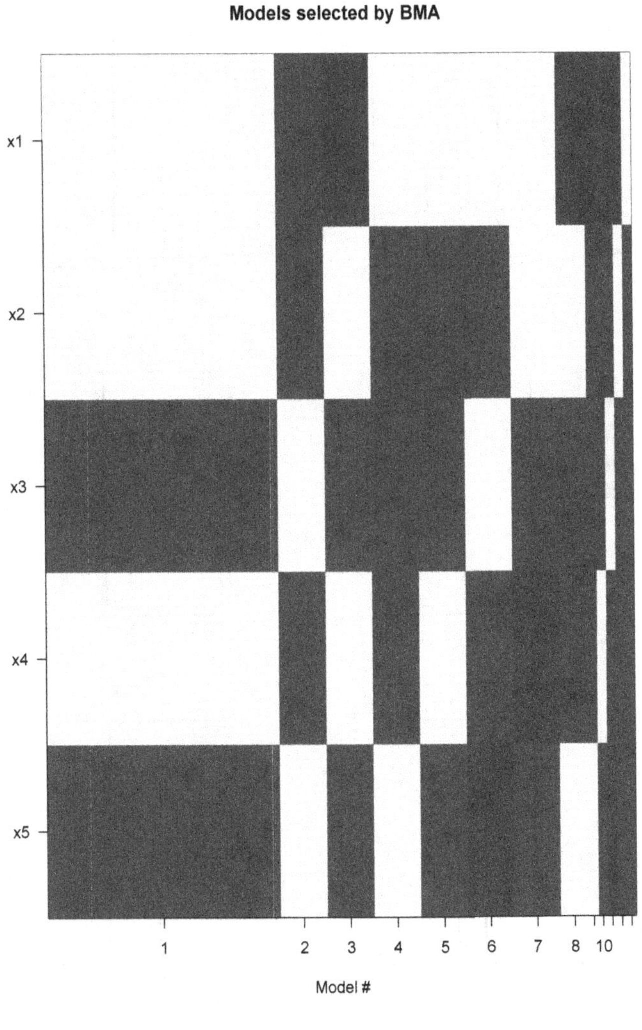

Fig. 9.3 The 13 Selected Models by BMA: the Quasar Data.

13 models, which have been arranged in order of their estimated posterior probability. The top five models, with cumulative posterior probability of 0.7136, are presented in Table 9.2.

According to BMA, the model that includes X_3 and X_5 is the best in the

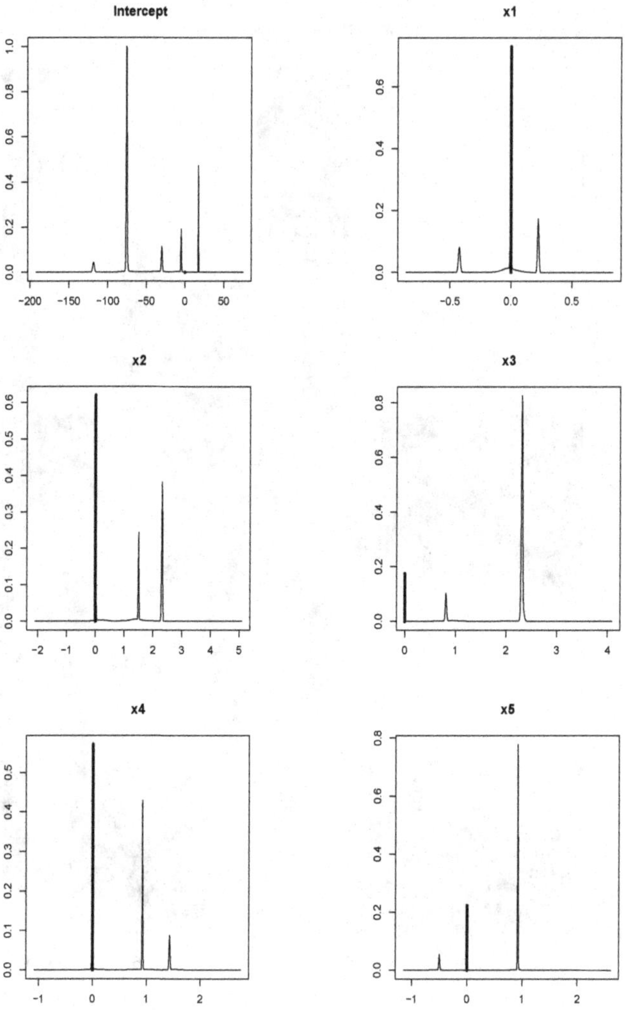

Fig. 9.4 Posterior Distributions of the Coefficients Generated by BMA: the Quasar Data.

sense that it has the highest estimated posterior probability 0.396. The first column of the table gives the estimated probability of being nonzero for each coefficient. According to the results, X_3 is most outstanding excluding the intercept term, followed by X_5 and then X_4. X_3 is selected by four out of the five best models. We can see that most of time its estimated slope is around 2.3, except in Model 4, which yields a quite different value 0.8136. Averaging over all 13 selected models, the estimated posterior mean is 1.7435 with estimated posterior standard deviation of 0.9247. Conclusions on other coefficient estimates can be made in the same fashion.

Plots of the estimated posterior distributions for each β are given in Figure 9.4. Again, one can draw some interesting observations. First, the posteriors in this example are very much like a discrete distribution, which is not necessarily true in general. The modes or bumps in each plot correspond to the varying estimated values in different models. This information is helpful in assessing various confounding effects among predictors. For example, the posterior of β_5 is essentially centered around one single number, which implies that the confounding effect of other covariates on X_5 is negligible. Secondly, one may also get better ideas about the final conclusion in testing $H_0 : \beta_j = 0$. For example, compared to others, the posterior of β_1 seems to fluctuate more between positive and negative values. Namely, the estimates of β_1 seems to encounter more frequent sign changes. The `BMA` package contains a few other interesting features, which have been omitted in this illustration.

Table 9.2 BMA Results for the Quasar Data.

	Estimated			Ordinary LSE of β in Model				
	$\text{pr}(\beta \neq 0 \mid D)$	$E(\beta \mid D)$	$\sigma(\beta \mid D)$	1	2	3	4	5
β_0	100.0	−58.5972	34.3667	−75.77	17.25	−75.77	−29.80	−75.40
X_1	27.1	−0.0033	0.1135	·	0.2240	−0.0015	·	·
X_2	38.1	0.5728	0.9265	·	2.308	·	1.508	0.0063
X_3	82.6	1.7435	0.9247	2.313	·	2.313	0.8136	2.307
X_4	42.9	0.3581	0.5243	·	0.9234	·	0.9257	·
X_5	77.7	0.5654	0.5222	0.9220	·	0.9221	·	0.9220
			Number of Variables	2	3	3	3	3
			R^2	0.999	0.999	0.999	0.999	0.999
			BIC	−.0166	−.0163	−.0163	−.0163	−.0163
			Estimated $\text{pr}(M_k \mid D)$	0.396	0.079	0.079	0.079	0.079

Problems

1. Show that the posterior joint density in (9.13) can be rewritten as the form given by (9.14).
2. Concerning the Bayesian linear model (9.11), construct a 95% Bayesian confidence interval for the mean response, $E(y_p|\mathbf{x}_p, \boldsymbol{\beta}, \nu) = \mathbf{x}_p^t\boldsymbol{\beta}$, of all individuals that have predictor values equal to $\mathbf{x} = \mathbf{x}_p$.
3. Show that the Bayesian estimator of $\boldsymbol{\beta}$ in (9.22) can be viewed as the generalized least squares estimator in the augmented linear model (9.23).
4. Concerning the semi-conjugate Bayesian linear model as given by (9.34), derive the conditional posterior distributions for $\boldsymbol{\beta}$ and ν.
5. For the quasar data in Table 5.1, apply a logarithm transformation on the response Y so that $Y \leftarrow \log(Y)$. Consider the semi-conjugate Bayesian linear model that includes all five predictors.

   (a) With specification

   $$\boldsymbol{\beta}_0 = \mathbf{0}, \ \mathbf{V} \approx \mathbf{O}, \ a = 0.0001, \text{ and } b = 0.0001,$$

   obtain the Bayesian estimator of $\boldsymbol{\beta}$ and σ^2, as well as their corresponding 95% confidence interval. Compare them to the results from ordinary least square estimation.

   (b) With the same prior specification in part (a), obtain the posterior densities of each β and ν using two methods, the analytic method and Gibbs sampler. Plot them on the same figure and comment on how the MCMC method approximates the true posterior densities.

   (c) Using the Gibbs sampler, construct a 95% Bayesian interval for predicting the response of a newcomer who has $\mathbf{x} = (3, -14, 45, 20, -27)^t$.

   (d) Redo the above analysis with a different prior specification

   $$\boldsymbol{\beta}_0 = \mathbf{0}, \ \mathbf{V} \approx 0.04 \cdot \mathbf{I}, \ a = 0.1, \text{ and } b = 0.1,$$

   and compare to see how an informative prior affects the results.

6. Verify equation (9.43).
7. Apply the BMA technique to the shingle sales data in Table 4.1. In particular, identify the best model according to the highest posterior model probability. Comments on the effect of each predictor on the response.

Bibliography

Agresti, A. (2002). *Categorical Data Analysis.* 2nd Edition. New York: John Wiley & Sons, Inc.

Akaike, H. (1973). Information theory and an extension of the maximum likelihood principle. In *Procedings of Second International Symposium on Information Theory,* (eds B. N. Petrov and F. Csaki). Budapest: Akademiai Kiado, pp. 267–281.

Aitkin, M. (1978). The analysis of unbalanced cross classifications (with discussion). *Journal of the Royal Statistical Society, Series A*, **141**, pp. 195–223.

Amemiya, T. (1985). *Advanced Econometrics.* Harvard University Press, Cambridge, USA.

Baltagi, B. H. (2001). *Econometric Analysis of Panel Data.* Wiley, John & Sons: New York.

Bard, Y. (1974). *Nonlinear Regression Estiamtion.* Academic Presss: New York.

Belsley, D. A., Kuh, E., and Welsch, R. E. (1980). Regression diagnostics. New York, John Wiley.

Berger, J. (1993). *Statistical Decision Theory and Bayesian Analysis*, 2nd edition. Springer-Verlag, New York.

Bishop, C. (1995). *Neural Networks for Pattern Recognition.* Clarendon Press, Oxford.

Box, G. E. P. (1979). Robustness in the strategy of scientific model building, Robustness in Statistics, R. L. Launer and G. N. Wilkinson, Editors. Academic

Press: New York.

Box, G. E. P. and Cox, D. R. (1964). An analysis of transformations. *Journal of the Royal Statistical Society, Series B*, **26**, pp. 211–252.

Box, G. E. P. and Tidwell, P. W. (1962). Transformations of the independent variable. *Technometrics*, **4**, pp. 531–550.

Breiman, Leo (2001). Random Forests. *Machine Learning*, **45** (1), pp. 5–32.

Breiman, L. and Friedman, J. H. (1985). Estimating optimal transformations for multiple regression and correlation (with discussion). *Journal of the American Statistical Association*, **80**, pp. 580–619.

Breiman, L., J. H. Friedman, L. J., Olshen, R. A., and Stone, C. J. (1984). *Classification and Regression Trees.* Monterey, Calif., U.S.A.: Wadsworth, Inc.

Breusch, T. S. and Pagan, A. R. (1979). A simple test for heteroscedasticity and random coefficient variation. *Econometrica*, **47**, pp. 987–1007.

Buse, A. (1982). The likelihood ratio, wald, and Lagrange multiplier tests: an expository note. *The American Statistician*, **36**, pp. 153–157.

Carlin, B. P. and Louis, T. A. (2000). *Bayes and Empirical Bayes Methods for Data Analysis*, Second Edition. Chapman & Hall/CRC.

Carroll, R. J. and Ruppert, D. (1988). *Transformation and Weighting in Regression.* New York, NY: Chapman and Hall.

Chambers, J. M., and Trevor J. Hastie, T. J. (eds.) (1992) *Statistical Models in S.* Pacific Grove, CA: Wadsworth and Brooks/Cole.

Chib, S. and Greenberg, E. (1995). Understanding the MetropolisCHastings algorithm. *The American Statistician*, **49**, pp. 327–335.

Cook, R. D. (1977). Detection of influential observations in linear regression. *Technometrics*, **19**, pp. 15–18.

Cook, R. D. and Weisberg, S. (1982) *Residuals and Influence in Regression.* New York: Chapman and Hall.

Cook, R. D., and Weisberg, S. (1999). *Applied regression Including Computing and Graphics.* New York: John Wiley & Sons, Inc.

Cox, D. R., and Hinkley, D. V. (1974). *Theoretical Statistics.* London: Chapman

& Hall.

Craven, P., and Wahba, G. (1979). Smoothing Noisy Data With Spline Functions. *Numerische Mathematik*, **31**, pp. 377–403.

Durbin, J., and Watason, G. S. (1951). Testing for serial correlation in least squares regression. II. *Biometrika*, **38**, pp. 159–178.

Efron, B. and Tibshirani, R. (1993). *An Introduction to the Bootstrap*. London, UK: Chapman & Hall.

Efron, B., Hastie, T., Johnstone, I., and Tibshirani, R. (2004). Least Angle Regression (with discussion). *Annals of Statistics*, **32**, pp. 407–499.

Farebrother, R. W. (1975). The minimum mean square error linear estimator and ridge regression. *Technometrics*, **17**, pp. 127–128.

Fernandez, G. C. J. (1997). Detection of model specification , outlier, and multicollinearity in multiple linear regression models using partial regression/residual plots. SAS Institute Inc., *Proceedings of the 22nd annual SAS users group international conference*, pp. 1246–1251.

Fox, J. (1997). *Applied Regression Analysis, Linear Models, and Related Methods*. Newbury Park, CA: Sage.

Fox, J. (2002). *An R and S-PLUS Companion to Applied Regression*. Thousand Oaks, CA: Sage Publications Inc.

Friedman, J. and and Stuetzle, W. (1981). Projection pursuit regression. *Journal of the American Statistical Association*, **76**, pp. 817–823.

Freund, R. J. and Littell, R. C. (2000). *SAS System for Regression*, 3rd ed. Cary, NC: SAS Institute.

Freund, Y. and Schapire, R. E. (1997). A decision-theoretic generalization of on-line learning and an application to boosting. *Journal of Computer and System Sciences*, **55**,119–139. `http://www.cse.ucsd.edu/~yfreund/papers/adaboost.pdf`

Frank, I. E. and Friedman, J. H. (1993). A Statistical View of Some Chemometrics Regression Tools. *Technometrics*, **35**, pp. 109–148.

Gail, M. and Simon, R. (1985). Testing for qualitative interactions between treatment effects and patient subsets. *Biometrics*, **41**, pp. 361–372.

Greene, W. H. (2000). *Econometric Analysis*, 4th Edition. Prentice Hall.

Grimmett, G. and Stirzaker, D. (2001). *Probability and Random Processes*, 3rd ed. Oxford. Section 7.3.

Hastie, T. J. and Tibshirani, R. J. (1986). Generalized Additive Models (with discussion). *Statistical Science*, **1**, pp. 297–318.

Hastie, T. J. and Tibshirani, R. J. (1990). *Generalized Additive Models.* New York: Chapman and Hall.

Hastie, T., Tibshirani, R., and Friedman, J. (2002). *The Elements of Statistical Learning: Data Mining, Inference, and Prediction.* New York, NY: Springer-Verlag.

Hastings, W. K. (1970). Monte Carlo sampling methods using Markov Chains and their applications. *Biometrika*, **57**, pp. 97–109.

Hoerl, A. E. and Kennard, R. W. (1970). Ridge regression: biased rstimation for nonorthogonal problems. *Technometrics*, **12**, pp. 55–67.

Hoerl, A. E., Kennard, R. W., and Baldwin, K. F. (1975). Ridge regression: some simulations. *Communications in Statistics*, **4**, pp. 105–123.

Hoeting, J. A., Madigan, D., Raftery, A., and Volinsky, C. T. (1999). Bayesian Model Averaging: A Tutorial (with discussion). *Statistical Science*, **14** (4), pp. 382–417.

Hosmer, D. W. and Lemeshow, S. (2000). *Applied Logisitc Regression.* New York: John Wiley & Sons, Inc.

Hurvich, C. M., and Tsai C.-L. (1989). Regression and Time Series Model Selection in Small Samples, *Biometrika*, **76**, pp. 297–307.

Johnson, R. A. and Wichern, D. W. (2001). *Applied Multivariate Statistical Analysis.* Upper Saddle River, NJ: Prentice Hall, Inc.

Kira, K. and Rendel, L. (1992). The Feature Selection Problem: Traditional Methods and a New Algorithm. *Proceedings of the Tenth National Conference on Artificial Intelligence*, MIT Press, pp. 129–134.

Koenker, R. (1981), A Note on Studentizing a Test for Heteroskedasticity. *Journal of Econometrics*, **29**, pp. 305–326.

Kononenko, I., Simec, E., and Robnik-Sikonja, M. (1997). Overcoming the myopia of induction learning algorithms with RELIEFF. *Applied Intelligence*, **17**(1), pp. 39–55.

Kotz, S. and Nadarajah, S. (2004). *Multivariate t Distributions and Their Applications.* Cambridge University Press.

Larsen, W. A. and McCleary, S. J. (1972). The use of partial residual plots in Regression analysis. *Technometrics*, **14**, pp. 781–790.

Lawless, J. F. and Wang, P. (1976). A simulation study of ridge and other regression estimators. *Communications in Statistics*, **5**, pp. 307–323.

Leamer, E. E. (1978). *Specification Searches.* New York: Wiley.

Li, K. C. (1991). Sliced inverse regression for dimension reduction (with discussions). *Journal of the American Statistical Association*, **86**, pp. 316–342.

Li, K. C. (1992). On principal Hessian directions for data visualization and dimension reduction: another application of Stein's Lemma. *Journal of the American Statistical Association*, **87**, pp. 1025–1039.

Lin, P.-E. (1971). Some characterization of the multivariate t distribution. Florida State University Statistics Report M200.

Loh, W.-Y. (2002). Regression trees with unbiased variable selection and interaction detection. *Statistica Sinica*, **12**, pp. 361–386.

Longley, J. W. (1967) An appraisal of least-squares programs from the point of view of the user. *Journal of the American Statistical Association*, **62**, pp. 819–841.

Madigan, D. and Raftery, A. E. (1994). Model Selection and Accounting for Model Uncertainty in Graphical Models Using Occam's Window. *Journal of the American Statistical Association*, **89**, pp. 1535–1546.

Maddala, G. S. (1977). *Econometrics.* New York: McGraw-Hill, pp. 315–317.

Madigan, D. and York, J. (1995). Bayesian Graphical Models for Discrete Data. *International Statistical Review*, **63**, pp. 215–232.

Mallows, C. L. (1986). Augmented partial residual Plots. *Technometrics*, **28**, pp. 313–319.

Mason, R. L., Gunt, R. F. and Hess, J. L., (1989). *Statistical design and analysis of experiments with applications to engineering and science.* New York: Wiley.

Mason, R. L., Gunst, R. F., and Webster, J. T. (1975). Regression analysis and problem of multicollinearity. *Communication in Statistics*, **4**, pp. 277–292.

McCullagh, P., and Nelder, J. A. (1989). *Generalized Linear Models.* London: Chapman & Hall.

McCulloch, C. E., Searle, S. R., Neuhaus, J. M. (2008). *Generalized, Linear, and Mixed Models.* New York: Wiley.

McQuarrie, A. D. R. and Tsai, C.-L. (1998). *Regression and Time Series Model Selection.* Singapore River Edge, NJ: World Scientific.

Mellows, C. L. (1973). Some comments on C_p, *Technometrics*, **15**, pp. 661–675.

Mellows, C. L. (1975). More comments on C_p, *Technometrics*, **37**, pp. 362–372.

Mendenhall, W. and Sinich, T. (2003). *Regression Analysis: A Second Course in Statistics*, 6th edition. Upper Saddle River, NJ: Prentice Hall.

Metropolis, N. and Ulam, S. (1949). The Monte Carlo method. *Journal of the American Statistical Association*, **44**, pp. 335–341.

Montgomery, D. C. and Peck, E. A. (1992). *Introduction to Linear regression analysis*, 2nd Edition. New York, NY: John Wiley.

Morgan, J. and Sonquist, J. (1963). Problems in the analysis of survey data and a proposal. *Journal of the American Statistical Association*, **58**, pp. 415–434.

Myers, R. H. (1990). *Classical and Modern Regression with Applications*, 2nd Edition. Pacific Grove, CA: Duxbury Press.

Nelder, J. and Wedderburn, R. W. M. (1972). Generalized linear models. *Journal of Royal Statistical Society, Series B*, **135**, pp. 370–384.

Osborne, M. R., Presnell, B., and Turlach, B. A. (2000). A New Approach to Variable Selection in Least Squares Problems. *IMA Journal of Numerical Analysis*, **20**, pp. 389–404.

Osborne, M. R., Presnell, B., and Turlach, B. A. (2000). On the LASSO and its dual. *Journal of Computational and Graphical Statistics*, **9**(2), pp. 319–337.

Papoulis, A. (1991). *Probability, Random Variables, and Stochastic Processes*, 3rd ed. McGraw-Hill, pp. 113–114.

Raftery, A. E., Madigan, D., and Hoeting, J. A. (1997). Bayesian Model Averaging for Linear Regression Models. *Journal of the American Statistical*

Association, **92**, pp. 179–191.

Rencher, A. C. (2000). *Linear Models in Statistics.* New York, NY: John Wiley & Sons, Inc.

Robnik-Sikonja, M., Kononenko, I. (2003). Theoretical and empirical analysis of ReliefF and RReliefF. *Machine Learning*, **53**, pp. 23–69.

Sall, J. (1990). Leverage plots for general linear hypothesis. *The American Statistician*, **44**, pp. 308–315.

SAS, SAS/GRAPH, and SAS/STAT (2009) are registered trademarks of SAS Institute Inc. in the USA and other countries.

SAS Institute Inc. (2004). *SAS/STAT 9.1 User's Guide.* Cary, NC: SAS Institute Inc.

Schwarz, G. (1978). Estimating the Dimension of a Model, *The Annals of Statistics*, **6**, pp. 461–464.

Smith, Petricia L. (1979). Splines as a Useful and Convenient Statistical Tool. *The American Statistician*, **33**, pp. 57–62.

Stine R. A. (1995). Graphical Interpretation of Variance Inflation Factors. *The American Statistician*, **49**, pp. 53–56.

Stone, C. J. (1985). Additive Regression and Other Nonparametric Models. *Annals of Statistics*, **13**, pp. 689–705.

Strobl, C., Boulesteix, A.-L., Zeileis, A., and Hothorn, T. (2007). Bias in random forest variable importance measures: Illustrations, sources and a solution. *BMC Bioinformatics*, **8**, article 25.

Tanner, M. (1998). *Tools for Statistical Inference: Methods for the Exploration of Posterior Distributions and Likelihood Functions*, 3rd Edition. New York: Springer-Verlag.

Thompson, B. (2001). Significance, effect sizes, stepwise methods, and other issues: Strong arguments move the field. *The Journal of Experimental Education*, **70**, pp. 80–93.

Tibshirani, R. (1988). Estimating optimal transformations for regression via additivity and variance stabilization. *Journal of American Statistical Association*, **83**, pp. 394–405.

Tibshirani, R. (1996). Regression shrinkage and selection via the Lasso. *Journal of the Royal Statistical Society, Series B*, **58**, pp. 267–288.

Wahba, G. (1983). Bayesian 'confidence intervals' for the cross validated smoothing spline. *Journal of the Royal Statistical Society, Series B*, **45**, pp. 133–150.

White, H. (1980), A heteroscedasticity-consistent covariance matrix estimator and a direct test for heteroscedasticity. *Econometrica*, **48**, pp. 817–838.

Wood, S. N. (2006). *Generalized Additive Models: An Introduction with R*. Chapman and Hall/CRC.

Yan, X. and Su, X. G. (2005). Testing for Qualitative Interaction. *Encyclopedia of Biopharmaceutical Statistics: Second Edition, Revised and Expanded*, Ed. By Dr. Shein-Chung Chow. Marcel Dekker, Inc.: New York, NY.

Index

ν-folder cross validation, 133

additive models, 207
additivity and variance stabilization (AVAS), 211
AIC, 282
AIC_C, 170
Akaike information criterion (AIC), 169
all possible regressions, 165
alternating conditional expectation (ACE), 210
analysis of covariance (ANCOVA), 76
analysis of deviance, 282
analysis of variance, 282
ANOVA, 23
autocorrelation, 266

backfitting, 208
basis, 43
Bayes factor, 310
Bayesian
  approach in general, 297
  linear regression, 297
Bayesian information criterion, 170
Bayesian model averaging (BMA), 297, 309
BIC, 170
Bonferroni, 134
Bonferroni CI for regression means, 71
Bonferroni CI for regression parameters, 71
boosting, 247
Box-Cox power transformation, 205
Breusch-Pagan-Godfrey test, 202

Chebyshev, 134
Chi-square distribution, 55
classical nonlinear regression model, 203
classification and regression trees (CART), 182
coefficient of variation, 25
collinearity, 81
confidence interval for regression mean, 69
confidence interval for regression prediction, 70
confounding, 72, 76
conjugate prior, 299
Cook's D, 149
corrected AIC, 170
covariance ratio, 150

cross validation, 132

data mining, 180, 182
DEBETAS, 149
degrees of freedom, 22
delta method, 263, 265, 278
deviance, 281
DFITTS, 148
dimension, 43
dot product, 44
dummy variable, 77, 288
Durbin-Watson test, 196, 266

effective dimension reduction (EDR), 186
effective number of degrees of freedom, 254
exploratory data analysis (EDA), 250
exploratory variable, 10
exponential distribution family, 272

Fisher scoring, 276
Fisherian approach, 297
fitted value, 12

gamma distribution, 300
Gauss-Seidel algorithm, 209
GCV, 253
Generalized Additive Model, 209
generalized cross-validation (GCV), 167, 209, 253
generalized inverse, 221
generalized least squares (GLS), 219
  F test, 231
  maximum likelihood estimation, 232
Generalized linear models, 269
  canonical link, 273
  deviance, 281
  Fisher scoring, 276
  Fisher's information matrix, 275
  likelihood equations, 274
  likelihood ratio test, 279
  Pearson residuals, 282
generalized variance (GV), 149
Gibbs sampler, 304
Gibbs sampling, 304
Glesjer test, 202
Goldfeld-Quandt test, 204
gradient, 276
Gram-Schmidt algorithm, 93

HAT diagonal, 148
HAT matrix, 96
heteroscedasticity, 197, 233
homoscedasticity, 195

idempotent matrix, 50
influential observation, 134
interaction, 72, 293
inverse matrix of blocked matrix, 68
iterative weighted least squares (IWLS), 233

jackknife, 167, 253

kernel method, 208

Lagrange multiplier test, 202
LASSO, 180, 219, 246
learning sample, 132
least absolute shrinkage and selection operator (LASSO or lasso), 246
least angle regression (LARS), 247
least squares
  nonlinear regression, 261

least squares method, 10
leave-one-out SSE, 108
likelihood ratio test (LRT), 279
linear hypothesis test, 66
linear model, 48
linear span, 43
linear subspace, 43
linearly independent vectors, 44
link function, 272
local polynomial regression, 208
log-linear model, 287
logistic function, 260
logistic regression, 272

Markov chain, 304
Markov Chain Monte Carlo (MCMC), 303
Markov chain Monte Carlo model composition (MC^3), 312
maximum likelihood (ML), 205, 274
mean shift outlier, 139
mean square error, 164
Mellows C_p, 167
MLE of simple regression parameters, 18
model diagnostics, 195
model misspecification, 157
model selection criterion, 165
  consistent, 170
  efficient, 170
multicollinearity, 219, 238, 250
multiple linear regression, 2
multiple regression, 48
multivariate normal density function, 54
multivariate t distribution, 301

neural networks, 187
Newton-Raphson method, 262, 276
non-full-rank linear models, 219
noninformative prior, 306
nonlinear least squares, 261
nonlinear model, 49
nonlinear regression, 3
normal density, 54
normal equation, 59

odds, 283
odds ratio, 283
One-Way Analysis of Variance (ANOVA), 219
ordinary least squares estimate, 197
orthogonal vectors, 44
outlier, 107, 134

parametric nonlinear regression, 259
Park test, 202
partial regression plot, 94
partial residual plot, 142
partial residuals, 208
Poisson regression, 287
prediction sum of squares (PRESS), 167
PRESS, 253
PRESS residual, 103, 135
PRESS statistic, 135
principal component analysis (PCA), 186, 243
principal components regression, 245
profile log likelihood, 206
project matrix, 46
projection, 45
projection pursuit, 187

QR decomposition, 95
quadratic form, 49

R, 6
R-student residual, 148
R-student statistic, 140
random forests (RF), 180
recursive partitioning, 181
reference cell coding scheme, 291
regression, 2
regression error, 12
regression prediction, 19
regression residual, 25
regression splines, 208
RELIEF, 182
residual, 12, 96
ridge estimator, 302
ridge function, 187
ridge regression, 219, 238
ridge trace, 253

S-PLUS, 6, 7
SAS, 5
selection bias, 184
shrinkage, 241
shrinkage estimator, 302
simple linear regression, 2, 9
simultaneous confidence region, 70
simultaneous inference, 25
sliced inverse regression (SIR), 186
smoothing, 208
smoothing splines, 208
splines, 208
SPSS, 6
standardized PRESS residual, 106
standardized residual, 133
statistical model, 2
studentized jackknife residual, 196
studentized residual, 103, 133
sum squares of residuals, 24
super-parameter, 300

test sample, 163
testing sample, 132
total variance, 22
trace, 50
training sample, 162
tree models, 181

variable importance, 180
  ranking, 180
variable screening, 179
variable selection, 157
variance inflation factor (VIF), 180, 250
variance of PRESS residual, 106

weighted least squares estimate (WLS), 233
White's test, 201